ADVANCES IN
CORROSION SCIENCE
AND TECHNOLOGY
VOLUME 1

T0212420

ADVANCES IN CORROSION SCIENCE AND TECHNOLOGY

Editors:

M. G. Fontana and R. W. Staehle

Corrosion Center, Department of Metallurgical Engineering
The Ohio State University, Columbus, Ohio

Contributors to Volume 1:

VITTORIO CARASSITI
Corrosion Study Center "A. Daccò"
Chemical Institute of the University
Ferrara, Italy

DUDLEY de G. JONES
Physical Chemistry Section
Central Electricity Research Laboratories
Leatherhead, Surrey, England

RONALD M. LATANISION
RIAS — Martin Marietta Corporation
Baltimore, Maryland

H. G. MASTERSON
Physical Chemistry Section
Central Electricity Research Laboratories
Leatherhead, Surrey, England

SAKAE TAJIMA
Tokyo City University
Tokyo, Japan

GIORDANO TRABANELLI
Corrosion Study Center "A. Daccò"
Chemical Institute of the University
Ferrara, Italy

A. R. C. WESTWOOD
Research Institute for Advanced Studies
Martin Marietta Corporation
Baltimore, Maryland

ADVANCES IN
CORROSION SCIENCE AND TECHNOLOGY

VOLUME 1

Edited by
Mars G. Fontana and Roger W. Staehle

SPRINGER SCIENCE+BUSINESS MEDIA, LLC • 1970

Library of Congress Catalog Card Number 76-107531

ISBN 978-1-4615-8254-0   ISBN 978-1-4615-8252-6 (eBook)
DOI 10.1007/978-1-4615-8252-6

© 1970 Springer Science+Business Media New York
Originally published by Plenum Press, New York in 1970
Softcover reprint of the hardcover 1st edition 1970

United Kingdom edition published by Plenum Press, London
A Division of Plenum Publishing Corporation, Ltd.
Donington House, 30 Norfolk Street, London W.C. 2, England

PREFACE

This series was organized to provide a forum for review papers in the area of corrosion. The aim of these reviews is to bring certain areas of corrosion science and technology into a sharp focus. The volumes of this series will be published approximately on a yearly basis and will each contain three to five reviews. The articles in each volume will be selected in such a way to be of interest both to the corrosion scientists and the corrosion technologists. There is, in fact, a particular aim in juxtaposing these interests because of the importance of mutual interaction and interdisciplinarity so important in corrosion studies. It is hoped that the corrosion scientists in this way may stay abreast of the activities in corrosion technology and *vice versa*.

In this series the term "corrosion" will be used in its very broadest sense. This will include, therefore, not only the degradation of metals in aqueous environment but also what is commonly referred to as "high-temperature oxidation." Further, the plan is to be even more general than these topics; the series will include all solids and all environments. Today, engineering solids include not only metals but glasses, ionic solids, polymeric solids, and composites of these. Environments of interest must be extended to liquid metals, a wide variety of gases, nonaqueous electrolytes, and other nonaqueous liquids. Furthermore, there are certain° complex situations such as wear, cavitation, fretting, and other forms of degradation which it is appropriate to include. At suitable intervals certain of the review articles will be updated as the demands of technology and the fund of new information dictate.

Another important aim of this series is to attract those in areas peripheral to the field of corrosion. Thus, physicists, physical metallurgists, physical chemists, and electronic scientists all can make very substantial contributions to the resolution of corrosion problems. It is hoped that these reviews will make the field more accessible to potential contributors from these other areas. Many of the phenomena in corrosion are so complex that it is impossible for reasonable progress to be made without more serious and enthusiastic interdisciplinary interest.

In addition to the discussion of scientific and technological phenomena the articles will also include discussions of important techniques which should be of interest to corrosion scientists.

Columbus, Ohio   R. W. STAEHLE
June 9, 1970   M. G. FONTANA

CONTENTS

Chapter 1

Techniques for the Measurement of Electrode Processes at Temperatures Above 100°C

D. de G. Jones and H. G. Masterson

Chapter 2

Surface- and Environment-Sensitive Mechanical Behavior

R. M. Latanision and A. R. C. Westwood

Chapter 3

Mechanism and Phenomenology of Organic Inhibitors

Giordano Trabanelli and Vittorio Carassiti

Chapter 4

Anodic Oxidation of Aluminum

Sakae Tajima

TECHNIQUES FOR THE MEASUREMENT OF ELECTRODE PROCESSES AT TEMPERATURES ABOVE 100°C

D. de G. Jones and H. G. Masterson

Physical Chemistry Section
Central Electricity Research Laboratories
Leatherhead, Surrey, England

INTRODUCTION

This paper is intended as an introductory guide to experimental techniques which have been found useful in electrochemical studies at temperatures high enough to pose problems of containment of the solution vapor pressure. There are compelling reasons, both fundamental and technological, for extending the temperature range over which meaningful electrochemical measurements can be made. On the one hand, many of the applications of electrochemistry in fields of technological interest are concerned with the properties of electrolyte solutions, and with the mechanism of electrode processes, at temperatures considerably higher than those at which electrochemical studies are usually carried out. The capital-intensive electrical power industries, both nuclear and conventional, and the chemical process industries provide frequent examples of problems where the basic electrochemical data available are inadequate for an understanding of the electrode–electrolyte interactions of interest.

On the other hand, there is considerable scientific advantage in developing adequate techniques for high-temperature aqueous electrochemistry. Most measurements have been made, to date, on cells whose temperatures are closely controlled to 25°C. At this temperature, measurement precision, both electrical and electrochemical, can be high, and thermodynamic data can be derived more accurately than from many other experimental techniques. The temperature dependence of electrode processes which require close thermostatic control during measurement is, itself, of interest, and frequent advantage has been taken of the limited temperature variation possible with cells designed for use at ambient pressure, to study temperature-dependent phenomena in this range. Examples which may be cited include[1,2]:

1. Standard electrode potential, activity coefficients, and Debye–Hückel equation constants.
2. Conductance.
3. Equilibrium constants.
4. Ionization constants.
5. Diffusion constants in polarography,[3] and electrodeposition of metals and alloys.[4]
6. Variation of electrocapillary maxima, and zero-charge point.[5]

The Gibbs–Helmholtz equation provides an expression relating the electrode-potential temperature coefficient thermodynamically to both the entropy change of a cell reaction

$$\Delta S = -nF\left(\frac{\partial E}{\partial T}\right)_P$$

and to the enthalpy:

$$\Delta H = nF\left[E - T\left(\frac{\partial E}{\partial T}\right)_P\right]$$

There is, however, a practical difficulty in making use of relationships between extensive properties for other than isothermal reactions. The temperature coefficients of standard electrode potentials are defined, arbitrarily, on the basis of a convention which sets the partial molal entropy of the hydrogen ion in solution to zero, so that electrode potentials may be compared over a range of temperature only on the basis that the potential of the standard hydrogen electrode does not vary with temperature. As Ives and Janz have stressed,[6] it is necessary, therefore, to take account of the specific chemical nature of the species involved in electrochemical reactions over a range of temperature, and to make use of the extended form of the thermodynamic relations, for example,

$$\Delta G = \Delta H_0 + \int_0^T \Delta C_p \cdot dT - T\int_0^T \left(\frac{\Delta C_p}{T}\right) dT - T\,\Delta S_0$$

rather than the simpler relationships between the extensive properties alone. This approach can, of course, be readily adopted for reactions for which the appropriate thermodynamic data are available over the temperature

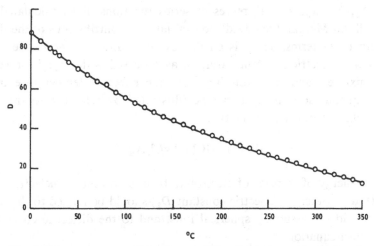

Fig. 1. The temperature dependence of the dielectric constant of water at saturation pressures.

range of interest. In this way, Lietzke and Stoughton[7] were able to compute the theoretical emf for cells combining pairs of electrodes (Ag/Ag₂SO₄ | Hg/Hg₂SO₄; Ag/AgCl | Hg/Hg₂Cl₂) and compare the results with experimental observation on high-temperature electrochemical cells. The dependence on temperature and on electrolyte composition, of deviations in the experimental results from the theoretical predictions, could then be interpreted, quantitatively, in terms of disproportionation and hydrolysis at higher temperatures of the less stable salts used in the reference electrodes.

Another effect of temperature on aqueous electrolyte solutions derives from the marked dependence on temperature of the dielectric constant. The data of Quist and Marshall,[8] given in Fig. 1, indicate that the value of this parameter for pure water in equilibrium with its vapor at 350°C is only about 16% of its value at 25°C. The DT parameter—of importance in the Debye–Hückel computation of the interionic-attraction-effect-on-activity coefficient—falls to about one third of its value at 25°C over the same range of temperature. One consequence of a decrease in the dielectric constant of the solvent, as illustrated by Harned with dioxane–water,[9] is the need to use the extended form of the Debye–Hückel equation, since the higher order terms are no longer negligible:

$$\ln f_{\pm} = - \frac{(\varepsilon z)^2}{2DkT} \frac{k}{1 + ka} + \left(\frac{\varepsilon^2 z^2}{DkTa}\right)^3 \left[\frac{1}{2} X_3(ka) - 2Y_3(ka)\right]$$
$$+ \left(\frac{\varepsilon^2 z^2}{DkTa}\right)^5 \left[\frac{1}{2} X_5(ka) - 4Y_5(ka)\right]$$

where X_3, Y_3, X_5, and Y_5 represent series functions of ka calculated by Gronwell, La Mer, and Sandved[10] to evaluate the contributions of the third and fifth order terms. There is an interesting contrast, however, between the solvent dielectric constant variation as observed with organic solvent–water mixtures, and that resulting from the effect of temperature on a purely aqueous solvent. This can be illustrated by reference to the emf data available from cells of the type

$$H_2(Pt) \mid HCl \mid AgCl/Ag$$

The free energy of transfer of electrolyte from a solvent of dielectric constant D_1 to another of dielectric constant D_2, should be related to the ionic radii, r_+ and r_-, assuming spherical ions, and to the dielectric constants, by the Born equation:

$$\Delta G_t^0 = \frac{Ne^2}{2} \left[\frac{1}{D_2} - \frac{1}{D_1} \right] \left[\frac{1}{r_+} + \frac{1}{r_-} \right]$$

The free energy of transfer can be determined electrochemically by comparison of the standard electrode potentials of cells, using solvents of different dielectric constants, with the standard electrode potential corresponding to a pure aqueous solution at 25°C. The standard electrode potentials are obtained by conventional extrapolation to infinite dilution of the data from cells covering a range of molality of hydrogen chloride solvent of each dielectric constant. The difference in standard potential is related to the difference in hydrogen chloride solvation energy in the two solvents by the equation

$$-\Delta G_t = F(E_1^0 - E_2^0)$$

The standard potential should then be linearly dependent on the inverse of the dielectric constant:

$$E_N^0 = \text{const} - \frac{Ne^2}{2F} \Sigma \frac{1}{r} \left(\frac{1}{D} \right)$$

But for organic solvent–water systems, this correlation is not good.[11,12] As may be seen for the methanol–water and dioxane–water examples given in Fig. 2, the decrease of E^0 with $1/D$, although of the same order as that predicted, can more often be described better by curved rather than linear plots, and there is a strong dependence on the nature of the organic solvent. As discussed in detail by Franks and Ives,[13] it appears that, in general, in water–organic solvent mixtures, the Born treatment is inadequate, and the

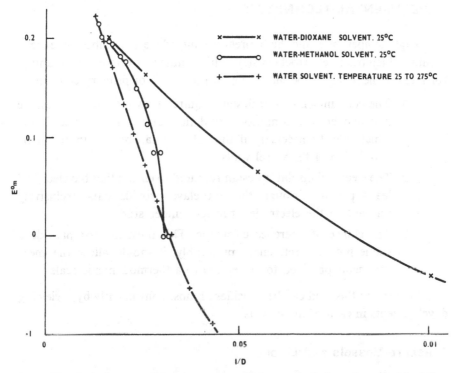

Fig. 2. The dependence on dielectric constant of the standard electrode potential of the cell $H_2(Pt)$ | HCl solvent | AgCl/Ag.

free energy of transfer ΔGt^0, must be largely determined by short-range interactions, i.e., ion–solvent interactions rather than the long-range coulombic forces appropriate to ion–ion interactions in dilute solutions. It is possible to apply the Born treatment to the extensive study of the cell

$$H_2(Pt) \mid HCl_{aq} \mid AgCl/Ag$$

carried out over the temperature range 25–300°C, by Lietzke and Greeley;[14] their values of the standard electrode potential $\varepsilon_m{}^0$ vs $1/D$ (using the Quist–Marshall values of Reference 8) are included in Fig. 2. Agreement with the Born requirement of a linear relationship is good, at least to 250°C, about which temperature electrochemical measurements, in fact, become increasingly difficult. Thus, although the concept of charged spheres interacting in a uniform dielectric may be still unduly simplified as a description of the electrochemical characteristics of high-temperature aqueous solutions, it at least provides an interesting starting point for their better understanding.

EXPERIMENTAL TECHNIQUES

High temperature and high pressure introduce a number of experimental difficulties to electrochemistry, which must be overcome if meaningful measurements are to be made. Three points are of primary concern:

1. The containment-vessel design required to withstand the pressure and temperature conditions, and the nature of the vessel lining which may be necessary if the solution can be contaminated by dissolution of the vessel walls.

2. The electrical insulator design required to ensure that the electrical leads passing through the autoclave provide data exclusively relevant to the electrode processes under study.

3. The choice of reference electrode. This must at least provide a stable potential reference; preferably, it should allow the measurements obtained to be related to a thermodynamic scale.

Progress in this field can be considered most conveniently by reviewing developments in each of these areas.

Pressure Vessels and Liners

It has not always been found necessary to use an autoclave. Some electrochemical studies have been carried out at high temperatures, but at moderate pressures, using vessels made of glass, plastic, or ceramic materials. Stene[15] made pH measurements at 150°C and 70 psig using a sealed glass apparatus; Titov and Agapov[16] studied the polarization characteristics of tantalum–niobium alloys in 90% sulfuric acid at 250°C and 8 psig, using a cell made of Pyrex. Pyrex vessels can be used under more rigorous conditions; Wanklyn and Britton,[17] for example, studied the corrosion of zirconium alloys in superheated steam at 500°C and 1000 psig, but, in general, conditions as rigorous as these have been more easily accommodated by autoclave techniques.

The principles involved in the design and construction of autoclave systems are well understood and there are many useful and detailed reviews.[18–28] However, autoclaves for high-temperature aqueous electrochemical application require special attention to the problem of avoiding contamination of the electrolyte solution. Autoclaves can be used without internal liners in some circumstances, but since the autoclave material must combine the necessary high-temperature strength with resistance to corrosion, the choice is limited to stainless steels and to the high nickel alloys, such as Nimonic, Hastelloy, and Monel. There must also be concern, when using an unlined autoclave, that corrosion may lead not only to contamination of the electrolyte solution but possibly to failure of the autoclave itself. Even the high nickel alloys are known to be subject to stress corrosion cracking in chloride and in hydroxide solutions.[29,30]

It is usually preferable, therefore, to carry out electrochemical experiments in lined autoclaves. The liner material, metallic, ceramic, or plastic, is selected primarily on the basis of high corrosion resistance and low solubility under the experimental conditions. The precious metals and their alloys give least concern for corrosion in most solutions under investigation, and therefore are commonly used as permanent liners to steel autoclaves. Franck, Savolainen, and Marshall[31] and Fogo, Benson, and Copeland[32] used platinum–iridium-alloy liners. Noyes,[33] Corwin, Bayless, and Owen,[34] and Ellis[35] used platinum, while Whitehead[36] plated the interior of his autoclave with rhodium. However, even the precious metals are not free from corrosion under certain circumstances. Dobson[37] noted that platinum corroded in $0.1M$ hydrochloric acid at 150°C while Reamer, Richter, and Sage[38] reported that platinum and gold–platinum alloys were less resistant to corrosion than pure gold, both in water at 315°C and in steam at 815°C. The metallurgical compatibility of liner metal with autoclave alloy at the required operation temperature must also, of course, be considered.

Glass and ceramics have often been used as autoclave lining materials but they are soluble in many high-temperature aqueous solutions, and provision must be made for replacing them regularly. Contamination due to the solubility of silica, silicates, and glasses is appreciable,[39,40] and this can give rise to misleading results during electrochemical or corrosion studies. Wanklyn and Britton[17] found, for example, that soluble borate species inhibit the corrosion of zirconium. A similar effect due to soluble silicate species has been observed by Videm[41] in the corrosion of aluminum and by Harrison, Jones, and Masterson[42] in the corrosion of iron.

Contamination from ceramic liners can also affect the corrosion behavior of metals in high-temperature aqueous environments. Although alumina and stabilized zirconia were found to be relatively stable in autoclave studies in which the test solution was static, alumina was found to dissolve in flowing distilled water, giving rise to a contamination level of 1.4 ppm in the test solution.[43] Booth[44] confirmed this unsatisfactory feature, and reported the formation of an inhibiting layer of boehmite on corrosion specimens. At the Central Electricity Research Laboratories, it has been noted[45] that the stability of alumina in the presence of high-temperature aqueous solutions improves as its purity is increased. We have also noted[46]

Table 1. Stability of Ceramic Materials in Aqueous Solutions at 300°C (24-h tests)

Solutions	Magnesia-stabilized zirconia	High-purity alumina	Yttria-stabilized zirconia	Thoria	Magnesia
Distilled water	0.42% wt. gain	0.04% wt. gain	disintegrates	0.23% wt. gain	disintegrates
0.1N HCl	0.64% wt. gain	0.1% wt. gain	disintegrates	cracked	disintegrates
13% NaOH	2.22% wt. gain	disintegrates	disintegrates	cracked	disintegrates

that, of a range of ceramic materials exposed at 300°C under static conditions to 0.1N hydrochloric acid, to 13% sodium hydroxide, and to distilled water, magnesia-stabilized zirconia was the most resistant material under all three conditions (Table 1).

PTFE and other fluorocarbon liners have been used with aqueous solutions at temperatures up to about 315°C. Waggener and Tripp[47] found that PTFE gave off unidentified gases between 206°C and 316°C; these did not contaminate their solutions but fogged the optical windows which they used. Moore and Rau[48] reported that PTFE at 205°C and 235°C gave rise, respectively, to 0.05-ppm and 35-ppm fluoride ion in water after a 48-h exposure; lower fluorocarbons gave 33-ppm fluoride ion after 28 h at 205°C and lowered the solution pH from 8.2 to 3.3. Krenz[49] reported a similar effect. Although they found that their solutions picked up little contamination from PTFE in a temperature range of 263–310°C, Tolstaya, Gradusov, and Bogatyveva[50] rejected this material as a liner because the zirconium alloys which they studied were susceptible to enhanced corrosion in the presence of fluoride ions. Lietzke et al.[51] avoided contamination problems by preheating PTFE liners in the autoclaves to 275°C. For electrochemical studies at the Central Electricity Research Laboratories on the corrosion of iron in high-temperature aqueous solutions, Harrison and Jones[46] used a pretreatment for PTFE based on heating it in air to 380°C; this reduced the subsequent fluoride and organic contamination from PTFE liners to low proportions.

The majority of electrochemical experiments carried out so far at high temperatures have been with static solutions. However, stirrers may be inserted through the sealing glands in the autoclave, or a rocking autoclave may be used if agitation of the solution is required.[21] An alternative method, which overcomes the mechanical difficulties of the gland system, is to couple the internal stirrer magnetically to an external drive, through an autoclave wall constructed of nonferromagnetic material. In this way, vertical[52] or horizontal[46] agitation of the solution may be arranged.

Autoclaves can also be designed in which the solution is continually replenished by recirculation. Examples of experimental studies in which this technique was used are described by Draley and Ruther,[53] Troutner,[54] and Bacarella.[55] Draley, Ruther, and Greenberg[56] modified their earlier apparatus to include a totally enclosed canned-rotor pump, working at 300°C and 3000 psig, thereby converting it to a recirculating system. To study the electrochemical behavior of aluminum in pure water, McMillan[57] devised a distilling recirculating autoclave capable of operating to 350°C and 2400 psig. At the Central Electricity Research Laboratories, we have used a somewhat similar system to measure directly the temperature dependence of the conductance of very pure water. The conductance cell was constructed in a platinum-lined autoclave (Fig. 3); an elaborate water purification train was provided, the pure water was flowed continuously through the cell on a once-through basis to minimize contaminant accumulation in the cell, and provision was made for a final distillation in the autoclave, if required.

In certain studies, it is instructive to observe the specimen while it is at working temperature through windows in the autoclave wall. A bibliography covering designs available and summarizing the use of windows in autoclaves has been compiled by Biggers and Chilton[58]; they recommended windows of single-crystal sapphire for use with aqueous solutions up to 250°C and 1000 psig, in view of the low solubility and high strength of this material. Franck[59] has, in fact, used sapphire windows up to 500°C and 52,500 psig.

For studies in which only a single aqueous phase is involved, a duplex autoclave system can be used. In this arrangement the test solution fills a sealed cell within the main autoclave and the two are connected by a pressure-transmitting diaphragm,[60-62] or by a liquid barrier.[63-65] The cell pressure is counterbalanced by using an isotonic solution in the main autoclave, or an applied pressure of inert gas, such as nitrogen or argon.

It should be noted that when the whole autoclave system is heated to the temperature of the experiment, the creep resistance of the metal limits the maximum temperature. Nickel-based alloys are the most satisfactory for high temperature use, but even these are limited to a maximum operating temperature of about 800°C. To work at higher temperatures, an internal heater can be used, so that the autoclave walls are kept at a temperature lower than that at which measurements are being made. In this way,

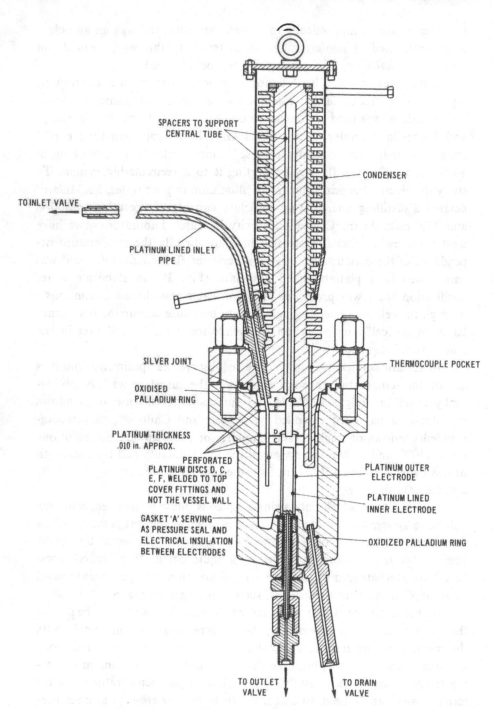

SPACERS TO SUPPORT
CENTRAL TUBE

CONDENSER

TO INLET VALVE

PLATINUM LINED INLET
PIPE

SILVER JOINT

OXIDISED
PALLADIUM RING

THERMOCOUPLE POCKET

PLATINUM THICKNESS
.010 in. APPROX.

PERFORATED
PLATINUM DISCS D, C,
E, F, WELDED TO TOP
COVER FITTINGS AND
NOT THE VESSEL WALL

PLATINUM OUTER
ELECTRODE

PLATINUM LINED
INNER ELECTRODE

GASKET 'A' SERVING
AS PRESSURE SEAL AND
ELECTRICAL INSULATION
BETWEEN ELECTRODES

OXIDIZED PALLADIUM RING

TO OUTLET
VALVE

TO DRAIN
VALVE

Fig. 3. The CERL high-temperature conductivity cell.

Vodar and Johannin[66] made thermal conductivity measurements at pressures up to 15,000 psig and temperatures of 1000°C, while keeping the autoclave walls at room temperature.

Insulation and Sealing of Electrode Leads

There are three methods in common use for sealing insulated electrode leads into autoclave heads or walls:

1. Metallized ceramic seals.
2. Compression seals.
3. Line or surface seals.

Metallized Ceramic Seals

The simplest form of metallized ceramic seal—a sparking plug—was used by Roychoudury and Bonilla[67] at pressures up to 600 psig at room temperature, but a sparking plug has also been used by Corwin, Bayless, and Owen[34] to study the conductivity of aqueous solutions above the critical point. However, the long-term pressure stability of sparking plugs is low; Parbrook[68] reported a leak rate of 15 cm³/min at 1200 psig. Kondrat'ev and Gorbachev[63] devised a more complex metallized seal and used it satisfactorily in conductivity and potential measurements above 340°C and 21,000 psig. It should be noted, also, that power-station boiler water level indicators, working on a differential conductivity principle, and incorporating metallized ceramic seals, have been used successfully at 350°C and 2500 psig.[45] Brenner and Senderoff[69] have used a combined metallized glass compression seal in electroplating studies at temperatures up to 300°C with total pressures (vapor and gas) of 2000 psig. Their seal consisted of glass-covered tungsten wire, which was coated with silver by chemical reduction, then plated with nickel, and fitted into a silver cone. The cone was compressed onto the nickel, making a seal which was pressure tight to 3000 psig at 300°C. These authors found that the glass resisted attack by acid and oxidizing agents, but was attacked by alkalis.

Compression Seals

The second class of insulated electrode seals to be considered—compression seals—rely on forming a pressure-tight connection by compressing the insulating material around the electrode lead. Care has to be taken in selecting a suitable insulating material. Parbrook[68] observed that polythene and PVC break down under pressure at temperatures up to 80°C in environments of water, transformer oil, or gases. Tolstaya et al.[50] found that mica and bakelite were unstable in water at 263–310°C but that PTFE was stable. Ellis[35] found that the conductivity of water at 200°C was higher in cells using quartz insulation than in those insulated with PTFE. He ascribed this difference to the appreciable solubility of quartz in steam. Similar objections have been made to the use of soapstone, mica, and other silica-containing insulating materials, and Harrison, Jones, and Masterson[42] have found that sufficient silica dissolved from soapstone in 1% aqueous solutions of sodium chloride at 250°C inhibits the corrosion of mild steel during anodic polarization.

There are several types of compression seal, some of which are commercially available. Four of these designs warrant a detailed description.

The Aminco seal, which is illustrated in Fig. 4, consists of a double-ended 60° cone through which the electrical lead passes. The seal is compressed by the follower onto the electrical lead, thereby making the insulating seal pressure tight. The most widely used insulating material, for reasons previously discussed, is PTFE. Walters and Eakin[70] used this seal to 10,000 psig and 150°C. Brenner and Senderoff[69] and Bacarella and Sutton[71] reported maximum useful seal temperatures of 225 and 260°C, the tempera-

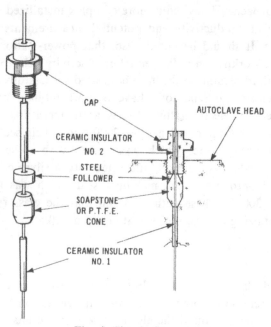

Fig. 4. The Aminco seal.

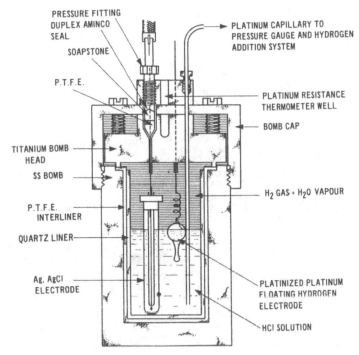

Fig. 5. Autoclave assembly used by Lietzke et al.[51]

ture being limited because of the extrusion of PTFE. Lietzke et al.[51] used a duplex cone of PTFE and soapstone, which is illustrated in Fig. 5, the PTFE being positioned nearer to the vapor phase. This seal was satisfactorily used to 275°C and 850 psig. Harrison and Jones[42] found that by containing the PTFE more closely in the autoclave head, as illustrated in Fig. 6, extrusion is prevented, and PTFE and zirconia-loaded PTFE can be used to 250 and 315°C, respectively. Bacarella and Sutton[71] have carried out polarization experiments at 295°C, using an air-cooling technique which kept the PTFE seal temperature to between 60 and 90°C.

A Conax seal can be used where the space available on the surface of the autoclave head is limited, or where there is an experimental requirement for a large number of electrode leads to be passed into the autoclave. The Conax seal[72] is illustrated in Fig. 7 and is a useful alternative to the Aminco seal. The problems in using PTFE in the Aminco seal have also been encountered[42] in using the Conax seal, and a similar approach to overcome them is being tried.

The Bridgman unsupported area seal has been used successfully both by Bridgman[19] and by Franck et al.[31] for passing insulated electrical leads

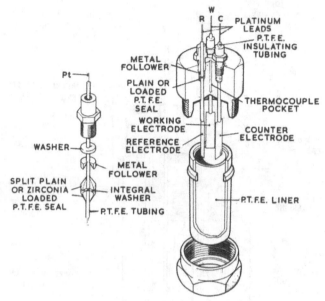

Fig. 6. CERL seal.

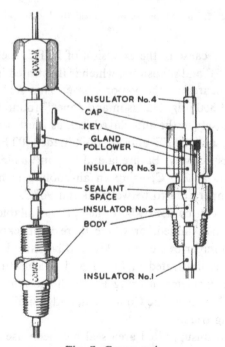

Fig. 7. Conax seal.

into pressure vessels for measurements of conductivity and other studies; these seals can withstand 60,000 psig up to temperatures of 150°C. By replacing the neoprene rubber and bakelite components by PTFE and mica, Draley and Ruther[53] were able to use this type of seal to 315°C and 1500 psig during studies on the corrosion behavior of aluminum alloys in high-temperature water.

Mineral insulated cable has been used by Distèche[73] in studies with glass electrodes at high pressure and ambient temperature. When sealed with Araldite, the cable served satisfactorily for insulated electrode leads at pressures up to 22,500 psig. Simon[74] has used magnesia-insulated, stainless steel-sheathed cables up to 100,000 psig at room temperature and to 60,600 psig at 300°C for studies with hydrocarbon fluids. Magnesia would not be suitable for use with aqueous solutions, for which alternative insulating ceramics, such as magnesia-stabilized zirconia, would be required.

Line Seals

A third class of seals—line or surface seals—in which glass and quartz are used as the insulating material, has been described in the literature.[75-78] Alumina, either in its sintered form as used by Johannin[78] or as synthetic sapphire as used by Fuchs[79] would be more suitable for use in high-temperature aqueous electrochemical work.

Reference Electrodes

Two approaches have been made to the problem of measurement of electrode potential in aqueous solutions at high temperatures and pressures. In one, the reference electrode is incorporated into the autoclave and operates at the same temperature and pressure as the rest of the electrochemical cell. The alternative approach is to maintain the reference electrode at room temperature and to connect it to the autoclave by a cooled electrolytic bridge. The bridge and the reference electrode can either operate at the same pressure as the autoclave electrochemical cell, or the bridge can be used to reduce the autoclave pressure so that the reference electrode operates at atmospheric pressure and temperature.

High-Temperature Reference Electrodes

Primary Electrodes. The *hydrogen electrode*, the primary standard, against which secondary electrodes are calibrated, has been used at room temperature under a wide variety of conditions, for example, alkaline hydroxide up to $4M$[80] and sulfuric acid up to $17.5M$.[81] It can be used with hydrogen pressures up to 60,000 psig[82] and as low as 0.15 psig.[83]

The hydrogen electrode has been used in a number of forms under a variety of conditions at elevated temperatures up to 275°C (Tables 2 and 3).

For studies where the solution vapor pressure does not exceed 1 atm, Pourbaix[84,85] and Bates and Bower[86] used a platinized platinum bubbling hydrogen electrode in concentrated sodium hydroxide at 105°C and $0.1M$ hydrochloric acid at 95°C, respectively.

Giner[87] has developed a platinum/hydrogen reference electrode, in which hydrogen is formed *in situ* on the electrode by cathodic polarization. This electrode has been used for fuel-cell studies in concentrated phosphoric acid up to 200°C.

At higher pressures, Stene[15] used a platinized platinum wire at 150°C with 15 psig partial pressure of hydrogen. Roychoudury and Bonilla[67] and Le Peintre[88] studied the effect of hydrogen partial pressure on a platinized platinum spiral up to 250°C. The most consistent series of experiments involving the use of a platinized platinum electrode, at elevated temperatures (up to 275°C), has been carried out by Leitzke *et al.*[51,89,90] using the platinized platinum electrode system already illustrated in Fig. 5. Harrison, Jones, and Masterson[42] have found that a platinum sheet of 1 cm² area acts as a reversible hydrogen electrode in $0.1M$ sodium hydroxide at 250°C; it shows no hysteresis during micropolarization (Fig. 8).

One objection to the use of the platinum/hydrogen electrode is the solubility of platinum in some high-temperature aqueous solutions. Dobson[37] found that dissolution gave rise to drifting emf measurements in $0.1M$ hydrogen bromide at 150°C, while Watson[91] found that sufficient platinum dissolved in 80% potassium hydroxide at 200°C to increase the activity of nickel fuel-cell electrodes in this solution.

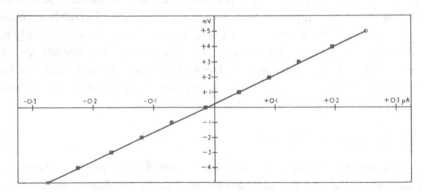

Fig. 8. Micropolarization curve of platinum electrodes in alkaline solution at 250°C.

Table 2. emf Measurements at High Temperatures and Pressures

Reference electrodes	Solution	Temperature range, °C	Application	Reference
Ag/Ag_2SO_4 | Hg/Hg_2SO_4	$0.05\text{-}0.57M$ H_2SO_4	25-250	development of high-temperature reference electrodes	Lietzke and Stoughton[111]
Hydrogen (Pt) | Ag/AgCl	$0.001\text{-}0.1M$ HCl	0-95	standard potential of the Ag/AgCl electrode and thermodynamic properties of dilute hydrochloric acid solutions	Bates and Bower[86]
Ag/AgCl | Hg/Hg_2Cl_2	$0.01\text{-}0.1M$ HCl	25-263	development of high-temperature reference electrodes	Lietzke and Vaughan[95]
Hydrogen (Pt) | Ag/AgCl	$0.001M$ and $0.01689M$ HCl	25-250, hydrogen pressure to 600 lb/in³	effect of temperature and hydrogen pressure on electrode combination for pH measurements	Roychoudury and Bonilla,[97] Lietzke[183]
$(Pt)PbO_2/PbSO_4$ | Ag/Ag_2SO_4	$0.3\text{-}2M$ H_2SO_4 in water saturated with Ag_2SO_4	25-140	activity of H_2SO_4 solutions and effect of Ag_2SO_4	Lietzke and Vaughan[148]
Hydrogen (Pt) | Ag/AgCl	$0.005\text{-}1M$ HCl	25-275	standard potential of the Ag/AgCl electrode	Lietzke et al.[51]
Hydrogen (Pt) | Ag/AgCl	$0.005\text{-}1M$ HCl	25-275	thermodynamic properties of HCl	Towns et al.[106]
Hydrogen (Pt) | Ag/AgBr	$0.005\text{-}0.5M$ HBr	25-200	standard potential of the Ag/AgBr electrode and the mean ionic activity coefficient of HBr	Towns et al.[109]
Hydrogen (Pt) | Ag/AgBr	total ionic strengths 0.01-3.8; ionic strength fractions of HBr 0.25, 0.5, and 0.75	25-150	activity coefficients of HBr and KBr in HBr-KBr mixtures	Lietzke and Stoughton[110]
Deuterium (Pt) | Ag/AgCl	$0.0102\text{-}1.27M$ DCl in D_2O	25-225	thermodynamic properties of DCl	Lietzke and Stoughton[101]
Hydrogen (Pt) | Ag/AgCl	total ionic strength 0.4 and 1, ratio of HCl to NaCl varied	25-175	thermodynamic properties of HCl–NaCl mixtures	Lietzke et al.[90]
Hydrogen (Pt) | Ag/AgI	$0.00315\text{-}0.2359M$ HI	25-200	calibration of Ag/AgI electrode	Körtum and Hausserman[181]

Table 3. pH Measurements at High Temperatures and Pressures

pH Electrodes	Solution	Temperature, °C	Application	Reference
Hydrogen (Pt) | Ag/AgCl or Hg₂Cl₂ (cold)	HCl-KCl, phthalate, phosphate, borate } buffers	150	calibration of pH scale at high temperatures	Stene[15]
Hydrogen (Pt) | Ag/AgCl	0.001 and 0.01689M HCl	250	comparison of measured pH at high temperature with calculated values	Roychoudury and Bonilla[67]
Glass electrode and pressurized calomel	HCl pH 2 and 3, and 25 commercial buffer solutions	160	pH measurements during wood pulping	Ingruber[97,113]
Glass electrode (internal electrode thallium chloride) vs thallium amalgam/thallium chloride	phosphate, acetate, phthalate } buffers	140	measurement of pH-temperature relationship of buffer solutions leading to development of buffer with zero pH temperature relationship	Fricke[88]
Hydrogen diffusion nickel electrodes vs Hg/HgO	0.01-10M KOH	240	measurement of pH of strongly alkaline solutions at high temperatures	Vielstich[93]
(a) Separated hydrogen electrodes	HCl, HCl/KCl mixtures, tartrate, acetate, phosphate and borate buffers	250	calibration of pH-temperature relationship of buffers	Le Peintre[86]
(b) Balanced pressure Pb amalgam glass electrode vs Ag/AgCl	buffer solutions	150	calibration of glass electrode in buffers measured in (a)	
Hydrogen (Pt) vs Ag/AgCl	0.001-1M HCl	275	calibration of pH scale at high temperatures	Greeley[184]
Balanced pressure Pb amalgam glass electrode vs Ag/AgCl	Similar to Le Peintre[86]	150	verification of Nernst's law up to 150°C	Fournie et al.[99]
Hydrogen (Pt) vs glass electrode (thallium ref.)	potassium, tetroxalate, phthalate, borates, acetic acid, benzoic acid, sulfonilic acid, salicylic acid. tartaric acid	150	calibration of pH scale at high temperatures	Kryuchov et al.[150]

Because of the corrosion problem with platinized platinum, other metals have been considered. Dobson[92] has examined the behavior of a palladium electrode which has been cathodically charged with hydrogen before use; Watson[91] and Vielstich[93] have both used sintered nickel electrodes through which hydrogen was bubbled. However, most metallic surfaces, including gold[59] and mercury,[94] have a significantly reduced overvoltage for hydrogen evolution at temperatures above 100°C, and the exchange currents for hydrogen evolution are correspondingly greater. Lietzke and Vaughan[95] and Roychoudury and Bonilla[67] have indicated that even the silver/silver chloride electrode tends to act as a hydrogen electrode at higher temperatures.

The glass electrode responds quantitatively to hydrogen ion over a wide range of concentration[96] and can therefore be considered as a primary reference electrode.

Table 4 summarizes the conditions under which balanced pressure glass electrodes have been used at room temperature at pressures up to 30,000 psig for measurement of the effect of pressure on pH and on the dissociation constants of electrolytes in aqueous solution.

For pH measurements at high temperatures and pressures (Table 3) the main problems with the glass electrode are the attack on the glass membrane by the aqueous solution and the stability of the internal reference electrodes.

Ingruber[97] found that the lifetime of commercial glass electrodes was severely limited at 180–200°C by the first of these factors and Fricke[98], Le Peintre,[88] and Fournie, Le Clerc, and Saint-James[99] had to use high-stability glasses for pH measurements in the range 150–200°C. An interesting development is the use by Dobson at the University of Newcastle of membranes of silica, quartz, fluorspar, or rutile as glass membranes. The effect of dissolution must still, however, be taken into consideration.

For the internal reference electrode, Fricke[98] developed a special thallium amalgam/thallium chloride electrode usable to 140°C, while Le Peintre[88] and Fournie *et al.*[99] used a 3% lead amalgam electrode up to 200°C.

Secondary Electrodes. The secondary electrodes which have been used successfully at high temperatures in aqueous environments can be grouped as follows.

Silver/Silver Compound Electrodes. Because of the low solubility of silver compounds at high temperatures[33,100] and their resistance to hydrolysis, these electrodes have been extensively used.

Silver/Silver Chloride Electrodes. The earliest reference to the use of the silver/silver chloride electrode at elevated temperatures was by Stene[15]

Table 4. pH Measurements at High Pressures and Room Temperature

pH Electrodes	Solution	Pressure range, lb/in²	Application	Reference
Balanced pressure glass and calomel electrode	water	2,000	measurements in dynamic corrosion experiments	Marburger et al.[185]
Balanced pressure glass and Ag/AgCl electrodes	HCl, acetic acid, carbonic acid, acetate buffer, bicarbonate buffer, phosphate buffer, sea water	22,500	effect of pressure on dissociation constants of solutions and measurement of pH of sea water at great depths	Distèche[73]
Balanced pressure glass and Ag/AgCl electrodes	buffer solutions	15,000	effect of pressure on pH as a preliminary to work on pH at high temperature (see Table 3)	Le Peintre[88]
Balanced pressure glass and Ag/AgCl electrodes	formic, phosphoric, acetic, carbonic acids, phosphate and acetate buffer, bicarbonate, sodium bicarbonate, sodium acetate, ammonia	15,000	effect of pressure on ionization constant	Distèche[186]
Balanced pressure glass and Ag/AgCl electrodes	$0.1M$ KCl + 0.001–$0.02M$ KOH	30,000	effect of pressure on ionization of water	Hamann[187]
Balanced pressure glass and silver/silver chloride electrode	acetic, formic, carbonic, phosphoric acid, adenosinetriphosphate, phosphoryloreatine	15,000	effect of pressure on pH and dissociation constants	Distèche[188]

who used a totally electrolytic electrode in conjunction with a hydrogen electrode for pH measurements at 150°C. Bates and Bower[86] measured the standard potential of thermal electrolytic silver/silver chloride electrodes to 95°C, while Lietzke and Vaughan[95] measured the emf up to 260°C of cells containing a silver/silver chloride electrode and a calomel electrode in hydrochloric acid solution. Roychoudury and Bonilla[67] measured the standard emf of the unsaturated electrolytic silver/silver chloride to 250°C. Lietzke, in a series of studies in chloride solutions at high temperature,[51,101] utilized a saturated thermal silver/silver chloride electrode, which was preheated by thermal cycling to 75°C in $0.1N$ hydrochloric acid. Prazak[102] observed that silver/silver chloride electrodes were more stable in saturated sodium chloride at 200°C than in $1N$ sodium chloride and used the former electrode for corrosion potential measurements. Fournie et $al.$[99] found the saturated silver/silver chloride electrode in $0.1M$ potassium chloride to be stable and reproducible for pH measurements at 150°C and 200°C. Booker,[103] in corrosion experiments in oxygenated alkaline solutions, used an electrolytically chloridized silver rod as a reference electrode to 180°C. Above this temperature, the electrode performed satisfactorily only in deoxygenated solutions (Booker, private communication, 1965). Ellis and Hills[94] used a similar electrode for dropping mercury and platinum microelectrode polarographic experiments up to 225°C in deoxygenated solutions, while Kolotyrkin, Bune, and Florianovitch[104,105] used an unspecified type of silver/silver chloride electrode for polarization experiments on iron–chromium alloys in $1N$ sulfuric acid at 200°C.

Lietzke, in recent studies on the thermodynamic properties of mixtures of hydrochloric acid–sodium chloride solutions[90] and hydrochloric acid–barium chloride solutions[106] up to 175°C, used the silver/silver chloride system described by him in 1960. Dobson, using thermal silver/silver chloride electrodes, is measuring the thermodynamic properties of both hydrochloric acid–sodium chloride and sodium chloride–sodium hydroxide mixtures up to 150°C.[92]

Silver/Silver Bromide Electrodes. The use of the silver/silver chloride electrode in aqueous solutions at higher temperatures becomes somewhat limited by the solubility of silver chloride. Silver/silver bromide electrodes are an attractive alternative since the solubility of silver bromide at room temperature and also at 100°C is an order of magnitude lower than that of silver chloride.[33] Following his use of the silver/silver chloride at 150°C, Stene[15] suggested the use of the silver/silver bromide electrode to extend the temperature of pH measurements to 200°C. Harned, Keston, and Donelson[107] used both thermal and thermal electrolytic silver bromide electrodes in 0.2M hydrobromic acid to 60°C. Their results differed from Bates, Robinson, and Hetzer[108] up to 50°C and the latter authors attributed this difference to variation in electrode-preparation technique. Towns et al.[109] and Lietzke and Stoughton[110] found that experimental errors with thermal silver/silver bromide electrodes at 200°C were higher than with thermal silver/silver chloride electrodes and for later thermodynamic studies have returned to the latter electrode.[90,101]

Silver/Silver Sulfate Electrodes. Lietzke and Stoughton[111] found that the silver/silver sulfate electrode was suitable for use in 0.5–0.05M sulfuric acid at temperatures up to 250°C. In a later study, a comparison was made between the silver/silver sulfate and the Pt | PbO_2/$PbSO_4$ electrodes up to 140°C in 0.3–2M sulfuric acid.

Mercury/Mercurous Compound Electrodes. Since both the calomel electrode and mercury/mercurous sulfate electrodes have been extensively used at room temperature,[6] attempts have been made to use them at higher temperatures.

Mercury/Mercurous Chloride Electrodes. Ingruber[112] and Leonard[113] found that the temperature limit for commercial saturated calomel electrodes was 100°C. Lietzke and Vaughan[95] examined this electrode to 263°C in 0.01–1M solutions of hydrochloric acid. The upper limit of temperature at which studies could be made varied with acid concentration, being 80° for 0.1M, 200° for 5M, and 230° for 1M. In 1M potassium chloride, the temperature limit was 70°C. They ascribed the erratic electrode behavior at temperature to the disproportion reaction

$$Hg_2Cl_2 \rightleftharpoons Hg + HgCl_2$$

followed by hydrolysis to HgO.

A similar temperature limitation at 200°C was reported by Prazak[102].

Mercury/Mercurous Sulfate Electrodes. Hydrolysis effects which limit electrode performance have also been observed by Lietzke and Stoughton[111] with the mercury/mercurous sulfate electrodes in sulfuric acid. The temperature limit in 0.05 and 0.2M sulfuric acid was 80°C, whilst in 0.5M solutions the limit was higher at 150°C. Van Ipenburg[114] preferred the mercury/mercurous sulfate electrode in 0.1N potassium sulfate for his experiments on the polarization of iron at temperatures up to 90°C because of its low temperature coefficient of emf.

Metal Electrodes. Platinum is inert in most solutions and for this reason has been used by a number of workers in high-temperature aqueous solutions.

Kondrat'ev and Gorbachev,[63] in experiments at 300°C, used a platinum plate 1 cm² in area in a separate compartment of the cell. Bacarella[55] used a platinum reference electrode in experiments in oxygenated sulfuric acid and found that the Tafel relationship for O_2 reduction on platinum and the exchange current were constant and independent of time, confirming its suitability as a reference electrode for these conditions. In studies of the corrosion of aluminum in high-temperature water, Videm[41] used a quartz-covered platinum wire as a reference electrode. Van Ipenburg (private communication, 1965) found that a platinum wire was not a stable reference electrode. This discrepancy was explained by Dobson[92] who found that platinum/hydrogen electrodes must have an area in excess of 1 cm² to operate satisfactorily.

Metal/Metal Compound Electrodes. The antimony/antimony oxide electrode has been used by Greer[115] for pH measurements at 100°C, and by Myers and Bonilla[116] to 300°C. It behaved irregularly between 100 and 300°C and this was ascribed by the authors to hydrate transformations. Draley and Ruther[53,117] have used the oxidized wall of a stainless steel autoclave as a reference electrode, in research on the electrochemical aspects of corrosion of aluminum alloys in water at 200°C. For corrosion studies in concentrated solutions, Every and Banks found that gold, platinum, and rhodium metal/metal oxide electrodes were reproducible reference electrodes up to 140°C in 100% sulfuric acid.[118] Subsequently, these electrodes, together with mercury/mercurous phosphate electrodes, were tested up to 115°C.[119] Le Peintre et al.[120] have calibrated the lead/lead sulfate electrode against the silver/silver chloride in 0.05M sulfuric acid to 250°C, where the former electrode was reproducible and reversible. Körtum and Hausserman[121] have calibrated the silver/silver iodide electrode vs the hydrogen electrode in hydroiodide up to 200°C.

External Reference Electrodes

Electrolytic bridges have been designed to connect the reference electrode, maintained at room temperature, to autoclave solutions at high temperature and pressure. Two types of design have been adopted, pressure reducing and pressure balancing.

Pressure Reducing Bridge Technique. Examples of electrolytic bridges of this type have been reported by Gerasimov et al.[122–124] who used a cooled birchwood plug in high-temperature studies (Fig. 9). In similar studies, asbestos string encased in PTFE tubing and a cooled cellulose plug in a

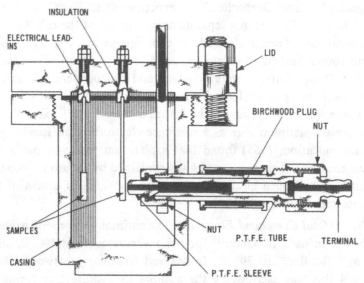

Fig. 9. Autoclave assembly used by Gerasimov.[124]

PTFE quartz capillary (Fig. 10) have been used by Bacarella and Sutton,[71] Hammar,[125] and Wilde.[126]

   Pressure Balanced Electrolytic Bridge. The balanced-pressure electrolytic bridge was first used by Stene[15] in the form of a tube filled with sand, to connect a calomel electrode at room temperature to a platinum/hydrogen

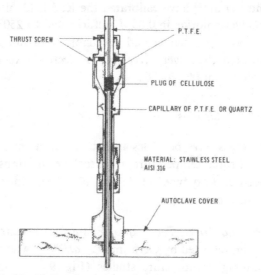

Fig. 10. Liquid junction with cellulose plug. (After Hammar.[125])

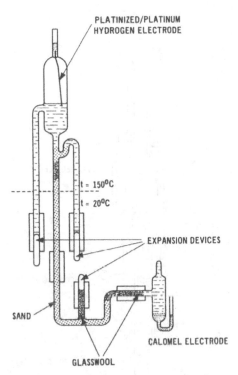

Fig. 11. High-temperature glass cell with external pressure-balanced reference electrode (After Stene.[15])

electrode at 150°C (Fig. 11). For similar pH measurements at high temperatures, Ingruber[97] used a balanced-pressure glass potassium chloride bridge system as shown in Fig. 12. Bacarella and Sutton[71] used a platinum electrode maintained at 60–90°C in an air-cooled extension to their autoclave. Jannsen (private communication, 1965) in a continuation of Van Ipenburg's work,[114] used a mercury/mercurous sulfate electrode in a separate cooled compartment, connected to the main autoclave via a glass-wool plug (Fig. 13). A similar system using a silver/silver chloride electrode has been developed by Postlethwaite.[127]

Care must be taken in the selection of construction materials for both types of bridge. Silica and silicate-containing materials are especially soluble and the wood and cellulose products release acetic, formic, and carbonic acids at elevated temperatures in highly humid conditions.[128]

Thermal Junction Potentials. Difficulties may arise when attempts are made to compare results of electrochemical experiments in widely differing

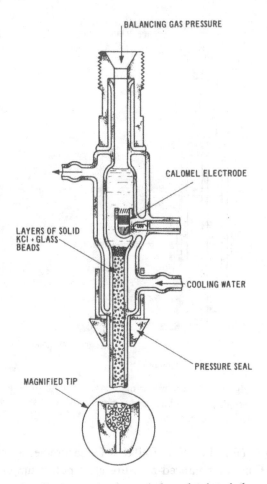

Fig. 12. External pressure-balanced calomel electrode. (After Ingruber.[112])

solutions over a range of temperature, because of the thermal junction potential across the electrolyte bridge when using an external reference electrode. Gerasimov, in descriptions of his autoclave system,[122–124] claims that
Bonnemay[129] has shown that the thermal junction potentials of solutions
are less than 10^{-6} V/°C. However, Bonnemay obtained this low value only
under carefully chosen conditions, selected to obtain the true temperature
coefficients of reference electrodes. de Bethune, Licht, and Swendeman[130]
have estimated the thermal junction potentials in solutions from 0.001 to
$1N$ solutions; salts such as potassium chloride, lithium chloride, and sodium
chloride gave values between $\pm$ 0.03 mV/°C, while acids gave values up to

−0.4 mV/°C, and alkalis up to +0.5 mV/°C. These values are consistent with measurements of the Soret effect, due to thermal diffusion, recent determinations of which Ikeda and Kimura[131] and Butler and Turner[132] indicate that acids and alkalis have higher coefficients than neutral salts, due to the higher mobilities of the H^+ and OH^- ions. Further, from the work of Agar and Breck[133,134] it is clear that the time taken to establish equilibrium due to the Soret effect is substantial. For example, Hübner,[135] in studies of thermal junction potentials between a silver/silver chloride electrode in an autoclave at 200°C in neutral chloride solutions and an external silver/silver chloride electrode (maintained at 25°C), observed a steady emf for the first 20 h followed by a drift of 50 mV in the next 5 days, to the ultimate steady value.

The situation with regard to the thermal junction potential therefore appears to be that values will vary widely with the solutions used and may drift substantially with time.

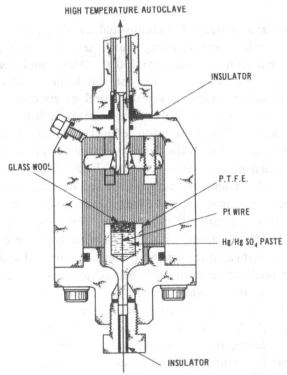

Fig. 13. External pressure-balanced reference electrode.
(After Jannsen.[114])

APPLICATION OF HIGH-TEMPERATURE ELECTROCHEMICAL TECHNIQUES

High-temperature electrochemical techniques have been found useful by workers in several scientific and technological fields, including corrosion studies, measurement of emf, measurement of pH, conductance measurements, electrodeposition and electrolysis, polarography, and fuel cells.

Corrosion Studies

The main interest in electrochemical studies of corrosion above 100°C is in relation to the power industry and covers not only ferrous alloys but other alloys, for example, zirconium and aluminum, which have advantages in the nuclear context. There is also a growing interest in nickel and titanium for use in high-pressure desalination plants.

Nonferrous Alloys

Aluminum and Alloys. Aluminum and its alloys, which have been considered for fuel-element cladding, suffer severe blistering attacks in water in the temperature range 200–350°C. Draley and Ruther[53] have examined the effect of various factors on this form of attack and have found that galvanic coupling to stainless steel or zirconium slowed the onset of blister corrosion. Subsequent experiments indicating that anodic polarization, reducible ions in solution, or nickel plating of the aluminum improved performance, led to the development of alloys containing nickel and iron. Potential measurements at 200°C in distilled water[117] indicated that only the nickel–aluminum eutectic showed an open-circuit potential more cathodic than aluminum, and also gave the lowest potential change under constant-current cathodic polarization. The alloy with silicon gave the greatest change, consistent with the corrosion behavior of the silicon alloy. Videm[41] investigated the effect of electrode potential and found that at pH 6 a potential difference of 0.2 V was sufficient to give complete protection to pure aluminum, while at pH 8, 0.6 V was required. In addition, he found that anodic prepolarization before the corrosion tests protected samples completely. This observation was explained in terms of protective film formation during the anodic polarization, both by Videm and by Greenblatt and Macmillan,[136,137] who examined the anodic and cathodic polarization behavior of aluminum and its alloys at 200–300°C in distilled water.

Similar effects have been observed in $0.1N$ KHCO$_3$ and $0.1N$ KNO$_3$ by Spaepen and Fevery,[62] who have investigated aluminum potential changes with temperature.

Nickel. In studies of the influence of temperature on the critical passivation current of nickel in $1N$ NiSO$_4$, Savchenkov and Uvarov[138] found that between 25 and 160°C there was an exponential increase in passive current, the value of the activation energy being about 10.5 kcal/mole over the range. At the highest temperature the corrosion-process rate was determined by the diffusion of nickel ions into the solution. Postlethwaite[127] has applied potentiokinetic and galvanostatic techniques to the study of passivity breakdown of nickel in $0.001 \rightarrow 5N$ sodium hydroxide, with sodium chloride additions, over the temperature range 25–275°C. The critical Cl$^-$/OH$^-$ ratio required for breakdown increases with OH$^-$ concentration and the induction time for breakdown was dependent on the potential, the Cl$^-$/OH$^-$ ratio and the OH$^-$ concentration. The effect of raising the temperature was to stimulate corrosion above 175°C, but above this passivity was encouraged. The critical potentials also moved to higher values and the induction time and critical chloride-ion level were raised. The breakdown of passivity gave rise to pitting at 25°C and general attack at higher temperatures.

Titanium. In relation to the crevice corrosion of titanium in desalination plants, Griers[139] has investigated the effect of temperature on the anodic and cathodic processes on titanium in $1M$ NaCl up to 150°C. The critical anodic current density increased with temperature at a rate corresponding to an activation energy of 11 kcal/mole. In oxygenated solutions, the cathodic process was the reduction of oxygen and the limiting diffusion current density increased with temperature, due both to greater oxygen diffusion rate and greater oxygen solubility.

Zirconium and Its Alloys. In studies of the corrosion of zirconium fuel-element cladding for the Steam Generating Heavy Water Reactor, Wanklyn and Hopkinson[17] examined, electrochemically, the role of hydrogen in the overall corrosion mechanism. It was found that cathodic polarization in water at 325°C decreased the formation of the protective film on both zirconium and its alloys, the effect being minimum in Zircaloy 2 (containing 1.5% tin, 0.12% iron, 0.1% chromium, and 0.05% nickel) and alloys with iron and platinum. Similar effects were observed if the specimens were subsequently cathodically polarized either at 325°C or at room temperature, after an initial unpolarized exposure at 325°C, and it is suggested that the oxide film had been modified by incorporation of hydrogen. In further studies, Wanklyn and Aldred[140] observed similar damage to Zircaloy 2 in alkaline solutions at 250°C caused by ac polarization.

Bacarella et al.,[55,71] in studies related to the corrosion of zirconium alloy for aqueous homogeneous reactors, have investigated electrochemical processes on zirconium in oxygenated sulfuric acid up to 300°C. In the initial work, instantaneous polarization measurements were used to follow the corrosion kinetics, which were hyperbolic, indicating a logarithmic oxidation law at 167 and 208°C. Between 200 and 300°C,[71] both logarithmic and cubic relationships were obeyed, and it was postulated that the anodic process was a field-dependent ionic transport of oxygen ions, and the cathodic process the reduction of oxygen. At 300°C a steady-state anodic rate was observed, which was independent of potential; it was suggested that this indicated equivalent film formation and dissolution. In later studies Bacarella and Sutton[141] have suggested that the film growth on zirconium between 174 and 284°C can be represented by a hyperbolic sine function relating anodic film growth to field strength.

Ferrous Materials

Iron and Mild Steel. In studies of iron corrosion in high-temperature concentrated alkalis, Pourbaix[84,85] utilized electrochemical techniques to determine passivation and corrosion potentials in the temperature range 80–100°C, where iron showed active–passive behavior, dissolving in the active region as ferroate. This transferred in solution and also on the counter- and reference electrodes to magnetite.

In studies specifically related to the corrosion of boiler tubes in conventional thermal power stations Van Ipenburg[114] at K.E.M.A., Arnhem, examined the effect of anions, such as chloride, sulfate, and phosphate, on the electrochemical reactions of iron in aqueous solutions at 90°C. These studies are at present being extended to 265°C. Similar studies at CERL by Harrison and Masterson[142] have shown critical ratios for phosphate/chloride ion at 250°C, and in more detailed studies at 250 and 300°C. Harrison, Jones, and Masterson[143] have examined the overall polarization characteristic of iron in acid and alkaline solutions. The results indicate a transformation from mixed electrode control in acid to anodic control in alkaline solutions. In alkaline solutions containing chloride ions, a thick nonprotective magnetite develops above the critical breakaway potential, while below this the film is thin and protective. In alkaline solutions free from chloride the anodic process changes with temperature in the range

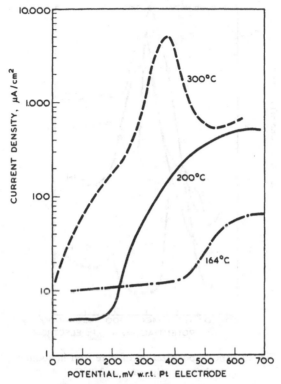

Fig. 14. Effect of temperature on anodic polarization
of mild steel in 3.25N NaOH.

164–300°C, as shown in Fig. 14. Above 200°C there appear to be two stages
in the anodic process.

1. Dissolution of iron to form porous magnetite and soluble iron
   species, probably dihypoferrite $HFeO_2^-$, which forms a secondary
   layer of crystalline magnetite on the electrode at potentials less
   than 200 mV to the rest potential by the reaction

$$3HFeO_2^- + H^+ = Fe_3O_4 + 2H_2O + 2e^-$$

2. Electrocrystallization of outer layer Fe_2O_3 from the dihypoferrite
   by the reaction

$$2HFeO_2^- = Fe_2O_3 + H_2O + 2e^-$$

   by kinetic processes in which the current is a cubic function of
   time at constant potentials up to 1000 mV positive to the rest
   potential indicating concurrent nucleation and three-dimensional

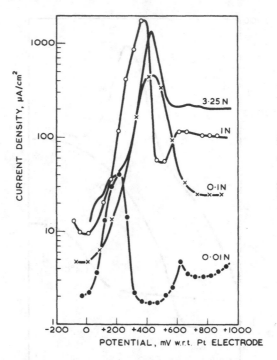

Fig. 15. Anodic polarization of pure iron in sodium
hydroxide solutions at 250°C.

growth processes. The effect of increasing the concentration of
sodium hydroxide, shown in Fig. 15, is to increase the height of
the current peak (due to the formation of α-Fe_2O_3), and it can
be shown that there is a relationship between the logarithms of
the current peaks and the solubility of iron in the varying concentrations of sodium hydroxide between 0.01 and 3.25N.

In related studies in KOH and LiOH (Fig. 16) it was shown that the
first stage, i.e., formation of two-layer magnetite films, was similar in the
three alkalis, but in the second stage α-Fe_2O_3 was only formed in KOH,
the lithium ferrite formed in LiOH decreasing the current density at high
potentials. In NH_4OH, due to the low solubility of iron, only solid-state
oxidation of iron to magnetite and hematite is observed.

Wilde[144] has applied linear polarization techniques to the study of the
corrosion of both mild and stainless steels in high-purity deoxygenated
water at 289°C, and found that a 99.6% correspondence was achieved
between electrochemical corrosion data and conventional weight-loss measurements. In a subsequent study[145] the corrosion of mild steel in water at

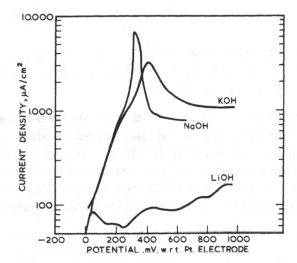

Fig. 16. Anodic polarization of mild steel in 3.25N NaOH, KOH, and LiOH at 300°C — not stirred.

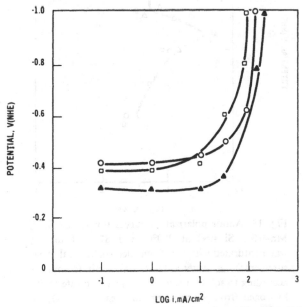

Fig. 17. Cathode polarization curves for 1% Cr–0.6% Mn–0.3% Si steel at 300°C and 87 atm in: ▲ distilled water saturated with air; ○ distilled water with addition of hydrazine; □ distilled water saturated with hydrogen. (After Gerasimov.[122])

289°C containing hydrogen, oxygen, and ammonia has been investigated electrochemically.

In relation to the corrosion of iron in high-temperature solutions, pH–potential diagrams have been constructed by Mason and Harrison[146] up to 300°C, and Lewis[147] up to 350°C.

Alloy and Stainless Steels. Electrochemical studies on alloy and stainless steels in relation to fuel-element cladding, reactor shells, and piping, have been concerned mainly with the effects of oxygen and chloride ion on the corrosion mechanism in high-temperature aqueous solutions.

Gerasimov,[122] in studies of 1% Cr–0.6% Mn–0.3% Si steel in distilled water at 300°C, found that oxygen affected the cathodic reaction (Fig. 17),

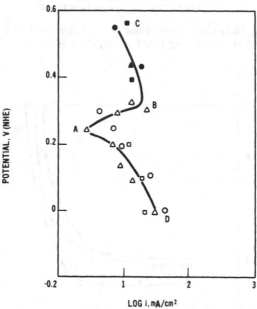

Fig. 18. Anode polarization curves for 1% Cr–0.6% Mn–0.3% Si steel at 300°C and 87 atm in: △ water saturated with air (from electrochemical measurements); ▲ water saturated with air (computed according to solution rate); □ water saturated with hydrogen (from electrochemical measurements); ■ water saturated with hydrogen (computed according to solution rate); ○ water with addition of hydrazine (from electrochemical measurements); ● water with addition of hydrazine (computed according to solution rate). (After Gerasimov.[122])

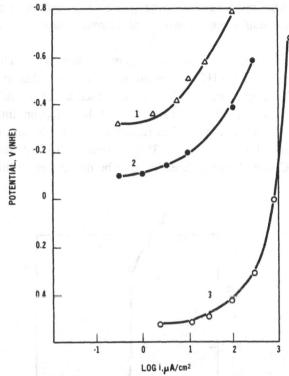

Fig. 19. Cathode polarization curves for 18% Cr–8% Ni–0.5% Ti steel at 300°C and 87 atm in distilled water and saturated with: 1) hydrogen (oxygen concentration, 0.05–0.1 mg/liter); 2) air (oxygen concentration, 8–10 mg/liter); and 3) oxygen at a partial pressure of 10 atm (oxygen concentration 400–430 mg/liter). (After Gerasimov.[122])

but had little effect on the anodic reaction (Fig. 18). Hydrazine, which is used as an addition to feedlines, had little effect on either electrode reaction. In further studies on 18% Cr–8% Ni–0.5% Ti stainless steel, oxygen again had an effect on cathodic reactions (Fig. 19), and little effect on anodic reactions (Fig. 20). The effect on the overall reaction, measured by the rest potential is significant, as shown in Fig. 21, where a positive shift of 700 mV is observed on the addition of oxygen to distilled water at 300°C. The effect of oxygen on the rest potential of iron–chromium alloys in $0.1N$ H_2SO_4 at temperatures up to 200°C has been studied by Kolotyrkin et al.[104] It can be seen from Fig. 22 that while at room temperature a partial pressure of 50 atm of oxygen does not significantly affect the passivation potential for alloys containing less than 13% Cr, at 200°C a significant effect is shown with 4% Cr. This must be due to oxygen and its influence on cathodic processes, since equivalent pressures of nitrogen cause no shift in potential.

Kolotyrkin *et al.* also showed that temperature had an effect on corrosion processes in 0.1*N* HCl, since alloys which corroded at 20°C were passive at 200°C; Spaepen and Fevery[62] observed a similar effect between 25 and 152°C with 18% Cr–8% Ni steel in 0.1*N* KHCO₃ containing 7.5-g/liter KCl. In contrast, Hübner[135] has found a decrease in pitting potential between 150 and 200°C on 20% Cr–35% Ni steel in distilled water containing 200-ppm Cl ion. These differences may be due either to the difference

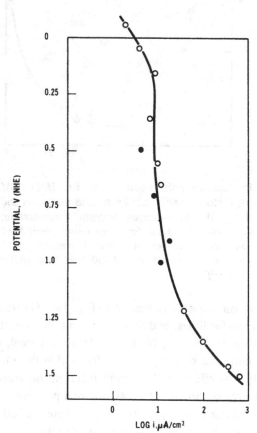

Fig. 20. Anode polarization curves for 18% Cr–8% Ni–0.5% Ti steel at 300°C and 87 atm in distilled water and saturated with: ○ air; ● oxygen at a partial pressure of 10 atm. (After Gerasimov.[122])

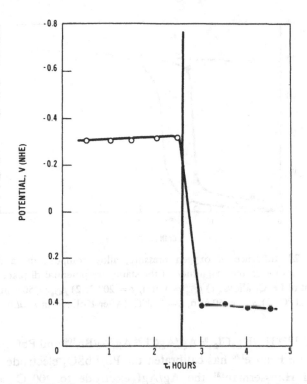

Fig. 21. Time variation of the stationary potential of 18% Cr–8% Ni–0.5% Ti steel at 300°C and 87 atm in distilled water and saturated with: ○ hydrogen; ● oxygen at a partial pressure of 10 atm. The moment of saturation of the water with oxygen is noted by a vertical line. (After Gerasimov.[122])

in alloy or the difference in chloride levels; the latter have a significant effect on the critical pitting potential of 18% Cr–8% Ni–0.5% Ti at 300°C, as shown in Fig. 23.

Measurement of emf

Considerable effort has been devoted to the measurement of the emf of cells at higher temperatures and pressures; this has contributed both to the development of reference electrodes for high temperature use, and to the derivation of thermodynamic data for the systems investigated. A list of studies is given in Table 2.

Notable among them is the extensive series of reference-electrode evaluations by Lietzke and his coworkers covering Ag/Ag_2SO_4 and

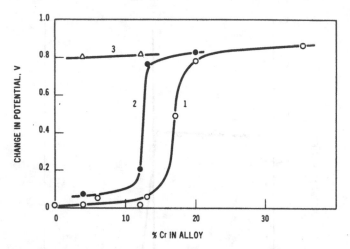

Fig. 22. Influence of oxygen pressure, alloy composition, and temperature on the magnitude of the stationary-potential displacement of Fe–Cr alloys: 1) $p_{O_2} = 1$ atm, $t = 20°C$; 2) $p_{O_2} = 50$ atm, $t = 20°C$; 3) $p_{O_2} = 50$ atm, $t = 200°C$. (After Kolotyrkin et al.[105])

Hg/Hg_2SO_4,[111,148] Hg/Hg_2Cl_2,[95] $Ag/AgCl$,[95] $Ag/AgBr$,[109] and $PbO_2/PbSO_4$.[148] In addition, Le Peintre[120] has calibrated the $Pb/PbSO_4$ electrode to 250°C, Körtum and Hausserman[121] the Ag/AgI electrode to 200°C, and Bates and Bower[86] the $Ag/AgCl$ electrode to 95°C.

The latter authors used the $Ag/AgCl$ electrode to derive the mean activity coefficient, the relative partial molal heat content, and the relative partial molal heat capacity of HCl. Lietzke[89] derived these functions, and the osmotic coefficient of HCl, up to 275°C, and reported that the extended Debye–Hückel equation represented the data as well at 275°C as at room temperature. Towns et al.[109] derived the mean ionic activity coefficients of HBr and the activity coefficient of HBr in HBr–KBr mixtures up to 150°C. At constant temperature and ionic strength, the logarithm of the activity coefficient of HBr in these mixtures varied linearly with the molality of KBr; the activity coefficient of KBr in these solutions varied less with changing ionic strength and temperature than the activity coefficient of HBr did in the same solutions. Körtum and Hausserman[121] have also derived the activity coefficient of HBr to 200°C. Thermodynamic properties of DCl solutions in D_2O, as well as the standard potential of $Ag/AgCl$ electrodes in these solutions, have been derived.[101] The standard potential was found to be lower in the dueterated system than in the corresponding protonated systems; the difference became greater the higher the temperature. Activity coefficient behavior in the two systems was consistent with

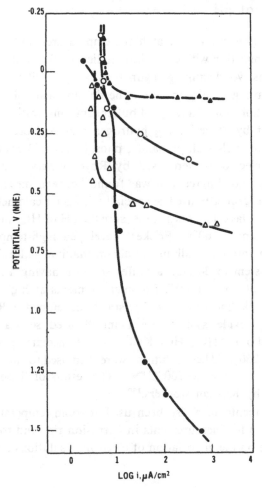

Fig. 23. Anode polarization curves for 18% Cr–8% Ni–0.5 Ti steel at 300°C and 87 atm. ▲ 0.1N sodium chloride; ○ 0.01N sodium chloride; △ 0.001N sodium chloride; ● distilled water. (After Gerasimov.[122])

a lower value of the dielectric constant of D_2O at all temperatures, going through a minimum at 100°C. The activity coefficients of HCl in HCl–NaCl mixtures have also been studied up to 175°C.[90] At constant temperatures and ionic strength, the logarithm of the activity coefficient of HCl in the mixture varied linearly with the molality of NaCl. There is a useful summary[149] of much of Lietzke's work, covering the period up to 1964; the monograph by Ives and Janz[6] provides a more general review.

Measurement of pH

Many measurements of pH at high temperatures and pressures have been made in connection with corrosion studies, or the control of technical processes such as wood pulping, hydrolysis, and esterification (Table 3). These studies have generally involved the development of reference electrodes and the calibration of a range of buffer solutions. Initial measurements were carried out by Stene[15] using hydrogen and Ag/AgCl electrodes to 150°C to calibrate HCl–KCl, phthalate, phosphate, and borate buffers. The same reference electrodes were used by Roychoudury and Bonilla[67] to 250°C in KCl and good agreement was found between measured and calculated values. Ingruber[112] used a specially designed calomel electrode in combination with glass electrodes to measure the pH of HCl and commercial buffer solutions up to 160°C. Fricke[98] developed high-temperature glass electrodes, using internal thallium amalgam/thallium chloride electrodes, and used this system to develop a buffer solution having a zero pH temperature coefficient up to 140°C. For measurements at high temperatures in concentrated alkaline solutions, Vielstich[93] developed a Raney nickel/mercury/mercury oxide system. Le Peintre[88] used separated hydrogen electrodes to calibrate HCl, HCl–KCl, tartrate, acetate, phosphate, and borate buffers to 250°C. These solutions were then used to calibrate specially developed glass electrodes to 200°C.[88,100] The results of these studies have been confirmed by Russian workers.[150]

pH measurements have also been used at room temperature and high pressure (Table 4) for measurements in corrosion rigs and to examine the effect of pressure on the dissociation of electrolyte solutions and the ionization of water.

Conductance Measurements

Franck[59,151] has made considerable progress in high-temperature conductance measurements, going well into the supercritical region, and he, Ellis and Fyfe,[152] and Hills and Ovenden[153] have reviewed the field extensively. Hamann[154,155] has reviewed the effect of pressure on electrolyte dissociation, and investigated the enhanced conductance of water and D_2O under shock-compression conditions. Pearson, Copeland, and Benson,[156] Corwin, Bayless, and Owen,[157] Quist and Marshall,[158–160] and Maksimova and Yushkevich[161] are prominent among contributors to measurement of electrolyte-solution conductance at elevated temperatures, and to the interpretation of electrolyte dissociation behavior.

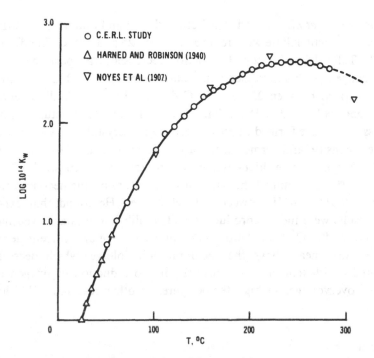

Fig. 24. The variation of log K_w with temperature.

The conductivity of water, of varying degrees of purity at 25°C, has been measured over the temperature range 25–300°C by Wright *et al.*,[162] Wilde,[163] and workers at CERL,[164,165] whose results are compared in Fig. 24 with earlier studies.[33,166] The conductivity maximum observed between 240 and 260°C has been discussed by Briere[167] on the basis of a theoretical equation, derived from the Born equation, which relates the dissociation constant and the dielectric constant of water over a range of temperature.

Electrodeposition and Electrolysis

Attempts have been made to increase the efficiency of electrodeposition of metals at temperatures up to 100°C; and to examine the possibility of depositing those metals which have not previously been deposited from aqueous solutions. Brenner and Senderoff[69] have examined the electrodeposition of a number of metals at 300°C. No satisfactory deposits were obtained under these conditions, due to hydrolysis and lowered solubility in the case of copper, nickel, zirconium, and titanium; reduction to lower valency states in the case of chromium and vanadium; and deposition of oxides of molybdenum and tungsten.

Vagramyan *et al.*[168] found that both cathodic and anodic overvoltages on cobalt in cobalt sulfate were reduced between 25 and 175°C. In addition, while the Tafel slopes for cobalt reduction decreased, transport coefficients and exchange currents increased with temperature up to 150°C. Activation energies derived between 25 and 150°C decreased with cathodic overvoltage.[169] Deposits formed at 150°C had a hexagonal crystal structure, a lower hardness than those formed at 25°C, decreased in crystal size with increase in current density, and formed with current efficiencies of 100%.[170] Similar effects were observed on high-temperature nickel electrodeposits.[171,172]

Franck[59] has examined the effect of temperature on the decomposition voltage of $0.01N$ NaOH between gold electrodes. He found that experimental results were more reproducible and equilibrium was achieved more rapidly above 425°C. The relationship between voltage and current density at 510°C was linear above the decomposition voltage, which decreased substantially with temperature, probably due to reduction of oxygen and hydrogen overvoltages at high temperature, as other researchers[51,94] have reported.

Polarography

Ellis and Hills[94] have examined the use of polarography in solutions at temperatures up to 225°C using both dropping mercury and platinum microelectrodes. The limiting diffusion currents for the reduction of uni- and divalent ions at a dropping mercury electrode increase by a factor of three to four in the temperature range 20–225°C. This is increased to 20–30 for reduction on platinum microelectrodes. The potential range of polarographic analysis in neutral and acid solutions was found to be limited by the rapidly decreasing overpotential for hydrogen evolution on mercury electrodes.

Conning, in studies at CERL, has used the dropping mercury electrode to 200°C in studies of double-layer capacities and reduction rates of cadmium and zinc ions in amalgams, using a single-pulse galvanostatic technique. Double-layer capacities as a function of potential for the mercury–hydrochloric acid interface have been measured in $0.1N$ HCl up to 200°C and in $1N$ HCl up to 100°C. The "capacity hump" observed at 20°C progressively disappears as the temperature is raised, and there is a small increase in the minimum value of capacity at potentials cathodic to the electrocapillary maximum; the increase is large between 20 and 50°C and very small thereafter to 200°C. Reduction rates of cadmium ion on cadmium amalgam in $0.1N$ HCl and H_2SO_4 were measured up to 200°C, and attempts were made to measure reduction rates of zinc ion in zinc amalgam in $0.1N$ HCl. Results could be obtained up to 100°C, but above this hydrogen over-voltage decreased sufficiently for hydrogen evolution to occur at the equilibrium potential of the zinc/zinc amalgam couple.

Fuel Cells

In order to increase the output of fuel cells based on aqueous electrolyte solutions, it has been found necessary to increase the temperature of operation. A number of fuel-cell systems using concentrated aqueous solutions have been studied, including:

1. Propane or hydrogen–oxygen systems in 105% phosphoric acid up to 200°C using catalyzed carbon electrodes.[173]
2. Hydrogen–oxygen systems in concentrated KOH at 200°C using hydrogen diffusion electrodes.[174]
3. Hydrogen or ethylene–oxygen systems in 50% KOH using nickel boride cathodes.[175]
4. Hydrocarbon–oxygen systems in cesium and rubidium carbonate and bicarbonate at 200°C using platinum black electrodes.[176]

By operating the fuel cell under pressure, Bacon et al.[177] were able to raise both the temperature of operation and efficiency, using 45% KOH as electrolyte and hydrogen or coal gas as fuel gas. The effect of pressure on the efficiencies of hydrogen–oxygen fuel cells based on potassium hydroxide, over a range of temperatures, has been investigated by Watson and Pearce[178] and Hartner et al.[179] Heath and Warsham[180] and Thacker and Bump[181] have examined the effect of hydrocarbon gases, which are less expensive than hydrogen, in a pressurized Bacon-type cell using KOH as electrolyte. Goloukin et al.[182] have examined in detail the processes taking place on the oxygen electrode in the Bacon cell. The electrode activity increased with pressure and temperature on both oxidized and lithium-dosed cathodes.

The Pratt and Witney modification of the Bacon cell consists essentially of increasing the concentration of KOH from 45% to 80%, which lowers the pressure of operation to 30 psig at a cell temperature of 250°C.

ACKNOWLEDGMENTS

The authors are grateful to their colleagues B. Case and G. Bignold for useful discussions, and the Central Electricity Generating Board for permission to publish this paper.

REFERENCES

1. H. S. Harned and B. B. Owen, *The Physical Chemistry of Electrolyte Solutions*, Reinhold, New York (1958).
2. D. A. MacInnes, *The Principles of Electrochemistry*, Dover Publications Inc., New York (1961), pp. 144, 201.
3. L. Meites, *Handbook of Analytical Chemistry*, McGraw-Hill, New York (1963).
4. A. Brenner, *Electrodeposition of Alloys: Principles and Practice*, Vol. 1, Academic Press, New York (1963), p. 139.
5. J. E. B. Randles and K. S. Whiteley, *Trans. Farad. Soc.* **52**, 1509 (1956).
6. D. G. Ives and G. J. Janz, *Reference Electrodes*, Academic Press, New York (1961).
7. M. H. Lietzke and R. W. Stoughton, *J. Chem. Ed.* **39**, 230 (1962).
8. A. S. Quist and W. L. Marshall, *J. Phys. Chem.* **69** (9), 3165 (1965).
9. H. S. Harned and B. B. Owen, Ref. 1, p. 331.
10. T. H. Gronwell, V. K. La Mer, and K. Sandved, *Z. Physik* **29**, 358 (1929).
11. D. Feakins and C. M. French, *J. Chem. Soc.* 2581 (June, 1957).
12. H. S. Harned and B. B. Owen, Ref. 1, p. 338.
13. F. Franks and D. G. Ives, *Quart. Revs.* **20** (1), (1966).
14. R. S. Greeley, Dissertation for Degree of Ph. D., University of Tennessee (May, 1959).
15. S. Stene, *Rec. Trav. Chim. Pays-Bas* **49**, 1133 (1930).
16. V. A. Titov and G. I. Agapov, *Zavod. Lab.* **26** (7), 839 (1960).
17. J. N. Wanklyn and C. F. Britton, *J. Nuclear Mats.* **5**, 326 (1962); J. N. Wanklyn and B. E. Hopkinson, *J. Appl. Chem.* **8**, 496 (1958).
18. D. M. Newitt, *The Design of High Pressure Plant and the Properties of Fluids at High Pressures*, Oxford University Press, Oxford (1940).
19. P. W. Bridgman, *The Physics of High Pressure*, Bell & Sons (1949).
20. B. F. Dodge, *High Pressure Technique, Perry's Chemical Engineers' Handbook*, 3rd Ed., McGraw-Hill, New York (1950).
21. E. W. Comings, *High Pressure Technology*, McGraw-Hill, New York (1956).
22. S. D. Hamann, *Physico-Chemical Effects of Pressure*, Butterworths, London (1957).
23. K. E. Bett and D. M. Newitt, The Design of High Pressure Vessels, *Chemical Engineering Practice*, Vol. 5, Butterworths, London (1958).
24. H. Tongue, *The Design and Construction of High Pressure Chemical Plant*, 2nd Ed., Chapman & Hall, (1959).
25. C. G. Brownell and E. H. Young, *Process Equipment Design–Vessel Design*, John Wiley and Sons, New York (1959).
26. W. R. D. Manning, High Pressure Engineering, *Bulleid Memorial Lectures*, Vol. 11, University of Nottingham (1963).
27. J. F. Harvey, *Pressure Vessel Design*, Van Nostrand, New York (1963).
28. R. S. Bradley, *High Pressure Physics and Chemistry*, Vols. 1 and 2, Academic Press, New York (1963).
29. W. L. Williams, *Corrosion* **13**, 539 (1957).
30. T. E. Evans, *Australian Corrosion Engineering* 25 (1960).
31. E. U. Franck, J. E. Savolainen, and W. L. Marshall, *Rev. Sci. Instr.* **33**, 115 (1962).
32. J. K. Fogo, S. W. Benson, and C. S. Copeland, *J. Chem. Phys.* **22**, 212 (1954).
33. A. A. Noyes, Publication 63, Carnegie Institute, Washington, U.S.A. (1907).
34. J. F. Corwin, R. G. Bayless, and G. E. Owen, *J. Phys. Chem.* **64**, 641 (1960).

35. A. J. Ellis, *J. Chem. Soc.* 2299, 4300 (1963).
36. G. Whitehead, *J. Appl. Chem.* **11**, 136 (1961).
37. J. B. Dobson, private communication (1964).
38. H. H. Reamer, G. N. Richter, and B. H. Sage, ASME Paper 63-WA-255, Philadelphia (1963).
39. M. A. Styrikovitch and K. Ya. Katkovskaya, *Teploenergetika* **2** (5) (1955).
40. G. C. Kennedy, *J. Econ. Geology* **45**, 629 (1960).
41. K. Videm, 2nd Int. Conf. on the Peaceful Uses of Atomic Energy, Vol. 5, United Nations, Geneva (1958), p. 121.
42. J. T. Harrison, D. de G. Jones, and H. G. Masterson, *A Simple Autoclave for Electrochemical Experiments* (to be published).
43. M. W. Boothroyd and D. H. Lee, A.E.R.E. Document HL61/1232 (1961).
44. G. C. Booth, Symposium on High Temperature Aqueous Electrochemistry, National Chemical Laboratory/National Physical Laboratory (1964).
45. P. E. Crosse, private communication.
46. J. T. Harrison and D. de G. Jones, unpublished (1967).
47. W. C. Waggener and A. M. Tripp, *Rev. Sci. Instr.* **30**, 677 (1959).
48. R. E. Moore and E. Rau, USAEC Report WAPD-BT-22 (1961), pp. 71–77.
49. F. H. Krenz, ASTM Special Technical Publication 368 (1963), p. 32.
50. M. A. Tolstaya, G. N. Gradusov, and S. V. Bogatyveva, *Corrosion of Reactor Materials–A Collection of Articles*, AEC Trans. 5219 (1962), p. 280.
51. M. H. Lietzke, R. S. Greeley, W. T. Smith, and R. W. Stoughton, *J. Phys. Chem.* **64**, 652 (1960).
52. R. K. Laird, A. G. Morrell, and L. Seed, *Disc. Faraday Soc.* **22**, 126 (1956).
53. J. E. Draley and W. E. Ruther, *Corrosion* **12**, 481t (1956).
54. V. H. Troutner, Hanford Atomic Products Operation Report HW 64111, TID 4500, USAEC (1960).
55. A. L. Bacarella, *J. Electrochem. Soc.* **108**, 331 (1961).
56. J. E. Draley, W. E. Ruther, and S. Greenberg, *Corrosion* **14**, 191t (1958).
57. A. F. McMillan, *Chemistry and Industry* 1279 (1959).
58. R. E. Biggers and J. M. Chilton, Review and Bibliography on the Design and Use of Windows for Optical Measurements at Elevated Temperatures and Pressures 1881–1959, USAEC Oak Ridge National Laboratory Report 2738 (1959).
59. E. U. Franck, CERL Symposium on High Temperature Aqueous Electrochemistry (1963).
60. J. K. Fogo, C. S. Copeland, and S. W. Benson, *Rev. Sci. Instr.* **22**, 765 (1951).
61. P. J. Ovenden, The Pressure and Temperature Coefficient of Electrolytic Conductance in Aqueous Solutions, Ph. D. Thesis, University of Southampton (1965).
62. G. J. Spaepen, M. J. Fevery, and J. Lincter, 9th Meeting of CITCE, Section 6, Brussels; *Corrosion Science* **7**, 405 (1967).
63. V. P. Kondrat'ev and S. V. Gorbachev, *Zh. Fiz. Khim.* **35**, 326 (1961).
64. D. A. Lown, A Study of Aqueous Solutions at High Temperatures, M. Sc. Thesis, University of Newcastle (1961).
65. D. A. Lown and W. F. Wynne-Jones, *J. Sci. Instr.* **44**, 1037 (1967).
66. B. Vodar and P. Johannin, *Ind. and Eng. Chem.* **49**, 2040 (1957).
67. R. N. Roychoudury and C. F. Bonilla, *J. Electrochem. Soc.* **103**, 241 (1956).
68. H. D. Parbrook, *J. Sci. Instr.* **30**, 26 (1953).
69. A. Brenner and S. Senderoff, *J. Electrochem. Soc.* **97**, 361 (1950).

70. R. Walters and B. E. Eakin, *Rev. Sci. Instr.* **28**, 204 (1957).
71. A. L. Bacarella and A. L. Sutton, *J. Electrochem. Soc.* **112**, 546 (1965).
72. M. W. Connell, U.S. Patent 2,625,573 (1953).
73. A. Distèche, *Rev. Sci. Instr.* **30**, 474 (1959).
74. I. Simon, *Rev. Sci. Instr.* **28**, 963 (1957).
75. H. J. Welbergen, *J. Sci. Instr.* **10**, 247 (1933).
76. G. M. J. Mackay, U.S. Patent 1,456,110 (1923).
77. A. Michels and C. Michels, *Proc. Roy. Soc.* **A231**, 409 (1953).
78. P. Johannin, *J. Rech. Centre Nat. Rech. Sci. Lab. Bellevue (Paris)* **26**, 324 (1954).
79. O. Fuchs, *Z. Electrochem* **47**, 101 (1941).
80. H. S. Harned and B. B. Owen, Ref. 1, p. 499.
81. H. S. Harned and B. B. Owen, Ref. 1, p. 574.
82. G. Tammann and E. Jenckel, *Z. anorg in allgem Chem.* **137**, 337 (1928).
83. S. Schuldiner and T. B. Warner, *J. Electrochem. Soc.* **112** (8), 853 (1965).
84. M. Pourbaix, Bulletin Technique Association des Ingenieurs sortés de l'Université de Bruxelles (1946), p. 67.
85. M. Pourbaix, Bulletin Technique Association des Ingenieurs sortés de l'Université de Bruxelles (1947), p. 109.
86. R. G. Bates and V. E. Bower, *J. of Res. of the National Bureau of Standards* **53**, 283 (1954).
87. J. Giner, *J. Electrochem. Soc.* **111** (3), 376 (1964).
88. M. Le Peintre, *Soc. Franc. Elect. Bull* **1**, 584 (1960).
89. M. H. Lietzke, R. S. Greeley, W. T. Smith, and R. W. Stoughton, *J. Phys. Chem.* **64**, 1445 (1960).
90. M. H. Lietzke, H. B. Hupf, and R. W. Stoughton, *J. Phys. Chem.* **69**, 2395 (1965).
91. R. G. H. Watson, private communication.
92. J. B. Dobson, Studies on High Temperature Aqueous Reference Electrodes, to be published.
93. W. Vielstich, *Zeit. Inst.* (6), 154 (1959).
94. A. J. Ellis ànd G. J. Hills, *Proc. S.A.C. Symposium, Nottingham*, Heffer & Sons, Cambridge (1965), pp. 430–445.
95. M. H. Lietzke and J. V. Vaughan, *J. Amer. Chem. Soc.* **77**, 876 (1954).
96. R. G. Bates, in *Reference Electrodes*, D. J. G. Ives and G. J. Janz, eds. (1961).
97. O. V. Ingruber, *Pulp and Paper Magazine, Canada* **55**, 124 (1954).
98. H. K. Fricke, in *Beiträge zur angewandten Glasforschung*, E. Schott, ed., Wissenschaftlicher Verlag, Stuttgart (1959), p. 175.
99. R. Fournie, P. Le Clerc, and M. Saint-James, *Silicates Industriels* **27**, 33 (1962).
100. R. J. Raridon, Measurement of Temperature Coefficients in Aqueous Systems up to 200°C using Packed Column Techniques, Ph. D. Thesis, Vanderbilt University, U.S.A. (1959).
101. M. H. Lietzke and R. W. Stoughton, *J. Phys. Chem.* **68**, 3043 (1964).
102. V. Prazak, *Werk. u. Korr.* 524 (1958).
103. C. J. Booker, Symposium on High Temperature Aqueous Electrochemistry, National Chemical Laboratory/National Physical Laboratory (1964).
104. J. M. Kolotyrkin, N. J. Bune and G. M. Florianovitch, European Symposium on Corrosion Inhibitors, University of Ferrara (1960), p. 493.
105. J. M. Kolotyrkin, N. J. Bune, and G. M. Florianovitch, AEC Translation 5219 (1960), p. 26.

106. M. H. Lietzke and R. W. Stoughton, *J. Phys. Chem.* **70**, 756 (1966).
107. H. S. Harned, A. S. Keston, and J. G. Donelson, *J. Amer. Chem. Soc.* **58**, 989 (1936).
108. R. G. Bates, R. A. Robinson, and H. B. Hetzer, *J. Phys. Chem.* **66**, 1423 (1962).
109. M. B. Towns, R. S. Greeley, M. H. Lietzke, and R. W. Stoughton, *J. Phys. Chem.* **64**, 1861 (1960).
110. M. H. Lietzke and R. W. Stoughton, *J. Phys. Chem.* **67**, 2573 (1963).
111. M. H. Lietzke and R. W. Stoughton, *J. Amer. Chem. Soc.* **75**, 5226 (1953).
112. O. V. Ingruber, *Industrial Chemist* **32**, 513 (1956).
113. J. F. Leonard, ASTM Special Publication 190 (1966), p. 16.
114. K. Van Ipenburg, 3rd European Congress on Corrosion, Brussels, *Electrotechniek* **41**, 445 (1963).
115. W. N. Greer, *Trans. Electrochem. Soc.* **72**, 153 (1937).
116. C. M. Myers and C. F. Bonilla, unpublished.
117. J. E. Draley and W. E. Ruther, Conference on Corrosion of Aluminium in High Temperature, Chalk River, Ontario, Canada (1956).
118. R. L. Every and W. P. Banks, *Electrochemical Technology* **4**, 275.
119. R. L. Every and W. P. Banks, *Corrosion* 153 (1967).
120. M. Le Peintre, *CR Acad. Sci.* **261**, 3389 (1965).
121. G. Körtum and W. Hausserman, *Ber. Bunsenges. Physikchem.* **69** (7), 594 (1965).
122. V. V. Gerasimov, Corrosion of Reactor Materials, a Collection of Articles, AEC Transl. 5219 (1962), p. 13.
123. V. V. Gerasimov and P. A. Akol'zin, *Method of Investigating Corrosion and Electrochemical Processes at High Temperatures and Pressures*, Moscow State Publishing House (1963), Chapt. 2.
124. V. V. Gerasimov, A. I. Gromova, and A. A. Sabinin, *Zavod Lab.* **24**, 1420 (1958).
125. L. Hammar, Progress Report S332, Aktiebolaget Atomenergi, Stockholm, Sweden (1965).
126. B. E. Wilde, *Corrosion*, to be published (1968).
127. J. Postlethwaite, *Electr. Acta* **12**, 333 (1967).
128. L. L. Shreir, *Corrosion*, Vol. 2, Sec. 19, Newnes (1963).
129. M. Bonnemay, 6th Meeting of CITCE, Poitiers, Butterworths, London (1955), p. 68.
130. A. J. de Bethune, T. S. Licht, and N. Swendeman, *J. Electrochem. Soc.* **106**, 616 (1959).
131. T. Ikeda and H. Kimura, *J. Phys. Chem.* **69**, 41 (1965).
132. B. D. Butler and J. C. R. Turner, *J. Phys. Chem.* **69**, 3598 (1965).
133. J. N. Agar and W. G. Breck, *Nature* **175**, 298 (1955).
134. J. N. Agar and W. G. Breck, *Trans. Farad. Soc.* **53**, 167 (1957).
135. W. Hübner, Progress Report S335, Aktiebolaget Atomenergi, Stockholm, Sweden (1965).
136. J. H. Greenblatt and A. F. Macmillan, First International Congress on Metallic Corrosion (1961), p. 429.
137. J. H. Greenblatt and A. F. Macmillan, *Corrosion* **19**, 146t (1963).
138. G. F. Savchenkov and L. A. Uvarov, *Protection of Metals* **1**, 569 (1965).
139. J. C. Griers, *Corrosion* **24**, 96 (1968).
140. J. N. Wanklyn and R. Aldred, *J. Electrochem. Soc.* **106**, 529 (1959).
141. A. L. Bacarella and A. L. Sutton, *Electrochemical Technology* **4**, 117 (1966).
142. J. T. Harrison and H. G. Masterson, 15th Meeting of CITCE, London (1964).

143. J. T. Harrison, D. de G. Jones, and H. G. Masterson, 16th Meeting of CITCE, Brussels (1965).

144. B. E. Wilde, *Corrosion* **23**, 379 (1967).

145. B. E. Wilde, *Corrosion* **24**, 338 (1968).

146. C. J. Mason and J. T. Harrison, CERL Laboratory Note No. RD/L/N71/66, Electrode Potential/pH Diagrams for the Iron–Water System at Elevated Temperatures (1966).

147. D. Lewis, pH–Potential Diagrams of the Iron–Water System up to 350°C, AB Atomenergie, Stockholm, Sweden (1966).

148. M. H. Lietzke and J. V. Vaughan, *J. Amer. Chem. Soc.* **79**, 4266 (1957).

149. M. H. Lietzke and R. W. Stoughton, in *Encyclopaedia of Electrochemistry*, C. A. Hampel, ed., Reinhold, New York (1964), pp. 505–511.

150. P. A. Kryuchov, V. D. Perkave, L. I. Stavortina, and B. S. Smolyakov, *Izvestia Akad. Nauk SSR Ser. Khim. Nauk* No. 7 (2) (1966), p. 29.

151. W. von Holzapfel and E. U. Franck, *Berichte der Bunsengesellschaft* **70**, 1105 (1966).

152. A. J. Ellis and W. S. Fyfe, *Reviews of Pure and Applied Chemistry (Australia)*, **7**, 261 (1957).

153. G. J. Hills and P. J. Ovenden, in *Advances in Electrochemistry and Electrochemical Engineering*, Vol. 4, Electrochemistry at High Pressures, P. Delahay, ed. (1966), p. 185.

154. S. D. Hamann, Annual Review of Physical Chemistry (1964).

155. S. D. Hamann and M. Linton, *Trans. Farad. Soc.* **62**, 2234 (1966).

156. D. Pearson, C. S. Copeland, S. W. Benson, *J. Amer. Chem. Soc.* **85**, 1044 (1963).

157. J. F. Corwin, R. G. Bayless, G. E. Owen, *J. Chem. Soc.* **64**, 641 (1960).

158. A. S. Quist, E. U. Franck, H. R. Jolley, and W. L. Marshall, *J. Phys. Chem.* **67**, 2453 (1963).

159. A. S. Quist, W. L. Marshall, and H. R. Jolley, *J. Phys. Chem.* **69**, 2726 (1965).

160. A. S. Quist and W. L. Marshall, *J. Phys. Chem.* **72** (9), 3122 (1968).

161. I. N. Maksimova and V. F. Yushkevich, *Zh. Fiz. Khim.* **37**, 903 (1963).

162. J. M. Wright, T. W. Lindsay, Jr., and T. R. Druga, WAPD-TM-204, USAEC (1961).

163. B. E. Wilde, *Electr. Acta* **12**, 737 (1967).

164. A. D. Brewer and J. E. C. Hutchins, *Nature* **210**, 1257 (1966).

165. G. J. Bignold, A. D. Brewer, and B. Hearn, unpublished (1968).

166. H. S. Harned and R. A. Robinson, *Trans. Farad. Soc.* **36**, 973 (1940).

167. G. Briere, *Electr. Acta* **13**, 119 (1968).

168. A. T. Vagramyan, L. A. Uvarov, M. A. Zhamagortsyan, *Soviet Electrochemistry*, **1** (1), 14 (1965).

169. A. T. Vagramyan, L. A. Uvarov, M. A. Zhamagortsyan, *Élektrokhimiya* **1** (6), 558 (1965).

170. A. T. Vagramyan, L. A. Uvarov, M. A. Zhamagortsyan, and Yu. M. Polukarov, *Élektrokhimiya* **3** (4), 363 (1967).

171. A. T. Vagramyan and L. A. Uvarov, *Trans. Inst. Metal Finishing* **39** (2), 56 (1962).

172. A. T. Vagramyan, L. A. Uvarov, and M. A. Zhamagortsyan, *Izv. Akad. Nauk SSSR Otd. Khim. Nauk* (2), 301 (1964).

173. W. T. Grubb and C. J. Michalske, *Nature* **201**, 287 (1964).

174. H. G. Oswin and S. M. Chodosh, in *Advances in Chemistry*, Vol. 61, Fuel Cell Systems, G. J. Young and H. R. Linden, eds., Amer. Chem. Soc., Washington, D.C. (1965).

175. R. Thacker, *Nature* **206**, 186 (1965).
176. E. J. Cairns and D. I. Macdonald, *Electrochemical Technology* **2**, 65 (1964).
177. F. T. Bacon, A. M. Adams, and R. G. H. Watson, in *Fuel Cells*, Mitchell, ed., Academic Press, New York (1963), p. 129.
178. R. G. H. Watson and L. J. Pearce, *Proc. 4th Int. Symp. on 'Batteries,'* Vol. 2, Pergamon Press, London (1965), p. 349.
179. A. J. Hartner, M. A. Vertes, V. E. Medina, and H. G. Oswin, Ref. 174, p. 141.
180. C. E. Heath and C. H. Warsham, *Fuel Cells*, Vol. 2, G. J. Young, ed., Chapman and Hall, London (1963), p. 182.
181. R. Thacker and D. D. Bump, *Electrochem. Technology* **3**, 9 (1965).
182. Yu. I. Goloukin, N. P. Vasilistov, and N. A. Fedotav, *Élektrokhimiya* **3** (7), 712 (1967).
183. M. H. Lietzke, *J. Amer. Chem. Soc.* **77**, 1344 (1955).
184. R. S. Greeley, *Anal. Chem.* **32**, 1717 (1960).
185. W. H. Marburger, J. Anderson, and G. G. Wigle, Argonne National Laboratory USA Report ANL5298 (1954).
186. A. Distèche, *J. Electrochem. Soc.* **109**, 1084 (1962).
187. S. D. Hamann, *J. Phys. Chem.* **67**, 2233 (1963).
188. A. Distèche and S. Distèche, *J. Electrochem. Soc.* **112**, 350 (1965).

SURFACE- AND ENVIRONMENT-SENSITIVE MECHANICAL BEHAVIOR

R. M. Latanision*

National Bureau of Standards
Washington, D. C.

and A. R. C. Westwood

Research Institute for Advanced Studies
Martin Marietta Corporation
Baltimore, Maryland

The influences of surface structure and environment on the mechanical behavior of crystalline inorganic solids are reviewed and possible mechanisms discussed. In particular, the various roles of such factors as the atomic, electronic, and defect structures of the near-surface regions, the presence of adsorbed surface-active species, alloyed layers, oxide films, gaseous or liquid environments, etc., are considered in connection with the Roscoe, Rebinder, and Joffe effects, liquid-metal embrittlement, complex-ion embrittlement, hydrogen embrittlement, and other phenomena.

INTRODUCTION

Many phenomena thought representative of solid-state behavior really involve only the surface or near-surface regions of the solid, the bulk not being physically involved. Friction, adhesion, corrosion, oxidation, and catalysis, for example, are predominantly related to the surface and its condition. The effects of environments on mechanical properties also belong to this class of phenomena, for it is at the surface that the critical interactions occur and where the greatest degree of control may be exercised

* NAS–NAE–NRC Postdoctoral Research Associate. Present address: RIAS, Martin Marietta Corporation, Baltimore, Maryland.

over them by such features as film formation, adsorption, local ion concentration, and the like.

In many early analyses of such phenomena, the atomistics of the surface were essentially disregarded, and the surface was considered simply as a sharp discontinuity between the solid and its environment. This approach, however, is completely inadequate. Moreover, a much clearer understanding of the atomic and electronic nature of surfaces and their environment-sensitive behavior than presently available will be required before any real understanding can be expected of such complex phenomena as the Rebinder effect, liquid-metal embrittlement, the electrocapillary effect, etc. Fortunately, surface physics is becoming an increasingly active field of research, and significant and useful advances are likely to be made in this area within the next few years.

The purpose of this paper is to review the phenomenology of certain types of surface- and environment-sensitive mechanical behavior. As far as possible, a mechanistic approach will be adopted, but it must be appreciated that much of what will be said in this regard should be considered current opinion, not established fact. The logical starting place for such a review is the surface itself, so a brief and elementary account of the nature and properties of surfaces is given in the second section. In the third section some of the effects of various environments on the mechanical properties of "clean" metals and nonmetals are reviewed, and in the fourth section the influences of solid-surface contaminants are described.

The premature failure of stressed metals in specific electrolytes, i.e., stress corrosion cracking, has been the subject of a recent symposium,[1] and though obviously related, will not be treated here. This review is also restricted to crystalline inorganic solids, thereby excluding vitreous nonmetallic and polymeric materials. However, the mechanical properties of the latter are also extremely sensitive to environmental conditions.[2,3]

THE NATURE OF CRYSTAL SURFACES

In many laboratory experiments, and in most industrial applications, the surface dealt with is a "technical"[4] surface that has been formed by some mechanical process. Even though such a surface may be clean from an engineering standpoint, it is inevitably physically damaged and chemically contaminated with oils, grease, and adsorbed gases. Further, most metals are covered with relatively stable oxides at room temperature, and when an oxidized surface is altered mechanically by cutting, grinding, or polishing, severe plastic deformation of the surface occurs and oxide particles are pushed under the surface. It follows that experiments on technical or engineering surfaces are of little value as far as mechanistic understanding of environment-sensitive mechanical behavior is concerned, because of the contaminated and unreproducible nature of such surfaces.

Clean Surfaces

Over the years, a number of studies have been made of the variations in electronic properties, bonding of surface atoms, and structural rearrangements which must take place to accommodate the "unsaturated" interatomic forces at "clean" surfaces. Reviews of this work may be found in References 5–11 and will not be discussed in detail here. It is appropriate, however, to describe some of the properties of clean surfaces at this time.

Surface Structure

Considerations of the crystallographic anisotropy of surface energy leads to the characterization proposed by Frank[12] and Cabrera[13] of surfaces as (a) singular surfaces, which correspond to a minimum in the Gibbs–Wulff plot of interfacial free energy as a function of orientation, and usually are low-index planes; (b) vicinal surfaces, which are of orientation close to those of singular surfaces, and formed of low-index facets separated by monatomic ledges; and (c) nonsingular or diffuse surfaces, which have interfacial free energies that are essentially independent of orientation. The model that is most often used for the structure of crystal surface stems from the ideas of Kossel,[14] Stranski,[15] and Burton, Cabrera, and Frank.[16] This model, which has been referred to as the terrace-ledge-kink (TLK) model (Fig. 1), has been used extensively in discussions of crystal growth, evaporation, crystal dissolution, and surface diffusion (see, for example, Reference 6). The structure in Fig. 1 is essentially that of a vicinal surface, showing low-index terraces and monatomic-height ledges which are occasionally displaced by monatomic distances at kink sites. With this model, the atomic arrangement of any particular plane can be deduced from three-dimensional crystallographic data.

An atlas of hard-sphere models of crystal surfaces has been compiled by Nicholas.[17] In considering hard-sphere surface models, however, it should be remembered that they represent ideal surfaces, no allowance having been made for the thermal motion of atoms at relatively high temperatures which can produce disordering of the surface, or for the relax-

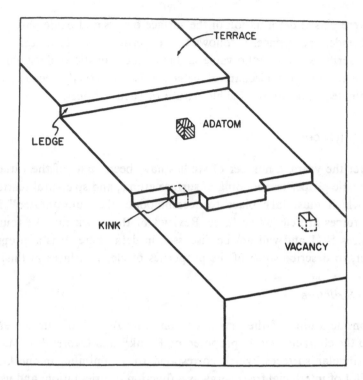

Fig. 1. Schematic of the terrace-ledge-kink (TLK) model for a crystal surface.

ation of surface atoms. Moreover, in some instances, continuation of the perfect lattice up to the interface does not necessarily provide the configuration of lowest energy—the presence of a certain concentration of defects may allow a lower energy.[18]

That atoms at the surface will in fact relax to positions different with respect to each other than those adopted in the bulk has been discussed by a number of workers, particularly Herring,[19] Shuttleworth,[20] and Mullins.[21] The occurrence of this effect, and the origin of surface stresses, is perhaps most simply appreciated from the argument provided by Shuttleworth,[20] as follows: consider a crystal at 0°K, and make the reasonable assumption that the interaction between nonnearest neighbor atoms *is* significant. Next consider an isolated plane of atoms. The equilibrium separation of atoms in this isolated plane will be different from that in the body of the three-dimensional crystal because the number of nonnearest neighbor atoms will be different in the two cases. For the isolated plane to have the same lattice spacing as that of the crystal, it will be necessary to apply external forces to the edges of the plane. These forces may be tensile or compressive, depending on the nature of the material. If the stressed plane is now moved towards the crystal until it becomes the surface plane, the external forces necessary to retain the bulk lattice constant in this plane will be reduced (to about 50%). However, because in equilibrium the distance between the outermost plane of atoms and the next also is different from that in the interior of the crystal, it will again be necessary to modify the tangential forces somewhat to take account of this difference. Similar, but smaller, tangential forces must also be applied to successive planes into the crystal.

On the basis of this argument, the surface stress S for a real crystal is defined as the total force per unit length that must be applied tangentially to the surface so that the surface planes have the same lattice spacing as the underlying crystal.

The parameter S should be distinguished from the reversible work γ required to create a unit area of surface. As Mullins[21] has commented, the most appropriate name for γ would be the specific surface work, but this term is rarely, if ever, used. More frequently γ is called the surface tension or specific surface free energy of the solid. Mullins[21] and Herring[18,19] prefer the former designation; Shuttleworth[20] prefers the latter. Workers in the field of fracture tend to utilize the term "fracture surface energy" when referring to the work done in creating unit area of surface by cleavage. In this text, we shall use the terms surface tension and surface free energy interchangeably.

According to Mullins,[21] the relationship between S and γ is

$$S_{ij} = \delta_{ij}\gamma + (\partial\gamma/\partial\varepsilon_{ij}), \qquad i, j = 1, 2$$

where δ_{ij} is unity if $i = j$ and zero otherwise, and ε is strain. Note that γ is always positive, whereas S may be either positive or negative.

Low-energy electron diffraction (LEED) experiments have revealed that the atoms in the topmost layers of many clean metal surfaces are arranged in the same manner as atoms in parallel planes in the body of the crystal, with only a slight adjustment in interatomic spacing. For nickel, for example, the spacing of atoms normal to the surface is reportedly increased by about 5% over similar atoms in the bulk of the crystal.[22] There are, however, exceptions to this, and occasionally the crystal periodicity and/or symmetry are significantly perturbed at the surface.[23] Surface structures which differ from that of the substrate have been observed on several fcc metals, although none have yet been found on bcc metal surfaces.[23] Temperature-sensitive structural rearrangements have also been reported to occur on clean platinum surfaces, while no such transformations occur in the bulk.[24] The surface structure of covalently bonded semiconductor crystals is sometimes considerably different than the structure of the bulk unit cell,[25] and this is so also for Al_2O_3.[26] Finally, there is ample evidence that adsorbed species can markedly alter surface structure.[27]

These and other observations imply that the surface of a crystalline solid can behave effectively as a separate phase, with physical properties that are distinctly different from those of the bulk.[23] The extent of such surface phases has been considered theoretically for fcc crystals,[28–31] molecular crystals,[32] and ionic solids.[33,34] These calculations[28,29] indicate that the mean-square displacement of surface atoms can be larger than that of atoms in the bulk by a factor of 30–200%, and that the component of the mean-square displacement perpendicular to the surface undergoes the greatest change with respect to the bulk value. The lattice spacing normal to the surface is predicted to be up to 13% greater near the surface than in the bulk, but this expansion is considered to decrease rapidly with distance into the crystal, being essentially zero by the fourth or fifth layer from the surface.[30,31] In general, the lattice spacing of the surface layer in ionic crystals is predicted to contract[33,34] and that of molecular crystals to expand, slightly.[32] LEED experiments, however, indicate a dilation of about 3.5% in the surface layers of LiF.[35]

Chemical Segregation at Free Surfaces

In addition to contamination by adsorption of gases, it is now recognized that segregation of impurities from the bulk may occur equally well at the free surface as at internal interfaces or dislocations.[2] Indeed, there is debate as to whether surface structures currently being reported for metals are truly representative of "clean" metal surfaces[36,37] or are associated with the presence of impurities.[38] LEED experiments by Fedak and Gjostein[38] indicate, for example, a hexagonal layer on the (100) surface of gold. This may well be caused by impurities originating in the bulk and accumulating on the surface by selective sputtering or diffusion along dislocation pipes. A surface or interfacial dislocation model has been suggested to account for this surface structure.

Recent developments in Auger-electron spectroscopy have also demonstrated that impurity segregation can occur at initially clean surfaces.[39] The surface region which contributes to the Auger spectrum is reportedly only a few atom layers thick,[37] and this powerful technique is sensitive to surface impurity concentrations on the order of tenths of a monolayer.[40] The segregation of sulfur to the surface of nickel, and of chromium, antimony, and sulfur in steel, have been clearly demonstrated by this technique.[39]

The driving force for solute segregation at free surfaces is probably the same as that leading to segregation at grain boundaries or dislocations. Relaxation of elastic distortion near the surface, for example, may be accomplished by solutes of appropriate size. Aust, Westbrook, and associates[2,41,42] suggest also that solute segregation arises because of the development of a gradient in vacancy concentration near free surfaces, the latter being an effective sink for vacancies. In those cases where solute–vacancy interactions are strong, vacancies will diffuse towards the surface dragging impurities along with them. The vacancy–solute complexes dissociate at the surface, vacancies are annihilated, and a net excess of solute remains at or near the free surface. In some cases, solute clusters may be formed near the surface. The driving force for this effectively uphill diffusion of solute atoms could be the decrease in free energy associated with the annihilation at boundary sinks of excess vacancies introduced during heat treatment.

By analogy with measurements on segregation at grain boundaries, the enriched region near the free surface is likely to extend several microns into the crystal. Although the existence of such an effect has not been experimentally established for metals, recent work by Jorgensen and Anderson[43] with ThO_2-doped Y_2O_3 indicates that effects of the segregation of ThO_2 at the free surface can be detected by means of indentation studies to depths of as much as $2\,\mu$, although calculations indicate that the excess solute is confined primarily to a region no more than 100 Å from the surface.

Space-Charge Effects

Because of the high density and mobility of the conduction electrons in a metal, interfacial electric fields cannot penetrate more than an angstrom or so into the bulk of a metal crystal. For nonmetals, however, because of the lower densities and mobilities of the charge carriers involved, the influence of such fields may be sizeable at much greater depths, for example, in excess of $1\,\mu$ for insulator materials. Space-charge effects in nonmetals will be considered in more detail in the next section.

Summary

The surface of a crystalline solid is far from being a simple discontinuity in an otherwise infinite continuum. The atomic arrangement and lattice spacing, both in the plane of the surface and normal to it, may be considerably different than that of atoms in the bulk. The surface may be distorted, may contain an above average concentration of point or line defects, and is unlikely to be atomically flat. Furthermore, solute and vacancy gradients are likely to exist in many metals and nonmetals, and the electronic properties of nonmetallic crystal surfaces will differ from those appropriate to the bulk.

Therefore, the near-surface region may to some degree be considered as a separate phase, the extent of which ranges from only a few atom layers in the case of structural rearrangements, to perhaps microns in terms of chemical and electronic properties. The mechanical properties of the surface layers may also be different from those of the bulk, particularly in the case of nonmetals.

ENVIRONMENTAL EFFECTS ON CRYSTALLINE SOLIDS WITH CLEAN SURFACES

The mechanical behavior of crystalline materials is largely determined by the generation, motion, and interaction of dislocations.* The fact that the presence and condition of the surface can have a major influence on plastic properties implies, therefore, that these factors must affect dislocation behavior. In particular, the surface may act as a region in which dislocations are readily formed, or as a barrier to the escape of dislocations from within the crystal. These effects are modified by the presence of surface films, and altered further if the solid is exposed to environments from which adsorption of surface-active species may occur.

The role of the surface in the plastic deformation of metals and non-metals has been reviewed recently by Grosskreutz and Benson[46] and Westbrook,[2] and can best be examined by deforming crystals with clean surfaces. Ionic crystals cleaved and deformed in a vacuum may meet this requirement, but metals exposed in any way to a laboratory atmosphere will be covered with adsorbed species such as water, and probably also by oxides or hydroxides, etc. One way in which this problem may be partially circumvented, however, is to remove the surface layers of a deforming metal crystal electrolytically. This approach can provide relatively clean surfaces, and has generated much useful information regarding environment-sensitive mechanical behavior.

---

* For texts on dislocation theory see Hirth and Lothe[44] and Nabarro.[45]

Metals

Clean Metals in Electrolytes

Surface-Removal Studies. There are numerous indications that plastic deformation begins in the surface layers of presumably clean metals. Heidenreich and Shockley[47] in 1947, and later Fujita et al.,[48] found that the Kikuchi lines produced by electron diffraction from aluminum monocrystal surfaces were extinguished by light plastic deformation. Following electrolytic removal of about 100–200 μ of metal from the surface, however, the lines reappeared, indicating the absence of plastic deformation at these depths. Likewise, the asterism of Laue diffraction spots from an extended monocrystal of α-iron was found by Sumino and Yamamoto[49] to be greatly reduced when a 200-μ surface layer was removed electrolytically.

Young[50] has shown also that dislocations are generated at or near the surface of highly perfect copper monocrystals in the first stages of plastic flow. Evidence for this is shown clearly in Fig. 2, which is a (111) Borrmann x-ray topograph of a lamella sliced from a lightly deformed crystal. It appears that some sources were introduced randomly at the surface by minor mishandling. However, no evidence of internal dislocation multiplication was found.

Considerations of the effects of surface removal on the flow properties of metal crystals has lead to the development of three mechanistic models which differ fundamentally in terms of the type and location of dislocation sources considered to be primarily involved in the deformation process. Each of these models will be described later and in detail in the context of the experimental evidence supporting them.* Briefly, however, the essential differences between each model are shown in Fig. 3. That utilized by Kramer and coworkers,[51–60] Fig. 3a, assumes that dislocations are generated by internal sources. As the dislocations leave the crystal, a fraction of them become trapped in the near-surface regions forming a zone of high dislocation density—the "debris layer"—which serves as a barrier to further dislocation egress. Other workers have suggested that certain surface-removal effects are best interpreted in terms of the generation of dislocations by surface sources.[61–64] The basis for this view is that a dislocation line terminating unpinned at the surface, but being pinned somewhere in the interior, behaves elastically as a dislocation line in a uniform medium which is anchored by the internal pinning point and its image above the surface.[65]

---

* Since, for polycrystals, grain boundaries probably are the most important source of dislocations, this discussion will be confined to observations on monocrystals.

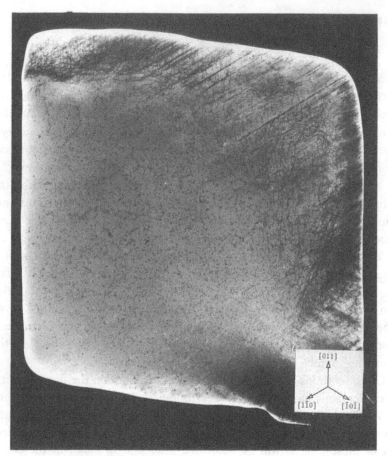

Fig. 2. A (111) Borrman x-ray topograph of a lamella from a copper monocrystal. Dislocations were generated at or near the surface. Edge length of lamella, 1 cm. (After Young.[50])

Thus, the effective length of a Frank–Read source at the surface may be twice its actual length L_S as shown in Fig. 3b. Since the stress required to operate such a source is inversely proportional to its unpinned length, a surface source should operate at approximately half of the stress required to operate an internal source of the same length. It follows that if surface sources of dislocations exist, they should be responsible for the yielding of clean metal crystals, and that dislocations should move inward from the surface in the early stages of plastic deformation. A third alternative, first suggested by Kuhlmann–Wilsdorf,[66] has been discussed recently by Latanision and Staehle.[67,68] They consider that the most likely dislocation generator in the early stages of deformation is a near-surface source. This

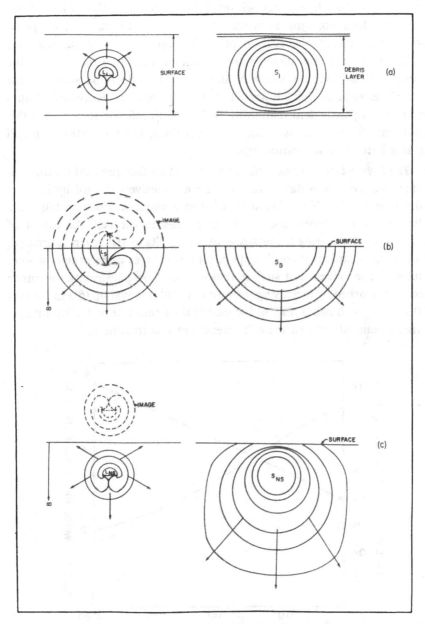

Fig. 3. Schematic illustration of (a) the "debris-layer" model, (b) a surface source of dislocations, and (c) dislocation generation by a near-surface source.

is pictured as a source of the Frank–Read type which happens to reside in the near-surface layers, say within a few microns of the surface. Such sources produce loops just as internal sources do, but because of the proximity of the surface, the portion of each loop nearest the surface slips readily out of the crystal while the remaining half-loop moves into the interior. Calculations by Sumino[69] indicate that dislocation sources near the surface, such as that illustrated in Fig. 3c, operate easily and should contribute to plastic deformation in preference to interior sources. On this hypothesis, yielding begins in the surface regions, but the surface can still act as a barrier to dislocation egress.

The Debris-Layer Hypothesis. The effect of surface removal on the mechanical behavior of metal crystals has been extensively studied by Kramer and coworkers.[51–60] They found that when aluminum monocrystals were pulled in tension during dissolution in an electrolytic cell, the extent of Stages I and II increased, the stress at which Stage III began decreased, and the work-hardening coefficients in all three stages of deformation were reduced.[51] The influence of surface removal on these parameters was made even more marked by increasing the rate of metal removal up to a maximum of 0.2 μ/sec, as shown in Fig. 4. Increasing the strain rate at a constant rate of metal removal reduced the effectiveness of this treatment.[51]

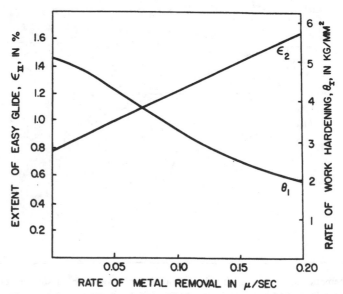

Fig. 4. Effect of surface-removal rate on the extent of easy glide and the rate of work hardening of aluminum monocrystals. Temperature, 3°C; $\dot{\varepsilon} = 10^{-5}$ sec^{-1}. (After Kramer and Demer.[51])

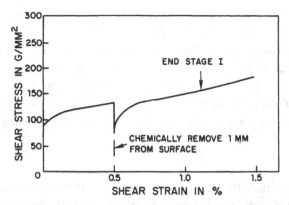

Fig. 5. Effect of electrochemically removing 1 mm from the surface of an aluminum monocrystal following 0.5% strain. Note recovery of the original yield stress upon reloading. (After Kramer and Demer.[51])

The critical resolved shear stress (c.r.s.s.) τ_0 was not affected by changes in the rate of metal removal. However, the initial yield stress and work-hardening behavior of an aluminum crystal deformed into Stage I could be recovered by surface removal (Fig. 5). Fabiniak and Kuhlmann–Wilsdorf[70] have also reported recovery of the initial yield stress by removing 50–100 μ from the surface of lightly deformed aluminum monocrystals, and Kramer[52] reports demonstrating this effect with gold monocrystals as well. The latter suggests that such effects are not necessarily related to the removal of an embrittling oxide layer. The general implication, in fact, is that work hardening in Stage I is concentrated in the near-surface region. In contrast, however, Nakada and Chalmers[71] were unable to detect any influence of electropolishing on the subsequent tensile behavior of gold crystals. They did, however, confirm that the flow stress of deformed aluminum crystals was lower after surface removal, and suggested, therefore, that electropolishing effects *were* associated with the presence of a surface-oxide film.

It has also been reported that the apparent activation energy for creep of aluminum, copper, and gold crystals at room temperature is significantly reduced when specimens are electropolished during deformation.[53] The decrease in activation energy is independent of strain, and dependent only on the rate of metal removal. At low strains, for example, the activation energy for aluminum monocrystals under ordinary testing conditions is about 4200 cal/mole. At a surface-removal rate of 0.2 μ/sec, the activation energy is reduced to 920 cal/mole.

Kramer[52,54] has interpreted these and other observations in terms of the formation of a dislocation-rich debris layer* in the surface regions of a plastically deformed crystal. The debris layer is considered to impede the motion of other dislocations, and serves as a barrier against which they pile up. Thus, the debris layer effectively introduces a back stress opposing the motion of dislocations moving into the surface region. Accordingly, the net stress acting on a mobile dislocation τ^* may be expressed as follows:

$$\tau^* = \tau_p - \tau_i - \tau_s$$

where τ_s is the back stress associated with the debris layer, and τ_i is the stress associated with the presence of other internal obstacles. τ_p is the applied plastic stress, i.e., $\tau_a - \tau_0$, where τ_a is the applied shear stress, and τ_0 the c.r.s.s.

τ_s has been measured by deforming specimens to a given strain, unloading, and then removing a predetermined amount of metal electrolytically from the surface. The difference between the final flow stress and the initial flow stress upon reloading is taken as a measure of the decrease in τ_s due to the removal of the debris layer in whole or in part. Such experiments indicate that the debris layer can extend to a depth of about 60 μ in aluminum,[54] and 100 μ in gold[56] or iron,[59] and that τ_s increases linearly with strain in Stages II and III.[54,56]

The interrelation between the stress components given in equation form above has been determined for aluminum monocrystals and is shown in Fig. 6. It has also been shown that the debris-layer stress may be eliminated by allowing deformed specimens to relax for appropriate periods of time.[54,57–60] Kramer considers the relaxation of τ_s to be very important in such deformation processes as creep,[57] cyclic creep, and work softening.[72]

The conclusion that work hardening in Stage I is confined predominantly to the surface layers implies that the transition to Stage II would occur at lower strains in small crystals. In general, it appears that the reverse is true.[73,74] It is clear, therefore, that not all surface effects can be described solely in terms of a debris-layer model.

The Possible Existence of Surface Sources. In experiments with silver monocrystals oriented for single slip (Fig. 7) Worzala and Robinson[61] found that by alternately removing a few microns of material from the crystal surface and then incrementally stressing, they were able to double

---

* The term "debris" has also been used to describe the distribution of point defects left in the wake of moving, jogged screw dislocations.

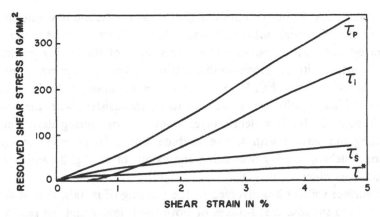

Fig. 6. Interrelation between τ^*, τ_p, τ_i, and τ_s for an aluminum mono-crystal. Temperature, 3°C; $\dot{\varepsilon} = 10^{-5}$ sec^{-1}. (After Kramer.[56])

the extent of Stage I. However, the slope of the stress–strain curve in Stage I, θ_I, was essentially identical to that of a similar crystal strained without the benefit of intermittent surface removal. They also found that the extent of easy glide could be increased if the surfaces were impacted lightly with silicon carbide particles. In Stage II, neither intermittent electropolishing nor particle impingement affected the flow curve.

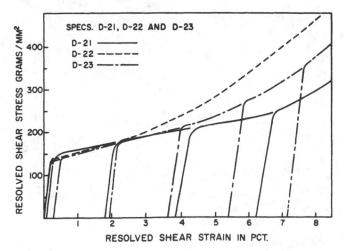

Fig. 7. Stress–strain curves for silver monocrystals tested in tension as follows: (i) D-21 was strained with incremental polishing; (ii) D-22 was strained directly; (iii) D-23 was strained and incrementally impacted with silicon carbide particles. (After Worzala and Robinson.[61])

Worzala and Robinson consider that these results are best interpreted in terms of preferential surface source operation. New surface sources may be "introduced" during polishing by intersection of the freshly produced, "clean" surface with the grown-in dislocation network, or during sprinkling by the introduction of half-loop sources such as those produced in ionic crystals.[75] These results are considered to be inconsistent with the debris-layer hypothesis, for if a debris layer does develop during deformation, sprinkling the surface with silicon carbide would be expected to create additional debris, thereby decreasing the extent of easy glide and not increasing it as observed in Fig. 7.

If surface sources are involved in the yielding of metals, then it seems reasonable to suppose that sources of both the primary and the secondary slip systems should be readily activated, possibly leading to the development of a layer close to the surface that is preferentially work hardened relative to the rest of the crystal. Results interpreted from such a viewpoint have recently been reported by Kitajima et al.,[63,64] who studied the orientation dependence of preferential surface hardening in copper and α-brass. They found that surface removal in Stage II leads to a decrease in the rate of preferential surface hardening, presumably related to the reactivation

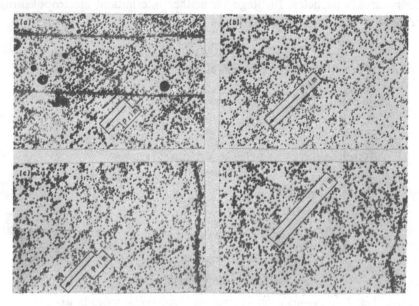

Fig. 8. Variation in etch-pit density (dislocation density) with distance from the $(1\bar{1}1)$ surface of deformed copper monocrystals. (a) at the original surface; (b) 40 μ beneath the surface; (c) 200 μ beneath the surface; (d) 350 μ beneath the surface. Magnification $\sim 70\times$. (After Kitajima et al.[64])

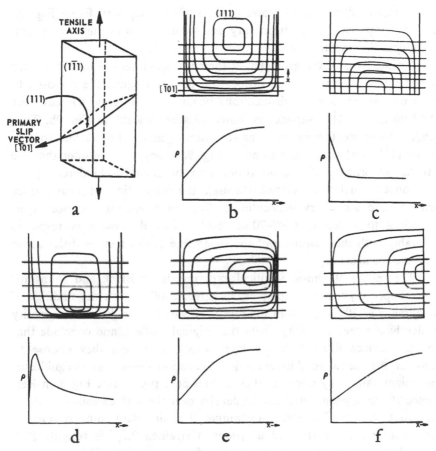

Fig. 9. Schematics of (a) crystal geometry and orientation and (b)–(f) variation of dislocation density ϱ as a function of distance x from the $(1\bar{1}1)$ surface of lightly deformed crystal, according to several models. (After Latanision.[77])

of surface sources and/or a decrease in the near-surface dislocation density. This is consistent with their observation[64] that for copper crystals deformed in Stage I, the density of dislocations is substantially higher within 40–70 μ of the surface than in the interior. The variation in the etch-pit distribution as a function of distance from the $(1\bar{1}1)$ surface of a deformed crystal is shown in Fig. 8. The geometry of this crystal (Fig. 9a) is such that the intersection of primary edge dislocations with the $(1\bar{1}1)$ surface would be revealed as etch pits along the primary slip direction, while slip steps would be created on the alternate pair of faces. The arrangement of etch pits in rows along the primary slip traces near the surface gives way to a more diffuse distribution in the interior, while the dislocations distributed more or less randomly among the rows at the surface disappear. From Fig. 8a, it is also apparent that both primary and secondary slip systems were activated near the crystal surface.

On the basis of the geometry of their test specimens, Kitajima et al.[64] consider that the variation in dislocation density is quite the opposite to what one would expect if dislocations produced by internal sources were piled up at the (1$\bar{1}$1) surface, as indicated schematically in Fig. 9b. They suggest, therefore, that dislocations are instead generated by surface sources at the (1$\bar{1}$1) surface, as illustrated in Fig. 9c. They also consider that the advance of primary glide dislocations into the crystal is hindered by the creation of barriers developed through the interaction of near-surface primary and secondary dislocations. A dislocation-dense surface layer, extending to a depth of 40–70 μ, results. This dimension corresponds favorably with their estimate of the mean free path of screw dislocations near the surface.

The results of almost identical experiments by Block and Johnson[76] with copper crystals are, however, in direct conflict with the findings of Kitajima et al. Block and Johnson report no variation in dislocation density to depths greater than 500 μ from the original surface, and conclude that no near-surface dislocation-rich layer exists. Although they choose to consider the measured differences in dislocation density as insignificant, one might cautiously interpret their actual data points (see Fig. 8 in Reference 76) to suggest a decrease in density near the surface followed by an increase toward the interior. For example, their first data point from measurements near the surface of a specimen strained 2.5% is roughly 50% lower than the next point, which is further from the surface. This is precisely the result one would expect to find if pileups are formed near the surface as indicated in Fig. 9b.

The reasons for the discrepancy in the measurements by Block and Johnson[76] and Kitajima et al.[64] are not clear. But it is important to appreciate that the interpretation of their etch-pitting results is not unambiguous, and that both sets of data may also be interpreted in terms of the operation of near-surface sources.[77] For example, the activation of near-surface sources at the (1$\bar{1}$1) surface could produce the results indicated in Fig. 9d. On the other hand, near-surface sources may also be preferentially activated on the alternate pair of specimen surfaces due to stress concentrations introduced by the notch effect of slip steps on these faces—a concept invoked earlier by Mitchell et al.[78] In this case, Fig. 9e, one would expect to observe a large number of edge dislocations of the same sign originating from these surfaces. Evidence suggesting this behavior has been provided by

Fig. 10. Etch-pit distribution on the cross-slip surface of an α-copper–aluminum alloy illustrating pileups of positive and negative dislocations in near-edge orientation on adjacent glide planes. Magnification ~1750×. (Mitchell et al.[78] Reproduced by permission of the National Research Council of Canada from the *Canadian Journal of Physics*.)

Mitchell et al.[78] (Fig. 10). Dislocations emanating from the surfaces exhibiting slip steps, and lying within a narrow band parallel to any particular glide plane, pile up in the same direction— suggesting that they are of identical Burgers vector. The dislocations within the two narrow bands ~3 μ wide located on either side of the principal band probably are piled up in the opposite direction, and probably are of opposite Burgers vector. The existence of pileups consisting of several hundred dislocations of the same sign, and the absence of pileups of equally large numbers of dislocations of opposite sign on the same glide planes, strongly suggests avalanching from sources at (Fig. 9f) or near the surface.

It will be appreciated that dislocation-density profiles based on the

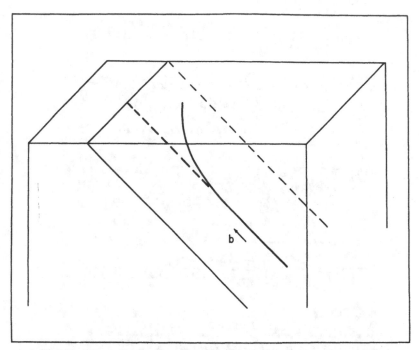

Fig. 11. Screw dislocation affected by image forces near free surface.

simplified models shown in Figs. 9b, 9e, and 9f may be much alike, the actual profiles depending upon the distribution of active sources. In these cases, the density of etch pits on parallel ($1\bar{1}1$) planes of a specimen deformed into Stage I should be lowest near the surface and increase with depth into the interior. Alternately, depending upon the distance of near-surface sources from the original ($1\bar{1}1$) face, the profiles in Figs. 9c and 9d may be very similar.

Despite evidence indicative of the operation of surface sources in the early stages of plastic deformation, however, a strong disclaimer to their existence in reality has been presented by Friedel.[79] Specifically, if the slip plane on which a potential surface source lies cuts the surface at any angle other than 90°, then there is a strong image force inducing the dislocation to cross slip into a position normal to the surface (Fig. 11). If the dislocation glides more easily on its initial plane, the cross-slipped portion is then likely to act as an effective pinning point, inhibiting any tendency for preferential operation of this source unless all other internal sources are heavily locked. Calculations by Lothe[80] show that there exist specific angles of incidence for dislocations with the free surface for which the forces vanish so that the dislocations can be truly straight approaching the surface.

However, considerations of dislocation core energy terms again lead to the conclusion that, within a few atom distances of the surface, dislocations will tend to align normal to the surface.[80] Moreover, in practical terms, metal surfaces are rarely clean, usually being coated with oxide films, and one would expect that the free end of the surface source shown in Fig. 3b would ordinarily be pinned. Furthermore, even if the surface were clean, the near-surface lattice distortions described earlier would tend to make operation of genuine surface sources difficult. It seems, therefore, that surface sources are likely to be involved in the yielding of metals only under very special circumstances—circumstances which are not likely to be ordinarily encountered in a metallurgical laboratory.

The Possible Existence of Near-Surface Sources. The results of studies by Latanision and Staehle[67,68] of the influence of continuous surface removal on the mechanical behavior of nickel monocrystals appear to be best interpreted in terms of the preferential operation of near-surface sources. In this work, crystals containing ~ 0.1 wt. % carbon were deformed in uniaxial tension in a $1N$ H_2SO_4 solution under potentiostatic control. Potentiostatic control allows precise description and control of the specimen's surface

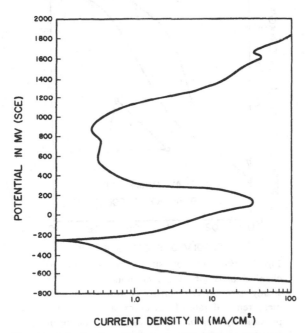

Fig. 12. Polarization diagram for nickel monocrystal in $1N$ H_2SO_4 at room temperature, determined potentiostatically. (After Latanision and Staehle.[68])

condition during each test, an advantage not extensively utilized in earlier work. The polarization diagram for the system is shown in Fig. 12.

The experiments of interest to this discussion were performed within the potential range -240 to $+100$ mV (SCE), i.e., the region of active dissolution in which nickel enters solution as a divalent ion. In this case, the air-formed film is quickly removed and the surface then maintained in an essentially clean state. It was found that the removal of approximately 3μ of material from the surface immediately before or during elastic deformation resulted in a substantial decrease in the value of τ_0 from that of identical crystals deformed in the laboratory atmosphere (Fig. 13). In addition, continuous slow surface removal during deformation significantly increased the extent of Stage I, while θ_{I} was slightly decreased. Likewise, Stage II was extended, and θ_{II} decreased. Surface replicas also revealed that slip lines formed while the surface was being actively dissolved were stronger and more widely spaced than those produced on crystals deformed in air (Fig. 14).

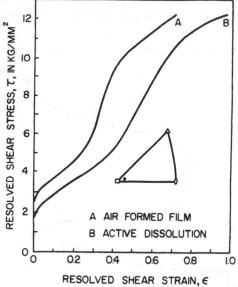

Fig. 13. Effect of slow surface removal on mechanical behavior of nickel monocrystals. The steady-state current density corresponded to a metal-removal rate of 7.5 atom layers/sec. Approximately 3μ of metal were removed prior to plastic deformation—curve B. Room temperature, $\dot{\varepsilon} = 3 \times 10^{-4}$ sec^{-1}. (After Latanision and Staehle.[68])

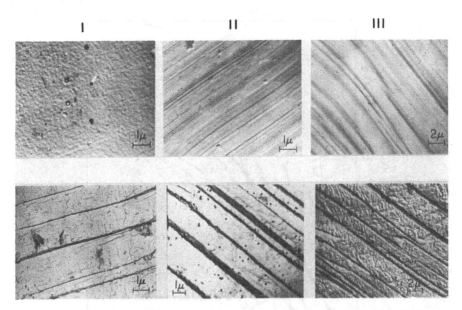

Fig. 14. Electron micrographs showing typical slip-line patterns produced during Stages I, II, and III on crystals strained (a) in laboratory air and (b) under active dissolution conditions. (After Latanision and Staehle.[68])

The observation that τ_0 may be reduced by surface removal immediately prior to deformation suggests that this parameter is not determined by the stress required to operate internal Frank–Read sources. This result also is not readily explainable in terms of the debris-layer hypothesis, since this layer is considered to form only after yielding.* From the standpoint of surface source operation, a decrease in τ_0 due to dissolution before straining would imply that the length of the surface sources is increased by surface removal. This could be achieved by the removal of a source-pinning natural oxide film. However, the slip line observations are not consistent with a surface-source explanation. Surface (oxide) removal would make more surface sources available for operation on many slip planes. This, however, should lead to a decrease in both the spacing and intensity of slip steps, contrary to observation.

An apparently consistent explanation, on the other hand, can be developed in terms of the preferred operation of near-surface sources which, by virtue of their position, are sensitive to surface conditions. A possible indication of this sensitivity may be obtained from other experiments by

---

* Kramer[72] has suggested, however, that microplastic yielding at stresses below the c.r.s.s. may create a debris layer which may subsequently be removed by dissolution.

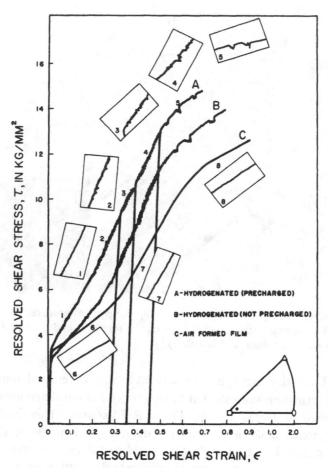

Fig. 15. Effect of hydrogen charging on the deformation of nickel monocrystals. Insets show detailed recorder tracings (load–extension). (After Latanision and Staehle.[67])

Latanision and Staehle[67] which show that serrated yielding occurs when nickel monocrystals are cathodically polarized during deformation (Fig. 15). At the potential used, −600 mV (SCE), hydrogen is produced on the surface of the specimen. Serrated yielding implies interaction between moving dislocations and solute atoms, and the important point to be made is that this occurs even though calculations indicate that the depth of penetration of hydrogen into the crystal could not be greater than 15 μ, and must be considerably less in the early stages of deformation.

It is of interest to consider those characteristics of a metal surface which would permit its near-surface layers to perform in the dual role of dislocation source and barrier. Sumino's calculation[69] indicates that a dislocation source near the surface which produces loops that are under the influence of their own image forces, can emit a large number of dislocations more easily than an internal source. Fabiniak and Kuhlmann–Wilsdorf[70] also point out that the surface acts as a vacancy sink, and that the first few microns directly beneath the surface may thus be depleted of vacancies. Dislocations near the surface are thus likely to be free of vacancies, i.e., unjogged, while the motion of dislocations in the bulk is likely to be hindered by these defects. These factors would allow the surface layer of a metal crystal to yield before the interior.

On the other hand, solute segregation may also occur near a free surface, probably to depths similar to the vacancy depletion distance. If the resultant alloy hardening or atmosphere locking is not such that the stress required to operate near-surface sources becomes greater than that needed to activate internal sources, then yielding will still begin near the surface but, of course, at stresses characteristic of the relatively more concentrated alloy present there. Clearly, the removal of such a layer immediately before deformation would decrease the c.r.s.s. Such a process may have contributed to the decrease in τ_0 observed by Latanision and Staehle[68] (Fig. 13). Solute segregation and vacancy depletion near the surface, both dynamic processes, tend to work in opposite directions from the standpoint of surface hardening. Presumably, active near-surface sources would be found at a distance from the surface where conditions affecting their ease of operation, including notch effects, are optimum. Factors involved in the probably analogous phenomenon of solute-induced hardening (or softening) near grain boundaries in metals and nonmetals have been considered by Aust et al.[41,42] They find that solvent–solute systems which exhibit strong solute–vacancy interactions and positive thermodynamic deviations from ideality are likely to become hardened near grain boundaries by a solute clustering mechanism. Systems containing solute atoms which do not interact strongly with vacancies, and exhibit negative thermodynamic deviations from ideality, are likely to show grain-boundary softening effects.

Another possibility, based on a suggestion by Latanision and Staehle,[68] is that the reduction in τ_0 caused by surface removal could be due to the elimination of the back stress on the operation of near-surface sources arising from a dislocation-rich subsurface layer which may exist even on undeformed crystals. This layer is distinct from the debris layer envisioned by Kramer, which develops after plastic deformation has begun, and is considered to arise because image forces can pull grown-in dislocations into the distorted region of the lattice immediately below the surface, where they become trapped. Dislocations pulled off their glide planes by the image forces, and therefore sessile, might provide additional trapping sites. The degree to which such a dislocation-rich layer might develop would depend on a number of factors, including crystal structure, specimen orientation, original dislocation distribution, and severity of surface lattice distortion.

Once plastic deformation has begun, however, there are several reasons for expecting that a preferentially work-hardened layer may develop near the surface. Even an ideal clean surface provides a barrier to the egress of a dislocation whose Burgers vector contains a component normal to the surface, because its emergence requires the creation of a slip step, and this requires extra (surface) energy. It might be anticipated that if the elastic self-energy of the dislocation, which is reduced near the surface due to image forces, exceeds the energy required to create the new surface step, then the dislocation will reach the surface relatively easily, and vice versa. An approximate calculation by Nabarro[81] suggests that for most metals the elastic energy of perfect edge dislocations is the larger, so that in principle they can leave the surface of a clean metal crystal freely. The reverse is true for partial dislocations. Thus, dislocations in a metal of low stacking fault energy may experience some difficulty in breaking through the surface of even a clean crystal. Shchukin,[82] on the other hand, has approached this problem in a slightly different manner, and concludes that the surface resisting forces should always exceed the elastic energy of near-surface dislocations.

Grosskreutz[83] has interpreted the electron diffraction contrast associated with slip lines on the surface of cyclically deformed gold (intermediate stacking fault energy) as being due to dislocation trapping at the surface and associated lattice distortion (strain). He concludes that, for this metal, the surface resisting force is greater than the elastic energy of emerging dislocations. Thus, a possible source of surface hardening is the piling up of dislocations against the barrier created by such residual strains at the surface.

The formation of a work-hardened surface layer is also likely to be sensitive to the character of the dislocations passing through the surface. Specifically, one would expect that edge dislocations are likely to be more effective contributors than screw dislocations, and this result has indeed been obtained by Nakada and Chalmers[84] (Fig. 16). The geometry of their aluminum monocrystals was such that screw dislocations emerged from one pair of sides (S sides), and edge dislocations from the other pair (E sides). They found from alternate electropolishing and incremental strain-

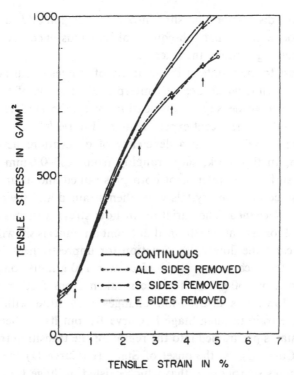

Fig. 16. The influence of the character of emergent dislocations on surface hardening in aluminum monocrystals. (After Nakada and Chalmers.[84])

ing tests that the flow curve of a crystal whose S sides only were removed was almost identical to that of a crystal which was deformed without electropolishing. On the other hand, the crystal whose E sides were polished behaved similar to one that was polished on all four sides. In this case, therefore, essentially all of the surface hardening appears to be associated with the emergence of edge dislocations. Other workers have also shown that the shape of the flow curve of fcc monocrystals is especially sensitive to the length of the glide path of edge dislocations.[73,85] Thus, the work-hardening behavior of a monocrystal is not only a function of its orientation, but also[54] of the crystallographic relationship between its principal slip directions and surfaces.

Another possible source of surface hardening has been discussed by Fleischer.[86] As mentioned earlier, it is known that the near-surface lattice spacing of a crystal is different from that of the interior. Fleischer has shown that dislocations crossing into a region of different lattice spacing along the slip direction must create sessile dislocations at the interface.

Each successive dislocation that slips into such a region adds a new sessile interface dislocation, so that subsequent dislocations encounter increasing difficulty in passing through the interface.

In contrast to the results of the majority of studies of surface-sensitive mechanical behavior, which are best interpreted in terms of the existence of an above-average density of dislocations near the surface of slightly deformed crystals, some recent experiments by Fourie[87-89] have been interpreted by him to indicate that a deficiency of dislocations actually exists in this region. In this work, slices ranging from 0.03–0.6 mm thick were cut from the surface and interior of both prestrained and as-grown copper crystals. These component crystals were then strained and their dynamical flow stresses determined. The variation in flow stress with position in the original crystal for crystals prestrained different amounts is shown in Fig. 17. Curve A indicates the flow stress variation for unprestrained crystals. The flow stress of the slice nearest the surface was 110 g/mm^2 compared to a peak flow stress of 150 g/mm^2 for slices taken from further within the crystal. The flow stress distribution changes very little with increasing amounts of prestrain through Stage I (Curve B), but its gradient increases when the crystal is prestrained into the region of the transition from Stage I to Stage II (Curve C). At the onset of Stage II (Curve D) the flow stress gradient increases sharply over that which existed in Stage I, and becomes even more marked in later stages of deformation.

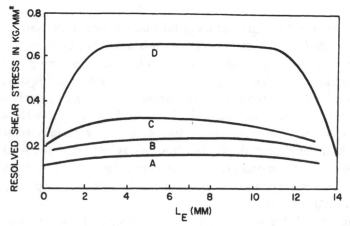

Fig. 17. The flow stress distribution in copper monocrystals plotted as a function of the length of the glide path of edge dislocations L_E, which has the value 0 and 14 mm at the original surfaces. Curve A is for an as-grown crystal; Curves B, C, and D are for prestrains of 0.02, 0.029, and 0.058, respectively. (After Fourie.[88])

Although these most interesting results are not entirely inconsistent with the preferential generation of dislocations in the near-surface regions, on the basis of transmission electron-microscope studies, Fourie has interpreted them in terms of (a) the operation of dislocation sources distributed uniformly throughout the crystal and (b) a net deficiency in edge dislocations of one sign extending over a distance from the original surface equal to the mean free path of edge dislocations.[88] Assuming that work hardening occurs by Hirsch's[90] model, such a deficiency at the surface would delay the onset of Stage II hardening in the surface regions. Thus a slice taken from near surface would exhibit an extensive Stage I, while one from the interior would exhibit only limited easy glide. In short, on this hypothesis, the interior of these crystals should work harden more rapidly then the surface region. This conclusion is, of course, at variance with that deduced from other studies on copper including the yielding behavior of monocrystals by Young,[50] the surface-removal and yield-stress recovery observations of Kramer,[51-60] and the etch-pitting work of Kitajima et al.[64]

It should be recognized that Fourie's work was performed on specimens thick in comparison to those of other workers, so that size considerations may be relevent. Another possible cause of this discrepancy is the fact that dislocation densities and structures observed in thin foils may not be representative of the bulk material from which the foils were prepared due to dislocation rearrangement and losses during electrolytic thinning.[91] On this point, the rapid relaxation of the surface layer stress on unloading noted by Kramer and coworkers,[57-59] and the accelerated relaxation which occurs with decreasing specimen dimensions,[57] imply that foils taken from slices near the crystal surface may well be denuded of dislocations by the time they are examined.[92,93]

Moreover, the dislocation-density measurements reported by Fourie[88] begin at a minimum distance of 100 μ from the unsliced crystal surface— which is just outside the range of thickness of the debris layer detected by Kramer on other fcc metals.

It should be added, however, that other attempts[94-96] utilizing conventional transmission electron microscopy have also failed to detect any above-average density of dislocations in foils prepared from the near-surface regions. Perhaps further work, utilizing high-voltage electron microscopy and foils on the order of a few microns thick, would produce dislocation-density measurements in which more confidence could be placed. Alternatively, attempts could be made to retard relaxation effects by neutron-irradiating still-stressed crystals so as to immobilize dislocations lying in the surface layers.

For the present, therefore, the existence or otherwise of near-surface debris layers must remain a controversial subject. The authors consider, however, that the weight of experimental evidence is in favor of their existence.

Possible Rationale for Surface-Removal Phenomena. In summary of the preceding discussions, the authors conclude that vacancy depletion, solute segregation, image-force effects, and notch effects due to surface ledges will be the primary factors influencing the yielding behavior of film-free metal crystals. Once the specimen has yielded, then energetic problems associated with the creation of surface slip steps, and various factors introducing near-surface lattice distortions, will tend to inhibit dislocation egress, and lead to the development of a preferentially work-hardened surface layer characterized by a relatively high dislocation density.

Based on this conclusion, a rationale for the early yielding behavior of film-free metal crystals may be suggested which appears to be consistent with most of the experimental evidence. Plastic deformation will begin when the applied shear stress is sufficient to activate near-surface sources. Surface removal before or during loading may or may not decrease the c.r.s.s. depending upon the relative importance to the specific crystal of such factors as the presence of trapped dislocations in the near-surface regions, vacancy depletion or solute-segregation effects, etc., and the rate of removal of the surface. Continuous surface removal during deformation will allow initially activated sources to continue to operate, leading to the formation of widely spaced, large offset slip lines.

Stage I work hardening is presumed to occur primarily as a consequence of the interaction of dislocation loops progressing into the metal from the surface with the grown-in dislocation forest. The transition to Stage II will occur when the stress concentration at internal barriers is sufficient to activate internal sources on secondary slip systems.

For a crystal deforming in the absence of a solvent environment, the surface layers are likely to constitute a barrier to dislocation emergence leading to trapping and the formation of a debris layer. In this case, near-surface source operation will tend to be inhibited, and alternate sources must then be activated. Consequently, slip will be finer and more homogeneously distributed in such crystals. The stress fields of dislocations piled up against the debris layer may also induce secondary slip from sources in the near-surface region at lower strains than for crystals undergoing surface removal. Thus, the extent of easy glide in these crystals will be less than that observed for those tested in solvent environments.

Electrocapillary Effects. The surface tension of an electrode is a function of the potential difference across the electrical double layer, i.e., the separation of charge which develops between the metal and the solution adjacent to it as dissolution occurs. A net surface charge q of either sign favors an increase in the area of the metal–electrolyte interface because the like charges of which it is composed repel one another.[97] One might expect, therefore, that any value of q other than zero would reduce the surface tension γ. The relation between γ, q, and the applied potential ϕ in the absence of specifically adsorbed ions was first derived by Lippmann[98] and is given by

$$(\delta\gamma/\delta\phi) = -q$$

In a plot of γ versus ϕ, the surface tension passes through a maximum when $q = 0$, i.e., at the point of zero charge. This point is generally referred to as the zero-charge potential or, more commonly, the electrocapillary maximum, because early work in this field was done with a Lippmann capillary electrometer.

Specifically adsorbed ions may tend to reduce the surface expanding tendency of the surface charge. In this case, the above equation becomes

$$-(\delta\gamma/\delta\phi) = q + (\delta(\bar{G})_{ads}/\delta\phi)$$

where $(\bar{G}_{ads})$ is the electrochemical free energy of specifically adsorbed ions. Adsorption of charged ions or organic molecules may, therefore, affect the shape and location of the entire electrocapillary curve. (For a review, see Chapter IV, Adamson.[10])

Various methods have been suggested for the determination of the electrocapillary maximum, but usually these are rigorous only for liquid metals. For this reason, mercury has been extensively studied, while information on solid metals is more meager. West[97] has compiled some of the available data on metal electrodes. For mercury the lowering of the surface tension produced by a potential difference of 1 V across the double layer is of order 100 erg/cm^2.[99] Presumably, similar changes would occur for solid-metal electrodes.

Rebinder and coworkers[100–102] have reported that the hardness (determined from the amplitude damping of an oscillating pendulum) and creep properties[103] of metals tested in electrolytes are sensitive to the potential difference across the double layer. For example, the variation in pendulum hardness of tellurium as a function of applied potential was found to go through a maximum at the zero-charge potential, i.e., at the position of maximum surface tension (Fig. 18). Moreover, the addition

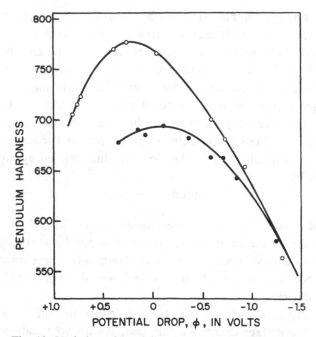

Fig. 18. Variation of pendulum hardness of tellurium with potential drop at solid–solution interface. Upper curve is for specimen immersed in $1N$ Na_2SO_4. Lower curve is for $1N$ Na_2SO_4 containing 0.5 mole/liter thiocarbamide. (After Rebinder and Venstrem.[100])

of surface-active agents reduced the hardness of tellurium, and displaced the maximum hardness to a different potential. Rebinder et al.[100–103] have interpreted these results in terms of adsorption-induced reductions in the surface free energy (tension) of the solid, these presumably facilitating the creation of the new surfaces produced during deformation.

There appear to be a number of reasons why this explanation is unlikely. First, a simple calculation shows that the energy expended in creating a hardness impression in a (ductile) metal surface is several orders of magnitude greater than that required to create the new surface area actually produced. Most of this energy is, of course, expended in generating and moving dislocations in the near-surface regions. Thus, even large changes in γ ($>50\%$) cannot significantly influence the measured hardness value. Also, more recent work has failed to produce the reported changes in the hardness of metal surfaces with environment when the hardness was determined by an indentation procedure, and not by the pendulum technique.[104,105] Bockris and Jones[104] have suggested, therefore, that the reported variations in pendulum hardness are in fact associated with surface-potential dependent changes in the coefficient of friction μ_F between the pendulum fulcrum and the surface. Supporting evidence for this view may be found in the work of Bowden and Young.[106] In addition, there appears to be a valid physical reason for suggesting that the hardness of a metal should *not* be adsorption sensitive—namely, that the influence of any adsorbed species is screened out by the highly mobile conduction electrons within a few atom distances of the surface. Thus, dislocation behavior in the near-surface region, and hence hardness, should not be significantly affected by their presence.

Finally, in many of the Russian studies, for example with tellurium, the applied potentials were varied over a wide range [+1.0 to −1.5 V (calomel)], and in some cases prolonged cathodic pretreatments intended to remove surface-oxide films before testing were utilized. It is conceivable, therefore, that surface-oxide or hydride films may actually have been present despite, paradoxically, their specific attempts to preclude complications arising from surface films.

It is considered that until more convincing evidence is available the reported effects of environments on the creep behavior of metals may also be better interpreted in terms of (a) surface-removal effects or (b) the influence of adsorbed species on the mechanical behavior of nonmetallic surface films, which in turn affect the behavior of the substrate metal (pseudo-Rebinder effects).[107]

Adsorption of Surface-Active Species

Even in the absence of dissolution reactions, the mechanical properties of clean crystalline solids can be significantly affected by liquid environments. For example, some metals are embrittled by certain liquid metals,[108] or weakened by adsorbed water molecules.[109] In the following sections we shall consider several of the more interesting examples of such adsorption-sensitive mechanical behavior. While the mechanisms of some of these are not yet clearly understood, it appears that certain examples of such behavior may be interpreted in a general, conceptual way from considerations of the type, concentration, mobility, and adsorption-induced redistribution of the charge carriers in the solid.[110,111] Adsorption-induced effects in metals will be considered in terms of this hypothesis next, and effects in nonmetals will be discussed later in the section.

Liquid-Metal Embrittlement. It has been known for some time that certain liquid metals embrittle certain pure or alloyed solid metals. Molten lithium, for example, severely embrittles polycrystalline copper and iron, but not aluminum; mercury embrittles zinc and brass, but not cadmium.[108,112] It appears that the prerequisites for the embrittlement of an otherwise ductile solid metal by an active liquid metal are (a) a tensile stress, (b) either a preexisting crack or some measure of plastic deformation and the presence of a stable obstacle to dislocation motion in the lattice, i.e., a concentrated tensile stress resulting in highly strained bonds across the potential fracture plane, and (c) adsorption of the active embrittling species specifically at this obstacle and subsequently at the propagating crack tip.[113,114] However, the factors that determine which liquid metal will actually embrittle which solid metal remain unclear. In general, it appears that to constitute an embrittlement couple, both the solid metal and the liquid metal should exhibit limited mutual solubility and little tendency to form intermetallic compounds.[108] Dissolution processes are not thought to be relevant to the mechanism, because a suitably prestressed metal specimen will fail instantaneously on being wetted with an appropriate liquid metal. Crack growth rates as high as 500 cm/sec have been reported.[108]

Embrittlement can occur in all degrees, and catastrophic embrittlement is only an extreme case. "Active" liquid metals usually affect the stress and strain at fracture, but do not affect the yield stress. The term embrittlement is used, therefore, to denote a reduction in strength or ductility, and is not intended to imply that the fracture process necessarily occurs in a completely brittle manner.

Since monocrystals can be embrittled, grain boundaries are not essential to the phenomenon. However, liquid-metal embrittlement usually is more severe in polycrystals than in monocrystals. Grain boundaries, of course, serve as stable obstacles to dislocation motion, and assuming a dislocation mechanism for crack initiation, this observation is not unexpected. Failure frequently occurs in an intercrystalline manner, presumably because less energy is required than for transcrystalline cleavage in most ductile metals. For highly anisotropic metals such as zinc, however, failure of polycrystalline specimens occurs predominantly by cleavage on basal planes.

In the past, liquid-metal embrittlement (L-ME) has been discussed in terms of a simple reduction in surface free energy associated with the adsorption of liquid-metal atoms.[101] Since surface free energy or surface tension is defined as the work necessary to form a unit area of surface by a process of division, such a hypothesis is undoubtedly correct. Mechanistically, however, it is not very satisfying because it does not provide any insight into the atomistic or electronic processes involved. Current thinking on the mechanism of embrittlement has been based on

the hypothesis that embrittlement results from an adsorption-induced reduction in the cohesive strength of atomic bonds at regions of stress concentration in the solid metal, e.g., at the tips of cracks or in the vicinity
of piled-up groups of dislocations.[112,115,116] To appreciate the general nature
of L-ME, it is useful to consider the nature of the interaction between an
adsorbed species and a metallic surface.[117,118] The high concentration of
mobile conduction electrons in the metal screens out the effect of adsorbed
species within a distance of a few atomic diameters of the surface.[119,120]
It follows, therefore, that adsorbates should not influence significantly the
bulk mechanical properties of oxide-free metals, and this conclusion is in
accord with the general observation that the yield or flow behavior of a
metal crystal is not affected when exposed even to highly surface-active
liquid metals. Adsorbates may influence fracture behavior, however, since
the propagation of a surface-initiated crack essentially involves the consecutive rupture of surface bonds.

Consider the possible effect of an adsorbing, surface-active, liquid-
metal atom B at a crack tip in a stressed solid metal A (Fig. 19). Chemisorption of atom B will cause some variation in the tensile strength σ of
the bond between atoms A and A_0 constituting the crack tip. If it is an
embrittling atom, σ will be reduced. Because of conduction-electron screening, however, it is unlikely that atom B will be able to influence the strength

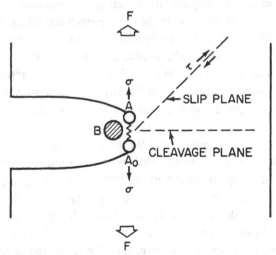

Fig. 19. Schematic of a crack in a solid, subjected to
an increasing force F. The bond A–A_0 constitutes the
crack tip, and B is a surface-active liquid-metal atom.
(After Westwood et al.[111])

of bonds across any slip plane for a sufficient distance from the crack tip to affect the ease of dislocation motion away from the crack. For this reason, the shear stress τ required to move dislocations on slip planes in the vicinity of the crack tip should not be significantly altered by the presence of atom B. Adsorption of an embrittling adsorbate leads, therefore, to a decrease in the ratio σ/τ. On the basis of arguments presented by Kelly et al.,[121] this decrease will be manifested as an increased tendency for the crack to propagate by cleavage rather than ductile shear.

The severity of embrittlement observed will depend upon the magnitude of the reduction of σ. Even for a normally ductile metal such as pure aluminum, an extremely active liquid metal can reduce σ/τ sufficiently to cause failure predominantly by cleavage.[117] Conversely, the ratio σ/τ can be increased by adsorbing some liquid-metal species which interacts at the A–A$_0$ bond so as to produce an increase in σ. An element known to form high melting point (strongly bonded) intermetallic compounds with the solid metal might be expected to act in this manner, and so serve as an inhibitor to L-ME.[118] Thus, for example, the addition of 0.4 at. % barium to mercury increased the fracture stress of polycrystalline zinc some 65% in this environment (zinc and barium form $Zn_{13}Ba$, mp $\sim 950°C$).[114]

The severity of embrittlement by mercury of a series of copper-base alloys increases as the stacking fault energy or ease of cross slip decreases.[122] In considering this variation, it might be suspected that alloying additions control embrittlement principally by affecting the parameter τ, and that the chemical nature of the solute has no particular effect.[122] However, Westwood et al.[118] note that the electron/atom ratio determines the stacking fault energy, and also affects the elastic constants of these alloys. Thus alloying can also affect σ, and it may be this factor which actually controls susceptibility in some instances. Significantly, the data shown in Fig. 20a exhibit an equally good correspondence with the electron/atom ratio (Fig. 20b).

In another study, Kamdar and Westwood[123] found that the susceptibility of polycrystalline zinc to embrittlement by liquid mercury is markedly increased by alloying the zinc with as little as 0.2 at. % copper or gold in solid solution. This result is shown in Fig. 21a. They also found that the critical resolved shear stress τ_0 for chemically polished zinc monocrystals stressed in air increased with copper and gold content, as shown in Fig. 21b. They concluded, therefore, that the increased susceptibility to embrittlement is not related to solute-induced changes in the stacking fault energy or bond strength—because of the low concentration of solute present—but to the variation in τ_0. Increasing τ_0 (decreasing σ/τ) inhibits the relax-

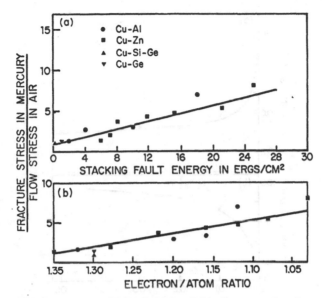

Fig. 20. Embrittlement of copper-base alloys as a function of (a) stacking fault energy (after Stoloff et al.[122]) and (b) electron/atom ratio (after Westwood et al.[118]).

ation by plastic flow of stress concentrations at grain boundaries, and facilitates crack initiation in the presence of mercury.

A factor which requires further attention is the variation in susceptibility to embrittlement with chemical composition of the liquid metal. A new approach to this problem, in which a potentially active metal is dissolved in an "inert-carrier" liquid metal of lower melting point, has made it possible to evaluate the embrittlement behavior of a number of new systems, and to compare the influence of several elements effectively in the liquid state at the same temperature.[111,118,124] The need for such a technique is evident when it is appreciated that direct investigation of many potentially interesting solid metal–pure liquid metal couples is not feasible because, at the required temperatures, the solid metal is either too ductile to maintain the stress concentration necessary to initiate and propagate a brittle crack, or is excessively soluble in the liquid metal resulting in crack blunting.

Recent studies indicate that L-ME is not as specific as had been previously thought. For example, polycrystalline pure aluminum is only slightly embrittled by mercury at room temperature, but additions of as little as 1–3 at. % of a number of elements to the mercury produce marked effects on the severity of embrittlement[118] (Fig. 22). The severity of embrittlement

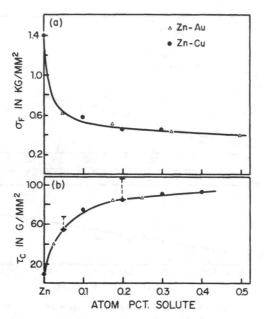

Fig. 21. (a) Effect of solute content on fracture
stress σ_F of polycrystalline zinc in liquid mercury.
The grain diameter of the zinc was $\sim$1 mm and
its engineering flow stress $\sim$1.9 kg/mm². The
strain rate was 4×10^{-5} sec^{-1} and the temperature was 25°C. (b) Variation of critical resolved
shear stress τ_C with solute content for zinc monocrystals. Range of values for asymmetric bicrystals
of Zn–0.05 at. % Cu and Zn–0.2 at. % Cu are
given by bars. (After Kamdar and Westwood.[123])

by liquid-metal solutions can be extremely temperature sensitive.[111,118,125]
Figure 23 illustrates, for example, the occurrence of brittle-to-ductile transitions in polycrystalline pure aluminum, the critical temperatures T_c for
which are determined by the composition of the mercury–gallium solution
environment. Similar effects have been observed for silver and brass.

It appears that previous analyses of apparently similar transitions
occurring in bcc or hcp metals tested in inert environments (e.g., that by
Petch[126]) are inappropriate for such environment-sensitive transitions, because the assumptions on which they are based are invalid for the latter case.
The assumptions are (a) that the temperature dependence of the yield
stress is the controlling factor and (b) that T_c is the temperature at which
the yield stress equals the fracture stress. For fcc metals such as aluminum,
however, the yield stress decreases only slightly with temperature over the

temperature range involved, and a significant amount of plastic deformation occurs before embrittlement results, even below T_c. Thus the yield stress does not equal the fracture stress at any temperature of relevance to this work.

A new analysis has been developed,[125] therefore, which assumes that such transitions are associated with the temperature-sensitive behavior of the ratio σ_e/τ, σ_e being the effective tensile fracture stress of atomic bonds in the presence of an adsorbed and embrittling liquid-metal species, and T_c being the temperature at which σ_e/τ increases above some critical value. While τ is a relatively temperature-insensitive parameter, σ_e is temperature sensitive because adsorption (desorption) is a dynamic and thermally activated process.

The strain-rate sensitivity of L-ME is not clearly established at this time.[118] Shchukin et al.[127] found that the strain at fracture, ε_F, for 1-mm-diameter zinc monocrystals oriented for single slip and coated with mercury, gallium, or tin is markedly dependent upon the strain rate, $\dot\varepsilon$. For $\dot\varepsilon \sim 10$–15% per min, all three liquid metals reduced ε_F from that in air, but at very low

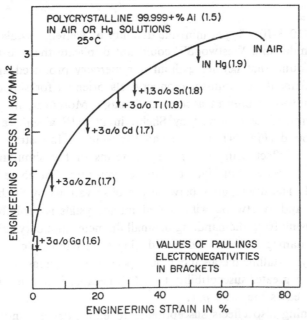

Fig. 22. Embrittlement of polycrystalline pure aluminum by various mercury solutions. Note apparent correlation between Pauling electronegativity of solute element and severity of embrittlement. (After Westwood et al.[118])

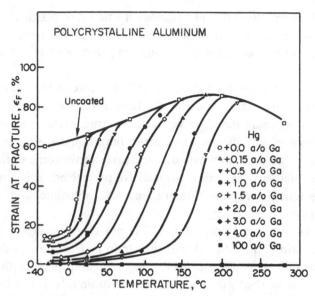

Fig. 23. Temperature dependence of strain at fracture for polycrystalline aluminum specimens in mercury, gallium, and mercury–gallium solutions. (After Preece and Westwood.[125])

strain rates (10^{-1}–10^{-3} % per min) the ductility of wetted crystals increased sharply. Kamdar and Westwood[113] could not duplicate these results, however. They found that neither gallium nor mercury produced any significant embrittlement of 6-mm-sq. monocrystals oriented for single slip and carefully handled to prevent accidental damage. Moreover, not only was the plasticizing effect observed by Shchukin et al.[127] absent, a converse effect was found (Fig. 24). For uncoated crystals at low strain rates, ε_F increased, this effect being interpreted in terms of the simultaneous deformation and recovery of zinc at room temperature. For coated crystals, ε_F decreased. The discrepancy between the observations by Shchukin and by Kamdar and Westwood with coated monocrystals is probably due to problems arising from the handling of small diameter monocrystals without accidentally damaging them and introducing kink bands, etc.

For polycrystalline metals it can be expected that increased strain rates will produce a greater susceptibility to L-ME because increasing $\dot{\varepsilon}$ raises τ, and this decreases the ratio σ/τ.

Prestraining also effects susceptibility to embrittlement, normally increasing the fracture stress and decreasing the strain at fracture.[128,129] However, the effects of prestrain are not always those expected, as indicated in Fig. 25 which presents some data obtained with a strain-aging aluminum alloy.[130] The data for the prestrained and amalgamated aluminum-alloy specimens clearly do not conform to a Petch-type relationship, since the slope of the data indicate a negative fracture energy—an unreasonable result! Other workers[131] have found that severely cold-worked (<50%) specimens of α-brass, iron, and Fe–Si alloys exhibit little susceptibility to L-ME, and fail transgranularly. For deformations of up to about 25%, however, increases in embrittlement were observed and failure was inter-granular. It has been suggested that fragmentation of the original grain boundaries by extensive cold working is responsible for the observed re-duction in susceptibility and change in fracture mode.

Rebinder Effects. According to the Russian literature,[101] the presence of certain adsorbed organic polar molecules (surface-active agents) can significantly affect the mechanical and electrical properties of metals. The yield stress, rate of work hardening, and stress and strain at fracture are found to decrease, and the creep rate and electrical resistance to increase, when metals are deformed in surface-active agents. Such effects have been

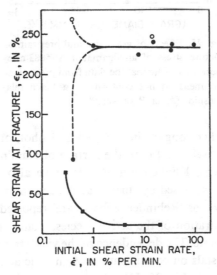

Fig. 24. Strain-rate dependence of the shear strain at fracture for (i) lower curve, 1-mm-diameter zinc monocrystals coated with gallium (after Shchukin *et al.*[127]), note "plasticizing" effect at low strain rates; (ii) upper curves, 6-mm-square zinc monocrystals partially coated with gallium (●) or uncoated (○) (after Kamdar and Westwood[113]).

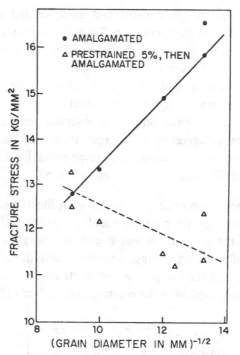

Fig. 25. Effects of grain size and prestrain on fracture stress of amalgamated Al-5083 alloy specimens. Note that the data from prestrained specimens do not conform to a Petch relationship. (After Rostoker.[130])

reported to occur when long-chain fatty acid, alcohol, or amine molecules are adsorbed from vaseline oil onto the surface of aluminum, copper, lead, tin, or zinc specimens. Klinkenberg et al.[132] have also reported that the creep rate of gold is increased by surface-active environments. These weakening effects, known as Rebinder effects, are reportedly dependent on temperature, concentration of the active species, chain length of the active molecule, and strain rate. The dependence of the yield stress and rate of work hardening of tin crystals on concentration of an oleic acid in a vaseline oil environment is shown in Fig. 26. The effect reportedly is maximized when a monomolecular layer of the active species is adsorbed. Much of the previous work on this phenomenon has been reviewed by Kramer and Demer,[133] and Machlin.[134]

Rebinder and coworkers[101] proposed many years ago that such effects result from a physical adsorption-induced reduction in the surface free energy of the metal which allows the easier creation and development of microcracks during deformation. Shchukin[135] has attempted to combine this early hypothesis with modern concepts of dislocation behavior.

At the present time, a considerable controversy exists regarding the actual existence of Rebinder effects in *clean* metals. The work of Andrade and coworkers[136] suggests that the reduction in strength of cadmium crystals occurring in dilute solutions of oleic acid in paraffin represents the removal of the strengthening effect of an oxide film, since it did not occur with crystals cleaned by thermal evaporation in vacuum. Nor could they find any influence of environment on the electrical resistance of cadmium or lead. Similar results have been obtained by other workers with monocrystals of tin[137] and zinc.[138] The latter consider that Rebinder effects in metals are due to disruption of a hardening surface film by the liquid, perhaps by dissolution or fragmentation. More recently, Westwood *et al.*[107] have suggested that such manifestations of the Rebinder effect may be a consequence of the influence of adsorbed molecules on the mechanical behavior of surface films, and that this effect in turn influences the behavior of the underlying metallic substrate. They consider that adsorbed organic molecules should not affect the flow behavior (a bulk property) of metallic specimens having clean surface because the influence of any adsorbate is screened out within a few atomic diameters of the surface, as discussed in connection

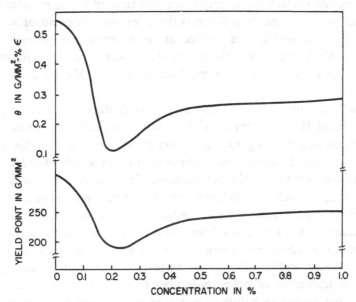

Fig. 26. Effect of oleic acid concentration on the yield point and work-hardening coefficient of tin monocrystals. (After Likhtman *et al.*[101])

with the phenomenon of L-ME. Only if the yielding behavior of a particular metal is controlled by the activation of surface sources might one expect that adsorbates would exert some influence. However, as mentioned earlier, unequivocal instances of such behavior appear to be the exception rather than the rule.

Although no Western worker has been able to reproduce the environmentally induced decrease in reduction in yield stress or electrical conductivity, variations in the work-hardening behavior of aluminum and copper crystals deformed in tension in solutions of paraffin oil or benzene containing various concentrations of stearic acid have been observed by Kramer.[139,140] These results show a marked similarity with others obtained by Kramer from surface-removal experiments (discussed earlier in this section), leading to the suggestion that fatty acid molecules react at the surface to form metal soap molecules which then desorb into the environment. The net effect is to remove the work-hardened surface layer which is considered to develop during deformation. The rate of surface dissolution is a function of the concentration of fatty acid molecules in the environment, the rate of reaction at the surface, and the rate of desorption of the soap molecules. Significantly, no effects were observed when the paraffin oil was presaturated with the appropriate metal stearate, and no effects were observed with gold. It is known that gold soaps are not normally formed because the free energy required is too large.[141] Kramer[140] also notes that metal soaps ordinarily will not form unless an oxide and/or water is present, but that working the surface of clean metals may promote their formation. Kramer suggests, therefore, that under certain conditions involving dissolution, effects on the mechanical behavior of clean metals may be observed.

Adsorbed water molecules also can affect (indirectly) the strength of metallic materials. For example, Nichols and Rostoker[109] have investigated the effects of adsorbed organic liquids on the fatigue life of high-strength steels. A systematic variation in lifetime with chain length of the organic molecules was noted (Fig. 27), but water was found to be the most effective embrittling agent evaluated. When a strong dehydrating agent ($CaSO_4$) was added to the organic environments, reductions in fatigue strength were no longer observed. On the basis of these and other observations, it was concluded that the reductions in strength in the organic liquids were in fact related to the adsorption of water at the surface of the test specimens, and not to the adsorption of polar molecules. Other workers[142–144] have also discussed the aggressive nature of water or water vapor in the premature failure of high-strength steels. Johnson and Willner[143] find, for example,

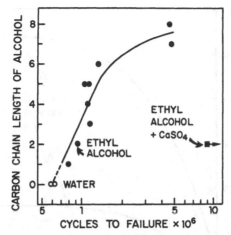

Fig. 27. Variation in fatigue life of a modified AISI-4340 steel when tested in several primary alcohols and in water. Note effect of dehydration with $CaSO_4$ on fatigue life in ethyl alcohol. (After Nichols and Rostoker.[109])

that the rate of crack growth in a high-strength steel reaches a maximum for relative humidities of about 60%, and this limiting rate is equal to that for specimens tested in liquid water (Fig. 28). This equivalence apparently results from the capillary condensation of water at the crack tip in high-humidity environments.[144] This type of embrittlement probably is related to the generation of hydrogen by some surface reaction at the crack tip, followed by its absorption into the lattice, and subsequent embrittling action.

In summary, we consider that Rebinder effects in metals probably are not a consequence of the adsorption of active species on clean metal surfaces as first postulated, but more likely involve either the dissolution of work-hardened surface layers, or adsorption-induced decreases in the hardness of oxide surface films which in turn lead to the easier egress of dislocations from the metallic substrate. Perhaps such effects should be termed pseudo-Rebinder effects to distinguish them from the genuine adsorption-induced reductions in strength observed with nonmetallic crystals.

Gaseous Environments and Vacuum Effects

Gaseous environments can affect mechanical behavior through a variety of mechanisms, some of which have been discussed by Westwood.[112,145]

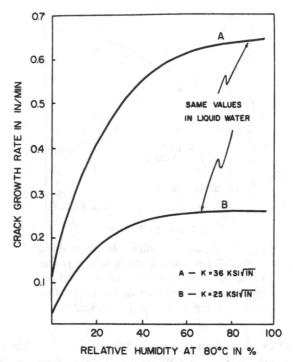

Fig. 28. Saturation of crack growth rates at relative
humidities in excess of about 60%. H-11 steel, 230-ksi
yield strength. (After Johnson and Willner.[143])

These include:

1. Chemical interaction with the surface layers to produce relatively
   hard and sometimes brittle surface films, such as oxides.

2. Diffusion into the surface layers of crystals (a) producing residual
   compressive stresses and a case-hardened layer, as in carburizing
   or nitriding, or (b) affecting the mobility of dislocations generated
   in the near-surface layers.[67]

3. Diffusion into the bulk of the crystal followed by segregation at
   dislocations thereby inducing changes in work-hardening and
   fracture behavior.[146–148]

4. Bulk diffusion followed by a reaction to produce a hard and
   embrittling second phase, for example, hydride platelets in α-
   titanium.[149]

5. Bulk diffusion followed by segregation or precipitation in voids,
   microcracks, or possibly regions of triaxial stress, as postulated

in various theories of hydrogen embrittlement (see References 112 and 150 for reviews).

6. Adsorption-induced effects on the binding of metal surface atoms, either at microcracks or on ostensibly perfect surfaces.

For the purposes of this discussion, we will consider only those phenomena related to surface rather than to bulk processes.

Hydrogen Embrittlement. In addition to the well known embrittlement of both bcc[150] and fcc[151] metals that occurs when an excess of hydrogen is present inside the metal (due to precharging, etc.), embrittlement may also occur when hydrogen is made available to the metal from an external environment during testing. The serrated yielding and increased rates of work hardening observed by Latanision and Staehle[67] in nickel monocrystals that were simultaneously deformed and cathodically charged is one example of this possibility (see Fig. 15). In this case, it has been suggested that hydrogen which diffuses into the surface layers of the crystal affects the mobility of dislocations generated by near-surface dislocation sources.

Embrittlement of high-strength steels has also been shown to occur in a hydrogen atmosphere.[152] As illustrated in Fig. 29, subcritical flaws will initiate at lower stress intensities and propagate at higher rates in 1 atm of purified hydrogen than in a fully humidified argon atmosphere. Oxygen, on the other hand, inhibits crack growth. This effect is shown in Fig. 30. An oxygen-arrested crack may be restarted only after oxygen is removed from the gaseous environment. The role of oxygen probably is to form an

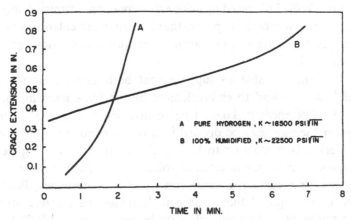

Fig. 29. Subcritical crack extension in hydrogen (1 atm) and humidified argon for H-11 steel of 230-ksi yield strength. (After Hancock and Johnson.[152])

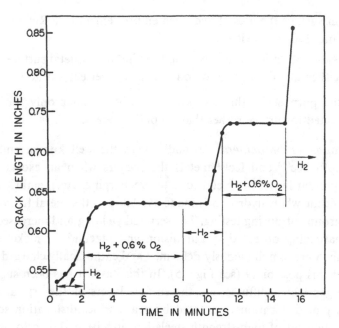

Fig. 30. Subcritical crack growth in an H-11 steel as function
of environment. (After Hancock and Johnson.[152])

oxide film which serves as a barrier to the passage of hydrogen into the
metal. When the oxygen supply is removed, hydrogen may reduce the
oxide and then penetrate the metal. Alternately, the film may crack on
stressing, allowing hydrogen entry. Since this type of hydrogen embrittlement is not markedly influenced by prior exposure of an unstressed specimen
to the hydrogen atmosphere, it appears that hydrogen can enter only through
a clean, oxygen-free surface, and that this prerequisite is provided at the tip
of a stressed crack.

It is not unreasonable to suppose that both external and internal
sources of hydrogen lead to embrittlement by the same mechanism, with
perhaps only differences in kinetics being involved.[150] Classical hydrogen
embrittlement theories may be divided into two general types. On the one
hand, brittleness is postulated to be associated with the development of
high pressures of hydrogen in internal voids.[150,153,154] Such pressures lower
the external stress necessary to cause crack growth. Alternately, Petch and
coworkers have proposed that hydrogen decreases the fracture stress of
steels via an adsorption-induced reduction in surface energy.[155-156] Troiano
and coworkers[157-160] have expanded on this hypothesis, and suggest that
hydrogen diffuses to regions of high triaxial stresses in front of cracks, and

reduces the cohesive or fracture strength of the material in this region. In support of this view, Johnson and coworkers[144,152] have found that the apparent activation energy for crack growth is close to that for hydrogen diffusion in high-strength steels. It may be important to note, however, that studies of the variation in electrical conduction, ferromagnetic moment, and ferromagnetic anisotropy of tungsten, iron, and nickel with chemisorption of oxygen, carbon monoxide, and nitrogen indicate that while these species weaken the binding of metal surface atoms, hydrogen does not.[161] This observation suggests, therefore, that the adsorption of hydrogen may not be an important factor in the hydrogen embrittlement of steels, and that perhaps the "pressure" theories are more relevant. This controversy is far from settled.

Mechanical Properties at Reduced Pressure and in Inert Environments. The effect of reduced pressures on the tensile-deformation characteristics of aluminum are shown in Fig. 31.[162] In some respects, the effects of

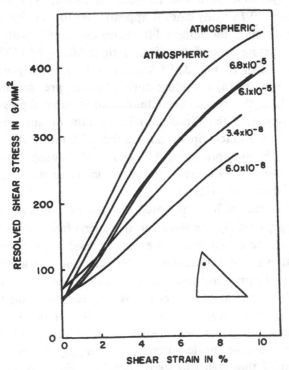

Fig. 31. Influence of air and vacuum environments on deformation behavior of aluminum monocrystals. (After Kramer and Podlaseck.[162])

reduced pressure on work-hardening parameters are similar to those obtained in surface-removal experiments: θ_I and θ_{II} are decreased, and ε_{II} and ε_{III} increased as pressure is lowered. It has been postulated, therefore, that the decreased work hardening rate of specimens deformed in vacuum is associated with a smaller surface-layer stress τ_s. This may result from a decreased rate of oxide-film formation on freshly exposed slip steps and hence a decrease in the effectiveness of the oxide coating (i.e., τ_s) as a barrier to dislocation egress.[162,163] Kramer[72] has recently found that the cyclic creep behavior of aluminum at reduced pressure also may be interpreted in terms of a decrease in τ_s.

Generally, the influence of environmental gases, particularly oxygen and water vapor, on the fatigue behavior of metals has been related to the crack-propagation stage.[164] Wadsworth and coworkers[165,166] have shown that the fatigue lives of aluminum, copper, and gold are extended when the pressure is reduced from 1 atm to 10^{-5} torr. Other workers have shown similar behavior in lead,[167] nickel,[168] type-316 stainless steel[168] and, more recently, polycrystalline pure and alloyed aluminum, Armco Iron, and titanium alloys.[169-171] In many cases it appears that there exists a threshold pressure beyond which the fatigue life increases rapidly with decreasing pressure. The pressure dependence of the fatigue life of Al 1100 (commercially pure) is shown in Fig. 32.[170] Vacuum effects have generally been interpreted in terms of (a) rewelding during the compression half-cycle of those crack surfaces that remain uncontaminated at reduced pressure during the tensile half-cycle,[172] (b) elimination in vacuum of any weakening of atomic bonds at the crack tip by adsorption of atmospheric gases[165-167] or atmospheric corrosion processes,[164,173] and (c), following the debris layer concept of Kramer,[169-170] a decrease in the surface-layer stress τ_s in vacuum due to a reduced rate of oxidation.

A number of studies have indicated that the rate of formation of slip markings is approximately the same for specimens fatigued in vacuum or in air, leading to the conclusion that environmental gases primarily affect crack propagation and not initiation. Grosskreutz and Bowles[174] point out, however, that at the pressures under which many of the above studies were conducted, $\sim 10^{-6}$ torr, a monolayer of oxygen can form on the surface in about 1 sec. The implication is that observation of possible effects on surface slip characteristics or the crack-initiation process would be difficult in such studies, because the test frequencies employed were such as to allow oxidation of the fresh slip steps formed during any one cycle. In accord with this view, Grosskreutz and Bowles found marked differences in slip behavior at 10^{-9} torr, namely, that the formation of slip bands on

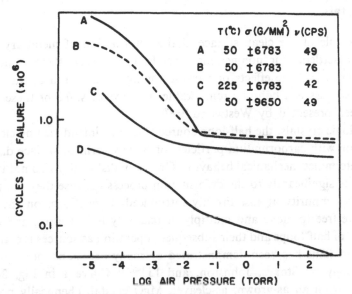

Fig. 32. Effects of frequency, temperature, and stress on the life-time vs pressure relationship for Al-1100 specimens. (After Shen and Kramer.[170])

fatigued aluminum monocrystals was suppressed. They predict, therefore, that the initiation of fatigue cracks at these pressures should be delayed, and there are some indications that this indeed occurs.[175] A more recent extension[176] of this work will be discussed later in this paper.

Few studies have been made of the effects of inert atmospheres on the mechanical behavior of metals. Thompson et al.[172] reported that the fatigue life of copper was increased by a factor of five in nitrogen as compared to air. Snowden,[167,177] on the other hand, found that the fatigue life of poly-crystalline lead was not much improved in nitrogen or hydrogen atmospheres over that in oxygen. Similar behavior has been reported by Achter et al.[168] for nickel. Shahanian and Achter[178] have found that the creep life and creep resistance of low-alloy steels and various high-temperature alloys is, in general, greater in air than in oxygen, nitrogen, hydrogen, or vacuum.

By and large, mechanistic understanding of the influence of gaseous atmospheres on fatigue behavior is poor, and further studies of this topic are necessary. To obtain interpretable results, however, attention should be given to the production of initially clean surfaces, and the composition of the environments should be carefully determined and monitored. These prerequisites have been virtually neglected in most of the work performed to date.

Nonmetals

The role of the clean surface in the deformation of metal crystals is far from settled. For nonmetallic crystals, in contrast, there is no such controversy, and the effects of varying the availability and operability of surface sources can be quite remarkable. Reviews of some of these effects have been presented by Westwood.[112,179]

In ionic crystals, the half-loop source (Fig. 33), introduced by cleaving, sprinkling with carborundum particles, or by any form of surface damage, often determines mechanical behavior. Grown-in dislocations do not usually contribute significantly to the deformation process because they are tightly locked by impurity atoms. Freshly introduced dislocations, on the other hand, are free to move and multiply at relatively low stresses, and so the injection of half-loops and their subsequent operation as sources is extremely significant. This has been demonstrated on many occasions, perhaps most convincingly by Stokes, Johnston, and Li.[180,181] Curve 1 in Fig. 34 was obtained from an as-grown, as-cleaved MgO crystal, chemically polished to remove half-loop surface sources introduced during cleavage, and into which a single fresh source was introduced by dropping a particle of silicon carbide onto the surface. When the crystal was stressed in tension this source operated, and dislocations from it completely filled the specimen. The yield stress was about 5 ksi and the strain at fracture ε_F, $\sim 7\%$. Curve 2 is from a similar specimen containing two artificially introduced dislocation sources. Dislocations from each source intersected and interacted to form a crack which immediately propagated, causing failure. For this specimen, the yield stress again was about 5 ksi, but ε_F was now only 0.2%. Curve 3 is from a similar specimen into which many sources were introduced. As before, intersecting slip bands caused cracks to form, but this time their propagation was hindered by other slip bands, so that a small but useful measure of ductility was achieved. Curve 4 illustrates the deformation

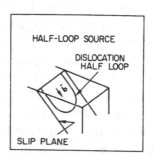

Fig. 33. Schematic of half-loop surface source.

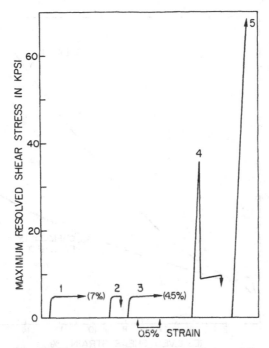

Fig. 34. Tensile stress–strain curves for carefully polished magnesium oxide crystals: (1) containing one artificially introduced surface source; (2) containing two surface sources; (3) containing many surface sources; (4) without surface sources; (5) annealed at 2000°C, cooled rapidly, and polished to remove surface sources. (After Stokes *et al.*[180,181])

behavior of a chemically polished crystal containing no artificially introduced and readily activated surface sources. This crystal was forced, therefore, to use harder interior sources (probably associated with impurity particles) and this raised the yield stress to about 35 ksi. Such specimens failed after about 0.1–0.4% strain. Finally, Curve 5 was obtained from a crystal annealed to 2000°C to dissolve precipitate particles (interior sources), cooled rapidly to prevent their reprecipitation, and then chemically polished to remove all subsequently introduced surface sources. In this case, yielding and catastrophic failure occurred simultaneously at stresses as high as 140 ksi ($\mu/150$). These experiments demonstrate that elimination of the usually present sources of dislocations can increase the yield and fracture stresses of oxide ceramic crystals by as much as 30 times.

Westwood[182] has found that as-cleaved crystals of LiF yield in compression at lower stresses but work harden more rapidly than chemically

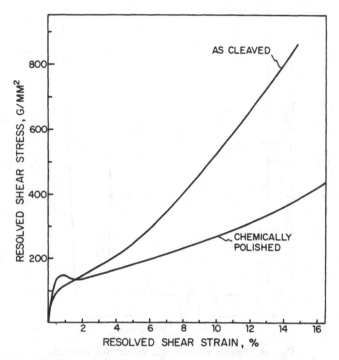

Fig. 35. Effect of surface condition on yield and flow behavior
of LiF crystals. (After Westwood.[182])

polished crystals (Fig. 35). A chemically polished crystal is free of surface
half-loop sources and thus forced to use harder interior dislocation gen-
erators resulting in an increased yield stress. When dislocation multiplication does begin, however, it proceeds at such a high rate that stress drops
often occur. An as-cleaved crystal, on the other hand, deforms by the
operation of many half-loop surface sources at a lower stress. However,
because many sources operate, dislocations on intersecting slip planes soon
meet and interact, resulting in a relatively high rate of work hardening.

Suzuki[183] has discussed similar observations in KCl in terms of the
activation of Fisher-type surface sources (discussed earlier in this section).
The increased yield stress observed in the case of crystals polished (on wet
silk) was presumed to be a consequence of a reduction in the length of
surface sources by the surface pretreatment. The slip distribution in these
experiments was determined by a birefringence technique using light polarized in the glide direction. Slip initiated near the free surface, and only
after strains of 0.3–0.4% was slip detected in the center of the crystal. The
birefringence pattern produced by glide on a secondary system at higher

strains indicated that dislocations of this system also were produced at the surface, and that strong interactions occurred between primary and secondary dislocations just beneath the surface. The result of this interaction was the development of a work-hardened surface layer. He found that the partial removal of this layer, by polishing off some 50 μ from the surface, reduced the subsequent flow stress considerably. Removal of $\sim$1 mm from the surface recovered the flow stress of the virgin crystal. These results suggest, therefore, that work hardening first occurs, and principally resides, in the surface layers of KCl crystals.

Mendelson[184] has discussed the role of the free surface in the plastic deformation of NaCl crystals in terms of the operation of, in our terminology, near-surface sources. His work has shown that the formation of a work-hardened surface layer occurs when alkali halide crystals are polished on wet silk prior to deformation (Suzuki's method), but not when crystals are polished in a solvent environment. The implication is that the formation of a work-hardened layer is dependent upon the presence of a multiplicity of dislocation sources injected by the pretreatment.

Solvent Environments (Joffe Effect)

The best known example of the effect of surface removal during deformation of nonmetals is the Joffe effect.[185] In the classic demonstration of this phenomenon, a salt crystal is shown to be weak and brittle if deformed in air, but up to 25 times stronger and considerably more ductile if deformed in water. This effect is demonstrated for irradiated KCl crystals in Fig. 36.[186] The upper crystal was bent in air; the middle crystal polished in water, carefully dried, and then tested in air; and the lower crystal deformed in water. It seems likely that the remarkable (but inherent) ductility revealed under surface-dissolution conditions results from a combination of factors,[112] namely, (a) the removal, by dissolution, of preexisting notches or cracks introduced by cleaving or mishandling (Joffe's original explanation), (b) the removal by dissolution, of preexisting and embrittling surface films, probably of potassium chlorate or polycrystalline potassium cloride formed by interaction of the crystal with moist air,[187,188] (c) the removal of surface barriers to dislocation emergence,[189] and (d) crack blunting by dissolution and a consequent reduction in propagatability. All of the above factors enhance ductility. High strength results from the elimination of surface-dislocation sources, and work hardening.[112]

Mendelson's[184] observations on NaCl crystals have lead him to conclude that the cross slip of screw dislocations at the surface is easier, and

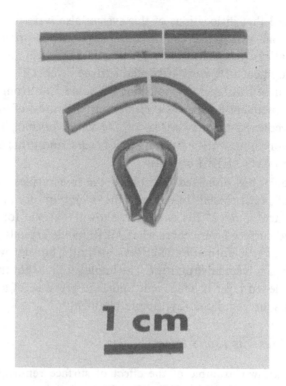

Fig. 36. Illustrating Joffe's effect with irradiated potassium chloride crystals. Upper crystal deformed in
air; middle crystal immersed in water for 60 sec, dried,
and then deformed in air; lower crystal deformed
under water. (After Westwood.[186])

plastic flow more homogeneous, when the surface is dissolved during deformation. In addition to the enhanced ductility which results, greater
strength is attributed to an increase in the density of dislocations in the
crystal, because multiplication via a double cross-slip mechanism is facilitated.

Hirth and Lothe[44] have suggested that the Joffe effect may, in part,
be associated with the influence of image forces on emergent dislocations.
As discussed earlier, image forces tend to pull screw dislocations approaching the surface into a perpendicular orientation. Such dislocations may thus
become kinked or jogged near the surface. Since jogs on screws are sessile,
the net effect is a pinning or relative immobilization of dislocations in the
near-surface region. Removal of this region by dissolution may, therefore,
facilitate yielding and plastic flow.

Some of the interesting effects of solvents on the fracture behavior of germanium crystals have been reviewed by Westwood.[112] Spectacular increases in strength are observed when germanium crystals are rapidly etched and deformed simultaneously. Fracture stresses approaching the theoretical cohesive strength have been recorded under these conditions, and this behavior is attributed to the removal of surface flaws by dissolution.

Effects of Surface-Active Species

As with metals, the effects of adsorbed species on the mechanical behavior of nonmetallic crystals may be understood in a general way by considering the type, concentration, mobility, and adsorption-induced redistribution of the charge carriers in the solid. Many of the phenomena to be discussed in this section have been recently reviewed in detail elsewhere.[110,111,114,190]

Complex-Ion Embrittlement. Because of the much lower concentration and mobility of the charge carriers in nonmetals than in metals, the influence of species adsorbed on a nonmetallic surface can be felt as much as several microns beneath the surface,[119] and their presence can, therefore, affect both flow and fracture behavior. For example, when polycrystalline AgCl is exposed at room temperature to aqueous environments containing highly charged complex ions, such as $6N$ sodium chloride presaturated with AgCl in which the predominant complex species is $AgCl_4^{3-}$, its fracture mode changes from ductile and transcrystalline to brittle and intercrystalline.[191] Both positively and negatively charged complexes can cause brittle behavior, and it has been found that the degree of embrittlement increases with concentration and charge of the critical complex species present in the environment, and is a function of the distribution of charge on the complex.

Metallographic studies on polycrystalline AgCl specimens stressed in complex-containing environments have revealed that cracking is initiated where slip bands are arrested at grain boundaries of large misorientation. In Fig. 37a, for example, intercrystalline cracks (arrowed), which appear dark in transmitted light, have been initiated at each of the blocked slip bands, A, B, and C. Once initiated, such cracks propagate along grain boundaries in a discontinuous manner, as evidenced by the "striae" on the intercrystalline fracture surface shown in Fig. 37b. Such striae are considered to mark arrest positions of the propagating crack. Tests on monocrystals have established that unnotched specimens are essentially immune to cracking in complex-containing solutions, but that notched monocrystals are severely embrittled, failing by discontinuous cleavage. It may be con-

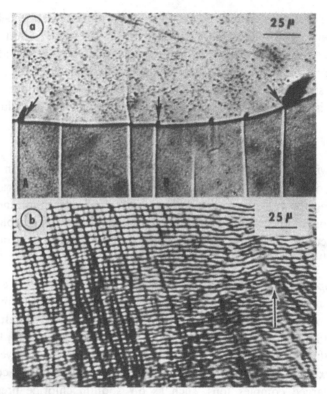

Fig. 37. (a) Polycrystalline AgCl strained in aqueous $6N$ NaCl containing $AgCl_4^{3-}$ complexes. Cracks are initiated where slip bands are arrested at the boundary, e.g., at arrows. Transmitted light. (b) Intercrystalline fracture surface reveals that crack propagation was discontinuous, apparently alternating between intercrystalline "cleavage" and plastic relaxation in the vicinity of the crack tip. The direction of crack propagation is denoted by the arrow. (After Westwood et al.[111,191])

cluded, therefore, that embrittlement is not a consequence of some inherent property of the grain boundary. Nevertheless, boundaries play an important role in the embrittlement of polycrystalline AgCl by acting as barriers to glide dislocations, introducing stress concentrations, and facilitating crack initiation.

It has been proposed[192] that this form of embrittlement may be associated with the repeated formation and rupture of point-defect-hardened charged double layers, as illustrated schematically in Fig. 38. The charge carriers in AgCl at room temperature are Frenkel defects,[193] silver interstitials Ag_i^+, and silver vacancies Ag_v^-, both of which are highly mobile at room temperature (the activation energy for motion of Ag_i^+ is 0.10 ± 0.05 eV, and for Ag_v^- 0.37± 0.10 eV). Following the arguments of Grimley and Mott[194] and others, therefore, it might be envisaged that the adsorption of complex ions of high negative charge at the surface of an AgCl crystal would result in the development of a compensating positive charge in the immediate vicinity of the surface. This charge, termed the "surface charge," would be produced by the presence of a sufficient concentration of Ag_i^+ ions. A potential balancing negative space charge, in the form of a more diffuse aggregation of Ag_v^- defects, would then form further into the crystal. This would represent the "near-surface charge," as defined by Lifshitz and Geguzin.[195] Such a situation is shown schematically in Fig. 38a. Adsorption of positively charged complex ions would be expected to produce a double layer of opposite sense.

The presence of a large concentration of point defects in the surface layers of a specimen would be expected to cause significant surface hardening. Thus, if a specimen exhibiting such a hardened layer were stressed, it is likely that stress concentrations, such as induced by a pile up of dislocations at a grain boundary, would cause it to crack (Figs. 38b and 38c). A crack so formed would propagate readily through the embrittled surface layer, but probably would blunt out when it entered the softer space-charge region of the double layer, or the more ductile matrix material.

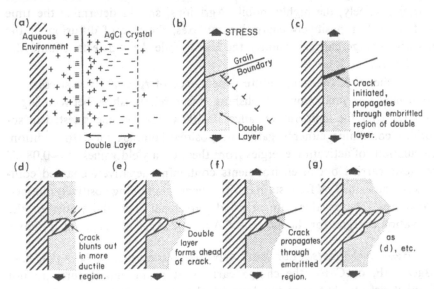

Fig. 38. Schematic of double-layer mechanism for the complex-ion embrittlement of AgCl. (Westwood *et al.*[192])

Meanwhile, however, the liquid environment would have entered the crack, the active species adsorbing on its walls and at its tip, and inducing thereby the formation of a new, defect-hardened region around and ahead of the crack. Then, when the region immediately ahead of the crack became sufficiently brittle, the crack would again propagate under the combined action of the applied stress and stresses set up by continued yielding in the plastic matrix, and again blunt out as it emerged from the severely hardened layer. This sequence, illustrated schematically in Figs. 38c through 38g, could be repeated until the specimen failed.

Convincing experimental evidence to support this mechanism for complex-ion embrittlement has been obtained.[192] For example, the discontinuous failure process described would produce striated fracture surfaces, and evidence for this is shown in Fig. 37b. In addition, the postulated surface hardening associated with the adsorption of complex ions has been revealed by microhardness studies. Finally, according to the model, the rate of crack propagation, and hence the time to failure t_F should be controlled by the rate of formation of the brittle layer ahead of the crack. Under certain conditions, this should be controlled by the rate of diffusion of the appropriate charge carriers to the vicinity of the crack tip, and thus be temperature dependent. In particular, the actual relationship with absolute temperature T should depend on the sign of the charge on the adsorbing species. When negatively charged species adsorb, the rate of diffusion of the positive charge carriers, namely, the highly mobile Ag_i^+ ions, should determine the time to failure. For positively charged complexes, the rate of diffusion of the negative charge carriers, namely the less mobile Ag_v^- defects, should determine t_F.

Studies[192] of the temperature dependence of t_F at a given stress have confirmed this prediction, establishing linear relationships between $\log t_F$ and $1/T$ over the temperature range 10–100°C for specimens tested in solutions containing either negative or positive complex ions. In addition, calculations of activation energies from these data yield values of $\sim$0.08 eV for tests carried out in environments containing negatively charged complexes, and $\sim$0.39 eV for tests in environments containing positively charged complexes. These values are in good agreement with the values of the activation energies for the motion of Ag_i^+ and Ag_v^- defects quoted above.

Rebinder Effects. For more typical ionic crystals than AgCl, e.g., MgO, LiF, or CaF_2, the charge carriers at room temperature are not Frenkel defects, but impurity-donated electrons or holes, and these are present in concentrations several orders of magnitude less than those of the carriers in a metal. The response of such solids to adsorbed surface-active species is, therefore, entirely different to that of either metals or AgCl. Such materials often exhibit the adsorption-induced reductions in microhardness known as "Rebinder effects."

Several possible explanations have been proposed for Rebinder effects, some of which have already been discussed in regard to Rebinder effects in metals. These include adsorption-induced variations in surface free energy,[101] surface cohesion,[145] surface stresses,[196] or the ratio of occupied to unoccupied dangling bonds at the cores of near-surface dislocations in semiconductor materials.[197] Another possibility, suggested by Westbrook and Hanneman,[198] is based on the premise that the coefficient of friction between an indenter and a crystal may be significantly altered by the presence of adsorbed species. Such an effect might be expected to cause some variation in the complex distribution of stress around the indenter. This could lead to a dependence of dislocation mobility on environment and hence microhardness. However, recent studies[199] of the relaxation behavior of near-surface dislocations in CaF_2 exposed to solutions of dimethyl formamide (DMF) and dimethyl sulfoxide (DMSO) have revealed that environments can influence dislocation mobility in the absence of stresses imposed by a loaded indenter. Moreover, for any particular DMSO–DMF environment, a correlation exists between the extent of dislocation motion induced by a loaded indenter and the amount of relaxation which occurs after the loader is removed. It has been concluded, therefore, that indenter-lubrication effects do not play a fundamental role in the occurrence of Rebinder phenomenon in CaF_2.[199]

The studies with CaF_2[199] and LiF[200] have also revealed that adsorption can lead to significant *increases* in hardness. This result is thermodynamically incompatible with the view that Rebinder effects are a consequence of adsorption-induced reductions in surface free energy.

It seems more reasonable to suppose, therefore, that these phenomena are a consequence of the influence of adsorbed species on the mobility of near-surface dislocations. On the basis of this assumption, an apparently consistent explanation for Rebinder effects in ionic crystals has recently been proposed by Westwood, Goldheim, and Lye (WGL).[107,201] It is known that in such solids, dislocation mobility is governed primarily by interactions with point defects and impurity atoms, rather than by the lattice friction stress (Peierls stress).[202,203] It is also considered that chemisorption on insulator materials results from electron or hole transfer between the adsorbate and near-surface point and line defects, rather than between the adsorbate and the lattice itself.[204] It has been suggested,[107,201] therefore, that such adsorption-induced changes in the state of ionization of near-surface point defects and dislocations introduce variations in their mutual interactions which then are manifested as environmentally induced changes in near-surface dislocation mobility, and hence in microhardness.

To examine the validity of this hypothesis, studies have been made of the effects of adsorbed ions, complex ions, and organic molecules on the room-temperature mobility of near-surface dislocation half-loops intro-duced by a diamond indenter into freshly cleaved surfaces of MgO[107,201] and CaF$_2$.[199] The experimental variables in this work were time of indenta-tion and environment. The indenter load was 10 g. The parameter measured was ΔL, where $\Delta L = L_t - L_2$, L_t being the extent of edge-dislocation glide occurring during an indentation time t, and L_2 that during a 2-sec indenta-tion, this being the minimum practical time for reproducible results. Values of L_t and L_2 were determined by optical measurement of the etch-pit distri-butions (rosettes) surrounding the indentations (see the insert to Fig. 39).

Figure 39 illustrates the variation of ΔL with indentation time for MgO exposed to several environments. When the surface was exposed to water-free toluene, no significant dislocation motion occurred following that produced within the initial 2 sec of loading. When the crystal was exposed

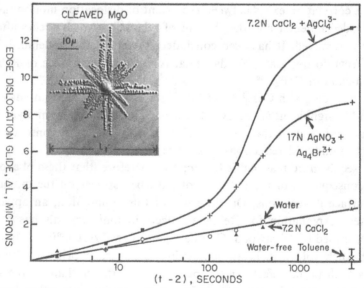

Fig. 39. Variation of ΔL with log (indentation time t minus 2 sec) for freshly cleaved, "as-received" Norton MgO crystals exposed to toluene or various aqueous environments, 10-g load, room temperature. (After Westwood *et al.*[107])

to moist air, water, or an aqueous salt solution, dislocation creep* occurred, the edge dislocations moving some $3\,\mu$ in 4000 sec. However, exposure to environments containing highly charged complexes resulted in significant increases in mobility. In a $7.2N$ $CaCl_2$ solution presaturated with $AgCl_4^{3-}$ complexes, edge dislocations propagated some $13\,\mu$ in 4000 sec. In $17N$ $AgNO_3$ containing Ag_4Br^{3+} complexes, dislocation loops propagated some $9\,\mu$ in the same period of time.

The actual variation of the extent of near-surface dislocation creep in some fixed time t, $\Delta L(t)$, with charge on the predominant complex species in the environment is quite complicated, as is evident from Fig. 40a. Note that an increase in ΔL is equivalent to a decrease in the microhardness of the crystal.

Other experiments[201] in organic environments have revealed that dimethyl formamide (DMF) or dimethyl sulfoxide (DMSO) molecules adsorbed on (100) MgO surfaces exert the same influence on the mobility of near-surface edge dislocations as negatively or positively charged silver chlorocomplexes, respectively. This correlation may be seen by comparing Figs. 40a and 40b. It appears from these data that DMF molecules act as electron-donor adsorbates with respect to (impure) MgO, as $AgCl_4^{3-}$ complexes would be expected to do, while DMSO molecules act as electron-acceptor adsorbates, as would Ag_3Cl^{2+} complexes.

The characteristic double minimum in the relationship between ΔL and environment can also be produced in water-free DMSO–DMF environments (Fig. 41, Curve I), and it has been shown that the exact locations of the minima at A and B can be controlled by changing the concentration or state of ionization of the impurities present in the material.[201] For example, in Fig. 41, Curve I is for "as-received" Norton MgO, containing, among other impurities, ~150 ppm of iron, predominantly as Fe^{3+}. Curve II was obtained using similar crystals which had been heat treated in hydrogen at $1650°$ for 24 h prior to indenting to reduce trivalent impurity cations to the divalent state. It is evident that these observations are in accord with the WGL mechanism for Rebinder effects in ionic solids.

Rebinder effects have also been observed in semiconductor materials. Westbrook and coworkers[196] have shown, for example, that several previously unrelated surface phenomena occur under ordinary laboratory testing conditions only because of the presence of adsorbed water. Such phenomena

---

* This is the "anomalous indentation creep" observed at room temperature when MgO, Al_2O_3, TiC, etc. are indented in a laboratory atmosphere, and shown by Westbrook and Jorgensen[105] to be associated with the presence of adsorbed water.

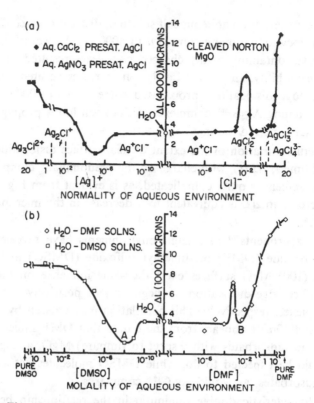

Fig. 40. (a) Variation of ΔL (4000 sec) with normality of aqueous solutions of either $CaCl_2$ or $AgNO_3$ presaturated with AgCl for "as-received" Norton MgO. The predominant complex species present in these solutions are indicated. (b) Variation of ΔL (1000 sec) with molality of aqueous solutions of DMSO or DMF. Compare with (a). Note similarity in form of curves. (Westwood et al.[201])

include (a) the photomechanical effect,[205] which is the decrease in micro-hardness of a semiconductor* material when subjected to illumination of some characteristic frequency; (b) the electromechanical effect,[206,207] which is the reduction in surface hardness of a semiconductor material which occurs when a small potential (~0.1 V) is applied between the indenter and the test surface, or a low current (~100 mA) is passed through the specimen; and (c) the apparent difference in hardness between A and B surfaces of III–V and II–VI compounds, B surfaces being softer in air.[208] This difference disappears when both surfaces are dry.

---

* A photomechanical effect has also been observed with MgO[107].

It is well known that chemisorption can lead to significant variations in surface charge and charge-carrier density in semiconductor materials,[204] and also that variations in bulk-carrier concentration can lead to variations in hardness (dislocation mobility).[209] Thus, Rebinder effects in such materials may be a consequence of direct electron transfer between the adsorbate and localized energy levels associated with the dislocation core and/or chemisorption-induced bending of energy bands in the vicinity of the surface leading to variations in the electronic structure of dislocation cores.[190]

Recently Westwood and Goldheim[210] have considered the possible relevance of their studies of environment-sensitive dislocation mobility in ionic solids to the practical problem of drilling nonmetallic solids. They

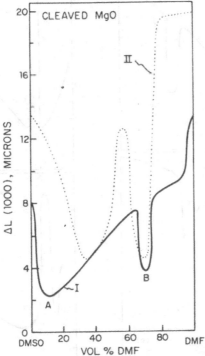

Fig. 41. Variation of ΔL (1000 sec) with composition of DMSO–DMF environment at room temperature for two types of MgO. Data points are omitted to improve clarity. Curve I is for "as-received" Norton MgO crystals. Curve II is for hydrogen-treated Norton MgO. Note shift in position of minimum A with state of ionization of impurities in MgO (cf. Curves I and II). (After Westwood et al.[201])

have determined the variation with environment of the depth of penetration
of a loaded carbide drill into MgO or CaF_2 monocrystals in 600 sec, D(600).
For comparison with their earlier work, the environments used were DMF–
DMSO solutions. The load on the drill was 6.7 kg when drilling MgO
crystals, and 1 kg for CaF_2 crystals. The drill speed was 1750 rpm, and
holes were drilled in a direction perpendicular to freshly cleaved surfaces
on the specimens.

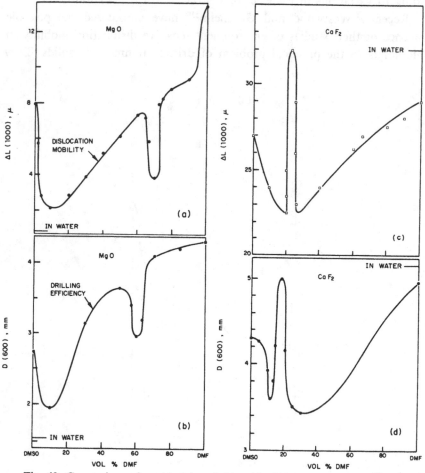

Fig. 42. Comparison of curves (a) and (c), with (b) and (d) suggests direct
relationship between drilling efficiency and dislocation mobility: (a) shows
the extent of edge-dislocation motion ΔL around an indentation produced in
a freshly cleaved {100} surface of MgO as a function of composition of
DMF–DMSO environment; (c) provides similar data for indentations in a
{111} surface of CaF_2; (b) and (d) illustrate the influence of DMF–DMSO
environments on the penetration of a carbide drill after 600 sec.

Some of their results are presented in Fig. 42. Note first that certain DMSO–DMF solutions which soften and facilitate drilling in MgO (with respect to its behavior in water), e.g., 50-vol. % DMF in DMSO (Fig. 42b), harden and make drilling more difficult in CaF_2 (Fig. 42d). It follows that the variations in drilling efficiency shown in Figs. 42b and 42d are not related to differences in the coolant properties of the solution environments. On the other hand, a comparison of these data with those in Figs. 42a and 42c reveals that $\Delta L(1000)$ and $D(600)$ vary in an essentially identical manner with composition of the crystals' environment. This suggests, therefore, that the efficiency of the drilling process in both MgO and CaF_2 may be related directly to the mobility of near-surface dislocations in these materials. A possible explanation of this correlation is that active environments can facilitate the initiation and propagation of cracks by dislocation processes in the region around and ahead of the drill tip—termed the "predestruction zone" by Rebinder.[211] Mechanisms by means of which microcrack growth in MgO can be assisted by the movement of dislocations have been discussed by Clarke et al.[212]

Liquid-Metal Embrittlement of Nonmetals. Liquid-metal environments can also produce significant embrittlement in semiconductor materials.[213] Figure 43, for example, illustrates the effects of liquid-gallium coatings on the bend strength of germanium monocrystals over the temperature range of 0–600°C. For temperatures between 100–350°C, the fracture stress in bending was reduced from about 120 kg/mm² in air to about 10 kg/mm² in gallium. However, it was found that this remarkable effect was not a genuine example of adsorption-induced liquid-metal embrittlement, but was in fact caused by a combination of selective dissolution[214] and the intrinsic notch brittleness of germanium, for cracks were observed to have initiated at crystallographic notches (left inset of Fig. 43) etched into the surface by the liquid metal. This observation has obvious implications for the ultrahigh-strength materials of the future. Such materials, while intrinsically notch brittle, are likely to be too hard to be notched mechanically under ordinary working conditions, but may be notched by chemical means—with potentially disastrous results.

At temperatures above 400°C, on the other hand, germanium crystals exhibited "true" liquid-metal embrittlement, sometimes fracturing below the upper yield stress following measurable plastic strain (Fig. 43, right inset). The variation in susceptibility to embrittlement with group of the liquid metal, and the variation with the solubility of germanium in the liquid metal were investigated. However, it was found that the severity of embrittle-

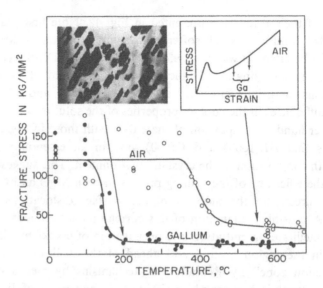

Fig. 43. Embrittlement of germanium by liquid gallium. Below about 350°C embrittlement is associated with the notch-etching effect illustrated in left inset. At about 450°C true liquid-metal embrittlement effects predominate, right inset. (After Pugh et al.[213])

ment was not related to the solubility of germanium in the environment, nor did extensive solubility prevent embrittlement. Group IV liquid metals were least embrittling, but groups II, III, and V liquid metals all induced approximately the same degree of embrittlement, and there was no correlation between degree of embrittlement and electron affinity of the liquid metal. Several other possible correlations were examined, but none was found to be significant.

EFFECTS OF SOLID SURFACE FILMS

Roscoe's discovery[215] in 1934 that thin oxide films on the surface of cadmium crystals caused a significant increase in their yield stress has been confirmed many times with various metal–film combinations. Indeed, there is now ample evidence that the presence of oxide layers, electrodeposited or evaporated films, or alloyed layers can profoundly affect the mechanical behavior of crystalline solids. Reviews by Kramer and Demer (for metals),[133] Westwood (for ionic crystals),[179] and others by Machlin,[134] Westwood[112,145] Grosskreutz and Benson,[46] and Westbrook[2] have clearly documented the influence of surface films in crystal plasticity.

Nonmetals

The effects of solid surface coatings on the mechanical behavior of ionic crystals appear to be reasonably well understood. For example, Westwood[182] found that when magnesium was diffused into the surface of a freshly cleaved LiF crystal to a depth of about 5 μ, the yield stress was increased by 50%, the rate of work hardening decreased, and the fracture stress was reduced by about 50%, as shown in Fig. 44. These effects can be accounted for as follows. An alloyed surface layer effectively restricts the operation of the half-loop sources introduced during cleavage either by (a) solute-ion locking or (b) preventing half-loop expansion and multiplication via increased lattice resistance caused by the presence of Mg ion vacancy pairs, or fine precipitate particles in the surface layers. As-cleaved and coated specimens are thus forced to use other sources of dislocations, namely, interior sources. This raises the yield stress. It also decreases the initial rate of work hardening by reducing the number of active sources and thereby reducing mutual interference effects between dislocations on intersecting slip planes. The sharp yield drop observed is probably due to the catastrophic breakthrough of the alloyed surface layer by piled up groups of dislocations. Indeed, groups of edge dislocations piled up just

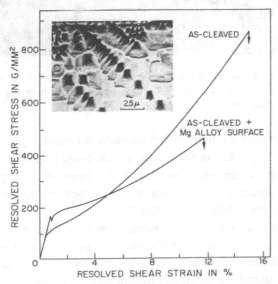

Fig. 44. Effect of diffusing magnesium into surface of an as-cleaved lithium fluoride crystal. Inset shows dislocations piled up beneath alloyed surface layer. (After Westwood.[182])

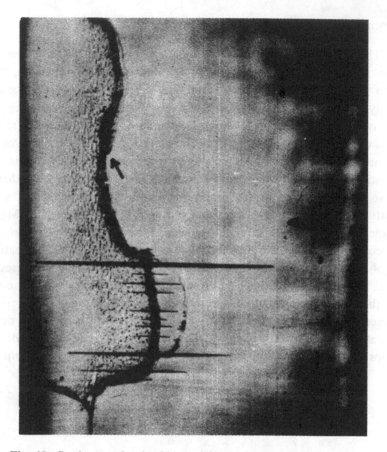

Fig. 45. Cracks associated with precipitate of polycrystalline sodium chloride on the tension surface of a bent sodium chloride crystal. Magnification about 25×. (After Stokes *et al.*[217])

beneath the surface may be revealed by etching coated crystals, as shown in the inset of Fig. 44. Premature failure of coated specimens probably results from the nucleation of cracks at the head of such pileups. The suppression of dislocation emergence at the surface of Mg-doped LiF has also been studied by Bilello and Cadoff.[216]

Figure 45 shows the effect of pileups beneath surface coatings when the bulk material is notch brittle.[217] In this case, cracks have formed at the interface between an NaCl monocrystal and a polycrystalline NaCl surface film produced by inefficient drying after polishing in water.

The influence of certain gaseous environments also may be understood in terms of their reaction with the crystal to produce embrittling surface films. Class *et al.*[187] report that both oxygen and ozone embrittle NaCl as a result of the formation of a sodium chlorate layer. In this case, the crystallography of the film–substrate interface is such that, assuming the formation of a coherent $NaClO_3$ layer, the film is placed in a state of compression. The substrate, of course, is placed in tension at the interface—a condition especially favorable for crack nucleation and propagation. Mendelson[184] also has observed a barrier effect associated with the ozonized surface of octagonal shaped NaCl monocrystals, and the air embrittlement of KCl[188] is likewise thought to be a consequence of the formation of a chlorate layer at the surface which acts as a barrier to dislocation egress and promotes crack nucleation.

The effect of surface segregation on the mechanical properties of ThO_2-doped Y_2O_3 has been demonstrated by Jorgensen and Anderson[43] and is shown in Fig. 46. Segregation of ThO_2 at the free surface increases the microhardness up to depths of as much as $2\,\mu$.

Metals

The controversy concerning whether a metal surface acts as a barrier to dislocation egress, a source of dislocations, or both, extends also into the interpretation of the effects of solid surface films on the mechanical behavior of metal crystals.

One of the most ingenious illustrations of the action of surface films

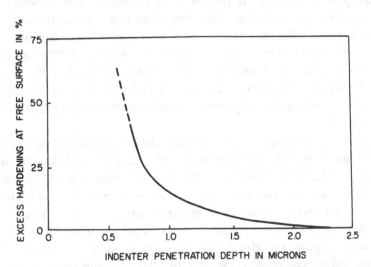

Fig. 46. Depth dependence of microhardness in ThO_2-doped Y_2O_3, suggesting surface segregation of ThO_2. (From Westbrook,[2] after Jorgensen and Anderson.[43])

as barriers to emergent dislocations is the abnormal aftereffect first reported by Barrett.[218] He found that the removal (by immersion in acid) of oxide films from the surface of twisted mono- or polycrystalline specimens of zinc or iron leads to an additional increment of twisting in the original twisting direction. The explanation proposed is that dislocations produced during the torsional deformation are unable to escape through the oxide film, and thus become piled up at the metal–oxide interface. Removal of the oxide releases these dislocations, and they then move out of the crystal, but in that direction in which they were moving during the deformation, causing the slight but detectable retwisting of the specimen.

A more detailed study of this phenomenon by Holt[219] indicates that the Barrett abnormal aftereffect occurs only in wires twisted through angles in excess of $\sim 10°$, and is associated with the presence of noncrystallographic cracks in the oxide film running at $\sim 45°$ to the wire axis. At small angles of twist, $\sim 0.5–1.5°$, a small normal aftereffect is observed.

Other workers have also interpreted the fatigue,[169,170] creep,[220] and tensile behavior[221–225] of film-covered metals in terms of the dislocation-blocking model.

Strengthening effects have been observed in metal monocrystals coated with evaporated or electrodeposited metal coatings,[220–222] thin oxide films,[169,170,223] and oxide or hydroxide layers developed in electrolytes in the absence of applied potentials.[138,223–225] In general, such films increase both the yield stress and the rate of work hardening, and in some cases the three-stage work-hardening behavior of fcc metal monocrystals is suppressed. Figure 47, for example, shows the effect of a 2-μ nickel–chromium coating on copper monocrystals oriented for easy glide.[226] For coated specimens, the measured yield stress is generally regarded as the stress at which dislocations piled up at the metal–film interface break through the surface film. Much of the early work on such phenomena has been discussed by Andrade.[136]

Another interesting example of the blocking effect of surface films has been noted recently by Latanision and Staehle.[68] In their experiment, the presence of a 150- to 200-Å NiO/Ni_3O_4 passive film on 99.8% pure nickel monocrystal straining electrodes was found to increase the yield stress of the electrodes and to produce a generally parabolic stress–strain curve (Fig. 48). The simultaneously measured change in current density which occurred in response to deformation, however, exhibited a series of abrupt increases and gradual decays (Fig. 49). In contrast to this behavior, actively dissolving specimens showed no such irregularities. Latanision and Staehle suggest, therefore, that the current-density surges are associated with rupture

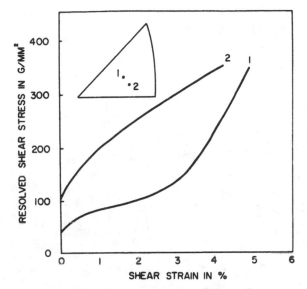

Fig. 47. Effect of nickel–chromium film on the stress–strain curve of similarly oriented copper monocrystals: (1) unplated, (2) plated. (After Garstone et al.[226])

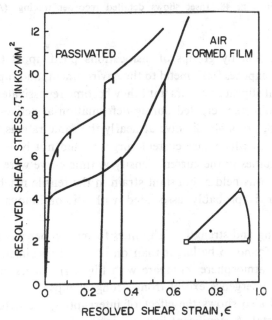

Fig. 48. Effect of passive film on the flow behavior of nickel monocrystal. (After Latanision and Staehle.[68])

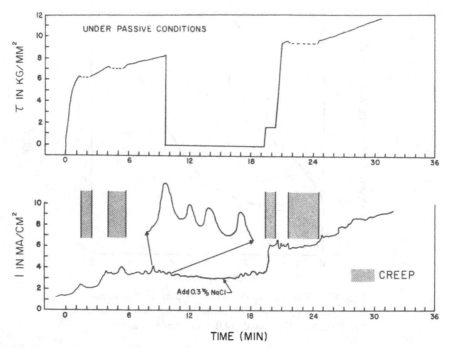

Fig. 49. Stress *vs* time and corresponding current density *vs* time curves for the passivated crystal in Fig. 48. Inset shows detailed recorder tracing. (After Latanision and Staehle.[68])

of the passive film by groups of dislocations piled up at the metal–film interface. This exposes fresh metal to the environment, and rapid dissolution of the emergent slip steps occurs until they become repassivated. The current then decays. Slip steps created during deformation at active potentials, on the other hand, probably dissolve at nearly the same rate as the remainder of the surface, in which case current surges would not be expected.

Discontinuities in the current density *vs* time curve were also observed when a crystal was held at constant strain in the tensile machine (Fig. 49). Such behavior is probably associated with discontinuous creep of the specimen.

The spacing and strength of slip lines formed under passive conditions (Fig. 50) was found to be larger than on identical specimens deformed in the laboratory atmosphere (compare with Fig. 14), suggesting suppression of dislocation emergence except from strong pileups.[68]

Figure 48 also shows the effect of interrupting the deformation of a passivated crystal. A significant increase in the flow stress occurs after the crystal is reloaded from almost zero stress following a delay during which the specimen was maintained at a passive potential. This effect is presumably the result of film growth and/or repair during the rest period.

Leach and coworkers[227] have studied the behavior of thin, anodically oxidized, polycrystalline aluminum wires, and find that the strengthening effect of thin surface films is removed during the process of further anodic oxidation. This implies that an initially embrittling film can become ductile in the presence of an ionic flux (anodic current).

Certain effects of alloyed surface layers on the mechanical behavior of metal crystals have been interpreted in terms of their influence on surface-source behavior by Adams.[228] He noted that while pure copper monocrystals do not exhibit yield point phenomena, the presence of small concentrations of zinc ($\sim$1%) in the surface layers produces sharp yield drops. When the zincified surface is polished off, specimens become softer, and the yield point disappears. When approximately 8% of zinc is present in the surface layers, the yield stress is significantly increased, but yield drops are no longer observed. These results have been interpreted as follows: assuming that slip begins via the operation of surface sources, introducing zinc into the surface layers should raise the c.r.s.s. by alloy hardening. However, providing that the strength of the surface layers is not doubled, surface sources should still operate before interior sources. (This hypothesis presumes, of course, that the glide plane on which the surface source operates is normal to the surface, and that the distribution of source lengths is uniform throughout the crystal.[79]) In heavily doped crystals, on the other hand, surface strengthening is sufficient to completely suppress the action of surface sources. Yielding begins, therefore, at interior sources and at stresses approximately twice that for pure copper crystals.

In rather similar experiments, Rosi[229] found that the c.r.s.s. and flow behavior of copper monocrystals was little affected by a silver-plated film, although the extent of Stage I in crystals oriented for easy glide was slightly

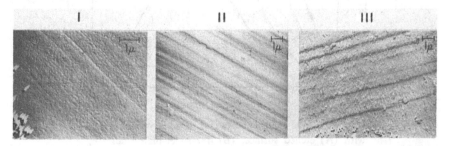

Fig. 50. Slip lines formed in Stages I, II, and III on passivated nickel monocrystals. (After Latanision and Staehle.[68])

reduced. However, if the crystals were heated to form an alloyed layer at the surface, then the c.r.s.s. was increased and the work-hardening behavior was similar to that of copper-alloy crystals containing a low, but homogeneously distributed, concentration of silver (Fig. 51).

While the results of Adams' and Rosi's work may be interpreted in part in terms of the surface source concept, they appear to be more consistent with the near-surface source hypothesis. This is because it is difficult to appreciate why any Fisher-type surface sources present are not readily converted to double-ended sources by the presence of a (presumably) polycrystalline surface film. The recent observation of Johnson and Block[230] that a 125-Å chromium coating does not affect τ_0 of copper monocrystals further suggests that surface sources do not play a major role in the observed strengthening effects. Such surface films, however, should not immediately affect the behavior of near-surface sources whereas alloying the surface layers should and, in the same sense as discussed by Adams,[228]

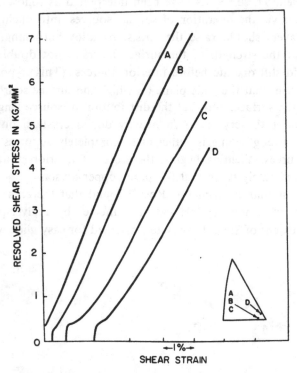

Fig. 51. Stress–strain curves for copper monocrystals with (A) etched surface, (B) 4-μ silver-plated surface, (C) copper–silver-alloy surface, and (D) an homogeneous Cu–0.5% Ag alloy. (After Rosi.[229])

sufficiently high solute concentrations should completely prevent the operation of near-surface sources. For relatively low surface-solute concentrations, however, slip should still begin by the activation of near-surface sources in the alloyed layer, and the resultant flow curve should be typical of that of the alloy, producing yield points, etc., as observed.[228,229]

The serrated yielding of nickel monocrystals during simultaneous hydrogen charging and plastic deformation (Fig. 15)[67] also suggests a dynamical interaction between solute atoms and mobile dislocation moving in the near-surface regions. Again, the implication is that slip begins in the near-surface layers and the resultant stress–strain curve is characteristic of the solid solution present at the surface.

As a general conclusion from the experimental work reported in this section, then, it appears that the most significant strengthening effects of surface films arise as a consequence of their action as barriers to the egress of dislocations from the crystal, and that the dislocations moving in the early stages of yielding are most probably generated in the near-surface layers. However, the operation of near-surface sources can be inhibited by doping with an appropriate solute element.

Mechanisms of Surface-Barrier Effects

The mechanisms by which films interfere with dislocation emergence from the substrate have been related to the elastic interaction of dislocations with the film-covered surface, and to the intimate geometrical and crystallographic relationships between the film and substrate. We will now consider the relevance of these parameters in light of recent experimental observations.

Elastic Theory

As indicated earlier, a dislocation near a free surface is less constrained elastically than a dislocation in the interior, and its self-energy is lower. The consequence of this is, in effect, a force attracting the dislocation to the free surface. The magnitude of this attractive force is roughly equal to that between a near-surface dislocation (at A in Fig. 52a), and an imaginary dislocation of the same strength but opposite sign at the mirror position outside the crystal (at B in Fig. 52a). For a screw dislocation parallel to the free surface, the attractive force per unit length of the dislocation line is given by[44]

$$F_L = - (Gb^2/4\pi r)$$

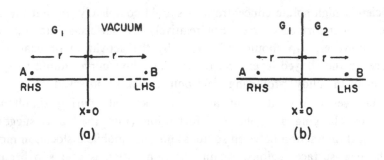

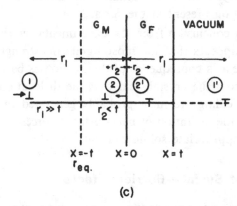

Fig. 52. Schematic of image-force effects on dislocations (a) near a clean free surface, (b) across the interface separating regions of different moduli, and (c) near a surface covered with a film of thickness t.

where G is the shear modulus of the crystal, b the Burgers vector, and r the distance of the dislocation from the surface. This is a long-range force which decreases ar r^{-1}.

When a change in modulus occurs across the interface, as indicated in Fig. 52b, the force between the dislocation and its image is given by [231]

$$F_L = - (G_1 b^2/4\pi r)(G_1 - G_2/G_1 + G_2)$$

where the new parameters are defined in the figure. An important variation now arises, however, in that the dislocation will be attracted to its image only if $G_1 > G_2$, but repelled if $G_2 > G_1$.

A film-covered metal may be considered a semiinfinite elastic medium of shear modulus G_M for $x < 0$, and modulus G_F for $0 < x < t$, where the metal–film interface is set at $x = 0$, and the film surface at $x = t$ (see Fig. 52c). In this case a dislocation of unit strength in the metal will induce a set of images, the strength of which will depend upon its position with respect to the metal–film interface. If $G_F < G_M$, the dislocation is attracted to the interface for all values of $x < 0$. If $G_F > G_M$, the dislocation will be repelled by the film if its distance from the interface, r, is much smaller than the thickness of the film—because its image will be in the film. If, on the other hand, the thickness of the film is negligible compared to r, the dislocation will be attracted toward the interface. The dislocation, therefore, feels a long-range attraction and a short-range repulsion. Consequently, there will be a position of stable equilibrium at some distance from the metal–film interface of the order of the thickness of the film. Calculations by Head[231] for screw dislocations and by Conners[232] for edge dislocations are in accord with this general conclusion.

For a dislocation to emerge from a crystal covered with a film which is of greater elastic modulus than the matrix metal, sufficient strain energy must be supplied by the applied stress to overcome the repulsive force discussed above (neglecting all other factors). If the applied stress is insufficient to do this the film acts as a barrier to slip, with the result that dislocations pile up near the metal–film interface, and the work-hardening behavior of the crystal is affected. Note also that as the thickness of the film increases, so does the distance over which dislocations must be moved against the repulsive image force. Thus a dependence of work-hardening behavior on film thickness may be anticipated.

The elastic repulsion of near-surface dislocations in the presence of an elastically harder coating has been invoked recently by Grosskreutz[176] for his observations on oxidized aluminum monocrystals cyclically deformed at reduced pressures ($\sim 10^{-9}$ torr). His work indicates that deformation in vacuum produced an increased tendency for entrapment of dislocation dipoles just beneath the surface, and that this tended to reduce the extent of surface slip-band formation. Such behavior is illustrated in Fig. 53. The increased density of trapped dipoles in specimens cycled at reduced pressure may be interpreted in terms of the above arguments since, from other experimental results, Grosskreutz suggests that the modulus of thin (3000 Å) Al_2O_3 films is increased fourfold under vacuum, possibly due to the removal of adsorbed water vapor. He has also found that the fracture strength of Al_2O_3 films is increased by about 50% in vacuum, so that the reduced tendency for slip-band formation at the surface may be due both to elastic repulsion effects and to the increased strength of the film.

Grosskreutz has also reported that the Young's modulus of thin Al_2O_3 films in (moist) air is less than that of bulk aluminum. If confirmed, this is a rather surprising result, since the modulus of bulk Al_2O_3 is several

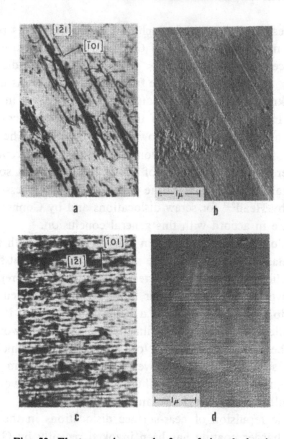

Fig. 53. Electron micrographs from fatigued aluminum specimens deformed (a) and (b) in air, and (c)
and (d) at 10^{-9} torr. (a) and (c) are transmission
electron micrographs showing dislocation arrangement near the surface; (b) and (d) are surface replicas
showing slip line patterns. (After Grosskreutz.[176])

times greater than that of aluminum.[233] Other workers, however, have also
shown that oxides which are normally brittle in bulk form may be appreciably ductile as thin films.[234] Whether this difference in behavior is related
to the greater influence of adsorbed species, such as water, on the mobility
of dislocations in thin films than in bulk specimens rather than to changes
in modulus has not been investigated.

The elastic repulsion concept cannot by itself account for the results
of some recent experiments[230] on the strengthening effects produced in
copper monocrystals by electroplated coatings of lead, gold, copper, chromium, rhodium, ruthenium, and rhenium. The modulus of these plating

materials in bulk form increases from lead to rhenium. According to elastic theory, one might expect that dislocations approaching the surface of a copper crystal would be attracted by lead or gold films, but repulsed by the others, and that this difference in behavior would be reflected by detectable differences in the yield stress. The results of these experiments are shown in Fig. 54. These data are complicated by the fact that some films are polycrystalline (chromium and ruthenium), and some are of different crystal structure than the substrate. However, there appears to be no systematic correlation between the observed strengthening and the elastic constants of epitaxial films of gold, copper, and rhodium. The fact that the copper plate itself increases the c.r.s.s. of the crystal is clear evidence that other factors are also of major importance.

Jemian and Law[235] have recently studied the abnormal aftereffect on polycrystalline wires coated with various metals. Figure 55 shows examples of the torsional relaxation and the abnormal aftereffect in copper and gold wires electroplated with coatings of higher or lower modulus than the substrate ($G_{Cr} > G_{Cu} > G_{Zn} > G_{Au}$). Each wire was given an initial twist, released, and allowed to unwind; and then after 16 min of recovery, the coatings were selectively removed by appropriate acids. Though these experiments are not unambiguous, in each case a transient winding occurred when the coatings were removed regardless of the relative magnitudes of the elastic moduli of the film and substrate materials.

It appears, therefore, that in real situations, the differential modulus effect plays a relatively minor role in determining the ease of egress of

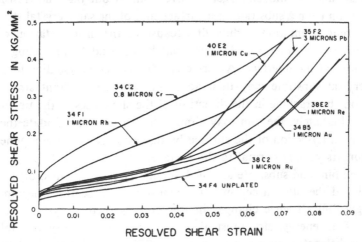

Fig. 54. Stress–strain curves from initially identical copper monocrystals coated with various electrodeposited metals. (After Johnson and Block.[230])

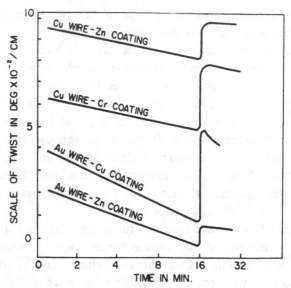

Fig. 55. The abnormal aftereffect in coated copper and gold wires. All coatings were dissolved in acid solutions after 16 min elastic recovery. (After Jemian and Law.[235])

dislocations from film-covered crystals. We shall now attempt to show that the atomistic structure of the interface is of more practical importance.

Atomistic Nature of the Film–Substrate Interface

Evans and Schwarzenburger[236] have pointed out that the transfer of a dislocation from a substrate to a surface film of the same crystal structure but different orientation is difficult because of the misorientation of slip planes across the interface. This mismatch will lead to the piling up of dislocations at the interface. A perfectly adherent polycrystalline film on a single-crystal substrate may then accommodate the resultant stress concentrations by cracking in the direction of the slip traces in the underlying crystal. Alternately, the film may simply detach to accommodate the step created by intersection of the surface by dislocations with a component of their Burgers vectors normal to the surface.

If the film and substrate are of different crystal structures, or if one is a metal and the other a ceramic, the transfer of dislocations becomes increasingly difficult since changes in the Burgers vector, crystal plasticity, stacking fault energy, etc., must also be accommodated. For example, an extended dislocation in a substrate of low SFE would be required to constrict before passing into a film of higher SFE.

The fundamentally more informative case of an epitaxial monocrystal film on a monocrystalline substrate of the same structure has been considered in detail by Brame and Evans,[237] Johnson and Block,[230] and Ruddle and Wilsdorf.[238] Brame and Evans investigated the deformation of thin (300–700 Å) monocrystal films of various elements evaporated onto monocrystal substrates of palladium or silver such that the film and substrate lattices were in parallel orientation. In this case, any misfit between the two lattices should be accommodated by a two-dimensional network of dislocations in the interface, the mesh size of the network decreasing as the degree of mismatch increases. An example of such accommodation dislocation networks is shown in Fig. 56. It might be expected that the effectiveness of an interfacial dislocation network as a barrier would diminish for small

Fig. 56. Transmission electron micrograph showing long, straight misfit dislocations in a deposit of platinum on gold. The plane of the figure is (001) and its vertical and horizontal boundaries are parallel to the ⟨110⟩ directions. Misfit, 3.9%; magnification, ∼250,000×. (Matthews and Jesser.[239])

lattice mismatches, that is, for small differences in the lattice parameters of the film and substrate. In this case, dislocations should be able to pass relatively freely from the substrate into the film. In agreement with this expectation, Brame and Evans have shown by means of transmission electron microscopy that dislocations do indeed transfer readily when the lattice misfit is less than about 3%, but are blocked at the interface when the misfit is much larger than this.

In general, the mode of deformation of an oriented film on a substrate depends upon the ease with which dislocations can be injected into it from the substrate. If injection occurs readily, the film will behave in a ductile manner; if not, the film will crack. The activation of dislocation sources in a thin film is not considered very likely, and there are indications from *in situ* electron-microscope studies of unattached monocrystal films formed by vapor deposition[240] that little or no dislocation movement occurs before fracture.

As mentioned earlier, any difference in lattice parameter between substrate and film will require a finite change in the Burgers vector of any dislocation crossing the interface. Thus, Brame and Evans[237] have suggested that a residual dislocation, having a Burgers vector of magnitude b_F-b_M, will be left behind at the interface for each dislocation which passes into the film (b_F and b_M are, respectively, the Burgers vector of glissile dislocations in the film and metal substrate). Such sessile dislocations should act as barriers to the passage of subsequent dislocations along active slip planes. Moreover, the influence of this effect should increase with strain, and the rate of increase of hardening due to this process should increase with degree of misfit between the two lattices.

Turning again to experimental observations, Ruddle and Wilsdorf[238] have recently reported the somewhat surprising result that oriented copper monocrystals coated with 600-Å epitaxial electrodeposits of nickel yield in tension at approximately half of the stress required for unplated crystals (Fig. 57). They interpret their results in terms of the theoretical prediction of van der Merwe[241] and Jesser and Kuhlmann-Wilsdorf[242] that for lattice misfits of ~4%, an electrodeposited coating will be elastically strained to match exactly the substrate lattice, at least in the early stages of growth. Thereafter, as the deposit thickens, an increasing fraction of the misfit will be accommodated by interfacial dislocations. Such behavior has been observed in thin-film nickel–copper bicrystals.[243-244] Ruddle and Wilsdorf propose, therefore, that on stressing the nickel-plated copper crystal, the presence of surface steps and the elastically strained condition of the dilated nickel surface film (misfit, +2.5%) allow relatively easy nucleation of

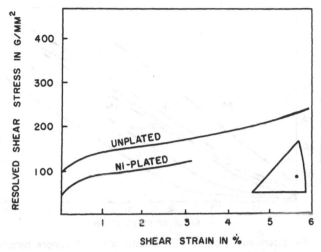

Fig. 57. Stress–strain behavior of unplated and nickel-plated copper monocrystals. (After Ruddle and Wilsdorf.[238])

dislocations in the film. As these dislocations move to the interface under the applied stress, dislocations present or generated at the interface are forced to glide into the substrate, inducing macroscopic yielding at a relatively low stress (calculated in terms of the cross section of the crystal).

Ruddle and Wilsdorf also report a similar effect for gold-plated films (misfit, −11.3%) on the yield stress of copper crystals. Johnson and Block's work with this system, on the other hand, shows no indication of any such reduction (Fig. 54).

In criticism of the Ruddle and Wilsdorf explanation, it seems difficult to envision the generation of a sufficient density of dislocations in the nickel film to account for the observed effect on the substrate material in view of the reported high strength[245] and brittle nature[240] of unattached films. It is clear that definitive work is required to decide whether, during the deformation of a metal–metal film composite, slip is transferred initially or preferentially from the substrate to the film, as suggested by Brame and Evans,[237] or vice versa, as proposed by Ruddle and Wilsdorf.[238]

The studies by Johnson and Block[230] demonstrate that the magnitude of the strengthening effect produced by a particular coating cannot be predicted on the basis of lattice mismatch alone (Fig. 54). For example, an electrodeposited gold coating strengthens copper crystals, and though the misfit is −11.3%, the gold film was found to behave in a ductile manner. Epitaxial rhodium also produce significant strengthening but, although the misfit was only 4.5%, exhibited brittle cracking. Epitaxial copper, with zero

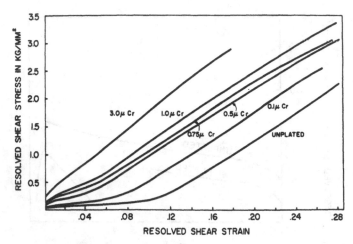

Fig. 58. Stress–strain curves for initially identical copper mono-crystals coated with polycrystalline chromium films of various thick-nesses. (After Block and Metzger.[246])

misfit, also produced significant strengthening. In this case, it is possible that purity of the film may be a relevent factor.

The marked film strengthening effects for chromium and rhodium noted in this study appear to be associated with cracking of these coatings, either prior to or during deformation. Etch-pitting studies revealed an increased density of dislocations near the interface, presumably generated by stress pulses accompanying film cracking. The depth of surface damage resulting from the cracking of 0.8-μ chromium coatings was $\sim 15\,\mu$, and this presumably interfered with the motion of dislocations in the near-interface regions producing the marked hardening observed. It is suspected that the extent of damage produced by film cracking increases with coating thickness, and also with strain—because of the intensification of film cracking. Such behavior possibly is reflected in the data of Fig. 58, which shows the effect of increasing the thickness of polycrystalline chromium coatings on the deformation behavior of specimens from the same monocrystal.[246] The c.r.s.s. and rate-of-strain hardening increase with coating thickness, while the extent of easy glide decreases.

It is well known that electrodeposited and evaporated metal coat-ings,[245,247] as well as oxide films,[227] often contain high residual stresses after growth. It seems likely that such stresses will also influence the mechanical behavior of substrate–film composites.

In summary, the influence of solid surface films on the mechanical behavior of metal substrates depends in a complex and interrelated manner on such factors as the crystal structure, composition, and mechanical properties of both components; the degree of atomic misfit at the interface; and the residual stresses in the film and the adhesion of the film. When significant differences exist between film and substrate, for example, in the case of a metal–ceramic composite, the resulting strengthening effects can be substantial. At the other extreme, however, e.g., metal–metal composites of the same crystal structure, the likelihood of strengthening or weakening effects will be determined by the sensitive and presently not well understood relationships between the above factors. At the moment, the behavior of such composites is largely unpredictable.

CONCLUDING REMARKS

From the preceding discussions, it is evident that the flow and fracture behavior of an inorganic monocrystal can be significantly influenced by the atomic, electronic, and defect structure of its near-surface regions, and by its environment. The extent to which such surface- and environment-sensitive mechanical phenomena are relevent to the behavior of more practical, engineering-type solids will depend on a number of factors, but principally on the complexity of their microstructure and the type of testing or deformation involved. For example, since fatigue failure usually is initiated at the surface, the fatigue life of a structural component is likely to be environment sensitive. On the other hand, the unidirectional tensile behavior of a metal-matrix composite material is not likely to be significantly influenced by the presence of an oxide film!

There are, however, a number of areas where further work is required before the mechanisms of certain surface- or environment-induced phenomena can be considered reasonably well understood even for simple solids. For example, we need to know more about the influence of adsorbed species on cohesion at the surface, particularly of metals, and the possible influence of lattice strain on the adsorption process. Such information is necessary for any real understanding of liquid-metal embrittlement phenomena, and could also be relevent to certain aspects of stress-corrosion cracking and hydrogen embrittlement.

Experiments should be devised to establish directly and unequivocally whether or not a dislocation-rich near-surface layer preexists or is established during the early stages of plastic deformation of "clean" metal monocrystals. If it does exist, what are the factors determining its existence, extent and influence?

The observation of significant flow stress gradients in the surface regions of large metal monocrystals[87-89] should be further investigated, and the controversy[248,249] of "hard" vs "soft" surface layers on deformed crystals settled.

It is important to establish the mechanism of the transfer of plastic strain across interfaces, for example, between a metal substrate and an epitaxial metal surface film. What are the factors determining whether dislocations are generated in the substrate and then injected into the film, or vice versa? Such studies would also be relevent to the behavior of composite materials.

The suggestion[110,111] that adsorption-sensitive mechanical behavior can be understood in a general way from considerations of the type, concentration, mobility, and adsorption-induced redistribution of the charge carriers in the solid seems to be consistent with observation. However, much more work is required before sufficiently detailed information will be available to allow predictability, and to explain, for example, why adsorbed DMSO molecules soften MgO but harden CaF_2.

Further studies of Rebinder-type phenomena in general could be of both scientific interest and technological value. Given improved understanding, adsorption-induced variations in surface hardness could be optimized for application to drilling, machining, compaction, comminution, and the reduction or enhancement of friction and wear.

The electrocapillary effect, and the influences of electrical fields on mechanical behavior, should also be further investigated. Machlin's[250] observation of a sevenfold increase in the plasticity of NaCl crystals in the presence of a 6-kV field should be pursued. Such work could lead to new technology for shaping otherwise "brittle" solids.

The mechanism by which absorbed hydrogen embrittles steel remains unclear. It is certainly about time the true nature of this important phenomenon was settled once and for all.

In short, a better understanding of surface- and environment-sensitive mechanical behavior could provide many important technological advances, ranging from easier methods of forming complex shapes from strong solids, to improved lifetime and reliability of performance of complex and expensive structures in aggressive environments.

ACKNOWLEDGMENTS

It is a pleasure to acknowledge the assistance of D. L. Goldheim and P. A. Boyer, and the helpful discussions with Drs. A. J. Melmed, J. Kruger, and A. W. Ruff, Jr. during the preparation of this review. We also wish to thank Professor S. Kitajima for making a copy of his manuscript available before publication. R. M. L., as an NAS-NAE-NRC Postdoctoral Fellow, is grateful to Dr. S. Silverman, Associate Director for Academic Liaison, National Bureau of Standards, for permission to pursue this review, and A.R.C.W. to the U.S. Office of Naval Research [Contract No. Nonr-4162(00)] for financial support.

REFERENCES

1. *Fundamental Aspects of Stress Corrosion Cracking*, NACE, Houston (1969).
2. J. H. Westbrook, in *Surfaces and Interfaces II*, Syracuse University Press, Syracuse, N. Y. (1968), p. 95.
3. W. H. Haslett, in *Environment-Sensitive Mechanical Behavior*, Gordon and Breach, New York (1966), p. 319.
4. L. E. Samuels, *The Surface Chemistry of Metals and Semiconductors*, John Wiley and Sons, New York (1959), p. 82.
5. *The Surface Chemistry of Metals and Semiconductors*, John Wiley and Sons, New York (1959).
6. *Metal Surfaces: Structure, Energetics, and Kinetics*, American Society for Metals, Metals Park, Ohio (1963).
7. *Solid Surface*, North Holland, Amsterdam (1964).
8. *Fundamental Phenomena in the Materials Science*, Vol. 3, Surface Phenomena, Plenum Press, New York (1966).
9. *Surfaces and Interfaces I*, Syracuse University Press, Syracuse, N. Y. (1967).
10. A. W. Adamson, *Physical Chemistry of Surfaces*, Interscience, New York (1967).
11. *Molecular Processes on Solid Surfaces*, McGraw-Hill, New York (1969).
12. F. C. Frank, in *Growth and Perfection in Crystals*, John Wiley and Sons, New York (1958), p. 3.
13. N. Cabrera, *Disc. Faraday Soc.* **28**, 16 (1959).
14. W. Kossel, *Nach. Ges. Wiss.*, *Göttingen* **135** (1927).
15. I. N. Stranski, *Z. Phys. Chem.* **136**, 259 (1928); **11**, 421 (1931).
16. W. K. Burton, N. Cabrera, and F. C. Frank, *Phil. Trans. Roy. Soc. London* **243A**, 299 (1950).
17. J. F. Nicholas, *An Atlas of Models of Crystal Surfaces*, Gordon and Breach, New York (1965).
18. C. Herring, in *Physics of Powder Metallurgy*, McGraw-Hill, New York (1951), p. 143.
19. C. Herring, in *Structure and Properties of Solid Surfaces*, University of Chicago Press, Chicago (1953), p. 5.
20. R. Shuttleworth, *Proc. Phys. Soc.* **A63**, 444 (1950).
21. W. W. Mullins, Ref. 6, p. 17.
22. A. U. MacRae and L. H. Germer, *Ann. N. Y. Acad. Sci.* **101**, 627 (1963).
23. G. A. Somorjai, *Ann. Rev. Phys. Chem.* **19**, 251 (1968).
24. H. Lyons and G. A. Somorjai, *J. Chem. Phys.* **46**, 2539 (1967).

25. R. E. Schlier and H. F. Farnsworth, in *Semiconductor Surface Physics*, Vol. 3, University of Pennsylvania Press, Philadelphia (1956); *J. Chem. Phys.* **30**, 917 (1959); H. F. Farnsworth, Ref. 5, p. 21.
26. J. M. Charig, *Appl. Phys. Letters* **10**, 138 (1967).
27. L. H. Germer, R. M. Stern, and A. U. MacRae, Ref. 6, p. 287; L. H. Germer, Ref. 8, p. 23.
28. B. C. Clark, R. Herman, and R. F. Wallis, *Phys. Rev.* **139A**, 860 (1965).
29. A. A. Maradudin and J. Melngailis, *Phys. Rev.* **133A**, 1188 (1964).
30. P. Wynblatt and N. A. Gjostein, *Surface Science* **12**, 109 (1968).
31. J. J. Burton and G. Jura, *J. Phys. Chem.* **71**, 1937 (1967).
32. J. Alder, J. R. Vaisnys, and G. Jura, *J. Phys. Chem. Solids* **11**, 182 (1959).
33. G. C. Benson, P. I. Freeman, and E. Dempsey, in *Adv. Chem. Series*, Vol. 33, Solid Surfaces and the Gas-Solid Interface, Am. Chem. Soc., Washington, D.C. (1961), p. 26.
34. G. C. Benson and K. S. Yun, *The Gas-Solid Interface* **1**, 203 (1967).
35. E. G. McRae and C. W. Caldwell, *Surface Science* **2**, 509 (1964).
36. G. A. Somorjai, *Surface Science* **8**, 98 (1967).
37. P. W. Palmberg and T. N. Rhodin, *J. Appl. Phys.* **39**, 2425 (1968).
38. D. G. Fedak and N. A. Gjostein, *Surface Science* **8**, 77 (1967); *Acta Met.* **15**, 827 (1967).
39. L. A. Harris, *J. Appl. Phys.* **39**, 1419, 1428 (1968).
40. R. E. Weber and W. T. Peria, *J. Appl. Phys.* **38**, 4355 (1967).
41. K. T. Aust and J. H. Westbrook, in *Lattice Defects in Quenched Metals*, Academic Press, New York (1965), p. 771.
42. K. T. Aust, P. Niessen, R. E. Hanneman, and J. H. Westbrook, *Acta Met.* **16**, 291 (1968).
43. P. J. Jorgensen and R. C. Anderson, *J. Am. Ceram. Soc.* **50**, 553 (1967).
44. J. P. Hirth and J. Lothe, *Theory of Dislocations*, McGraw-Hill, New York (1968).
45. F. R. N. Nabarro, *Theory of Crystal Dislocations*, Clarendon Press, Oxford (1967).
46. J. C. Grosskreutz and D. K. Benson, Ref. 2, p. 61.
47. R. H. Heidenreich and W. Shockley, in *Strength of Solids*, The Physical Society, London (1948), p. 57.
48. F. E. Fujita, D. Watanabe, M. Yamamoto, and S. Ogawa, *J. Phys. Soc. Japan* **11**, 502 (1956).
49. K. Sumino and M. Yamamoto, *J. Phys. Soc. Japan* **16**, 131 (1961).
50. F. W. Young, Jr., in *Dislocation Dynamics*, McGraw-Hill, New York (1968), p. 313.
51. I. R. Kramer and L. J. Demer, *Trans. AIME* **221**, 780 (1961).
52. I. R. Kramer, *Trans. AIME* **227**, 1003 (1963).
53. I. R. Kramer, *Trans. AIME* **230**, 991 (1964).
54. I. R. Kramer, *Trans. AIME* **233**, 1462 (1965).
55. C. Feng and I. R. Kramer, *Trans. AIME* **233**, 1467 (1965).
56. I. R. Kramer, Ref. 3, p. 127.
57. I. R. Kramer, *Trans. ASM* **60**, 319 (1967).
58. I. R. Kramer and C. L. Haehner, *Acta Met.* **15**, 199 (1967).
59. I. R. Kramer, *Trans. AIME* **239**, 520 (1967).
60. I. R. Kramer, *Trans. AIME* **239**, 1754 (1967).
61. F. J. Worzala and W. H. Robinson, Ref. 3, p. 183; *Phil. Mag.* **15**, 939 (1967).

62. B. Chalmers and R. S. Davis, in *Dislocations and Mechanical Properties of Crystals*, John Wiley and Sons, New York (1956), p. 232.
63. S. Kitajima, H. Oasa, and H. Kaieda, *Trans. Jap. I. M.* **8**, 185 (1967).
64. S. Kitajima, H. Tanaka, and H. Kaieda, *Trans. Jap. I. M.* **10**, 12 (1969).
65. J. C. Fisher, *Trans. AIME* **194**, 531 (1952).
66. D. Kuhlmann-Wilsdorf, Ref. 3, p. 681.
67. R. M. Latanision and R. W. Staehle, *Scripta Met.* **2**, 667 (1968).
68. R. M. Latanision and R. W. Staehle, *Acta Met.* **17**, 307 (1969).
69. K. Sumino, *J. Phys. Soc. Japan* **17**, 454 (1962).
70. R. C. Fabiniak and D. Kuhlmann-Wilsdorf, Ref. 3, p. 147.
71. Y. Nakada and B. Chalmers. *Trans. AIME* **230**, 1339 (1964).
72. I. R. Kramer, private communication.
73. J. T. Fourie, *Phil. Mag.* **15**, 187 (1967).
74. F. R. N. Nabarro, Z. S. Basinski, and D. B. Holt, *Adv. Phys.* **13**, 193 (1964).
75. J. J. Gilman and W. G. Johnston, *Sol. State Phys.* **13**, 147 (1962).
76. R. J. Block and R. M. Johnson, *Acta Met.* **17**, 299 (1969).
77. R. M. Latanision, *Scripta Met.* **3**, 465 (1969).
78. J. W. Mitchell, J. C. Chevrier, B. J. Hockey, and J. P. Monaghan, Jr., *Can. J. Phys.* **45**, 453 (1967).
79. J. Friedel, in *Electron Microscopy and Strength of Crystals*, Interscience, New York (1963), p. 605.
80. J. Lothe, in *Fundamental Aspects of Dislocation Theory*, Rept. of Conf. at the National Bureau of Standards, Gaithersburg, Md., April, 1969.
81. F. R. N. Nabarro, *Adv. Phys.* **1**, 332 (1952).
82. E. D. Shchukin, *Proc. Acad. Sci. USSR* **118**, 1105 (1958).
83. J. C. Grosskreutz, *Acta Met.* **13**, 1269 (1965).
84. Y. Nakada and B. Chalmers, *J. Appl. Phys.* **33**, 3307 (1962).
85. Y. Nakada, U. F. Kocks, and B. Chalmers, *Trans. AIME* **230**, 1273 (1964).
86. R. L. Fleischer, *Acta Met.* **8**, 598 (1960).
87. J. T. Fourie, *Can. J. Phys.* **45**, 777 (1967).
88. J. T. Fourie, *Phil. Mag.* **17**, 735 (1968).
89. J. T. Fourie, *Scripta Met.* **2**, 63, 629 (1968).
90. P. B. Hirsch, *Disc. Faraday Soc.* **38**, 111 (1964).
91. C. T. B. Foxon and J. G. Rider, *Phil. Mag.* **17**, 729 (1968).
92. I. R. Kramer and C. L. Haehner, *Acta Met.* **15**, 678 (1967).
93. I. R. Kramer and A. Kumar, *Scripta Met.* **3**, 205 (1969).
94. P. R. Swann, *Acta Met.* **14**, 900 (1966).
95. J. Krejci and P. Lukas, *Czech. J. Phys.* **B18**, 954 (1968).
96. U. Essmann, M. Rapp, and M. Wilkens, *Acta Met.* **16**, 1275 (1968).
97. J. M. West, *Electrodeposition and Corrosion Processes*, Van Nostrand, New York (1965).
98. G. Lippmann, *Ann. Chim. Phys.* **5**, 494 (1875).
99. D. A. Vermilyea, Ref. 1, p. 15.
100. P. A. Rebinder and E. K. Venstrem, *Dokl. Akad. Nauk. SSSR* **68**, 2 (1949).
101. V. I. Likhtman, P. A. Rebinder, and G. V. Karpenko, *Effect of A Surface Active Medium on the Deformation of Metals*, H.M.S.O., London (1958).
102. V. I. Likhtman, E. D. Shchukin, and P. A. Rebinder, *Physicochemical Mechanics of Metals*, Academy of Sciences, USSR, Moscow (1962).

103. E. K. Venstrem and P. A. Rebinder, *Z. Fizh. Khim.* **26**, 12 (1952).

104. J. O'M. Bockris and R. Parry-Jones, *Nature* **171**, 930 (1953).

105. J. H. Westbrook and P. J. Jorgensen, *Trans. AIME* **233**, 425 (1965).

106. F. P. Bowden and J. E. Young, *Research* **3**, 235 (1950).

107. A. R. C. Westwood, D. L. Goldheim, and R. G. Lye, *Phil. Mag.* **16**, 505 (1967).

108. W. Rostoker, J. M. McCaughey, and H. Markus, *Embrittlement by Liquid Metals*, Reinhold, New York (1960).

109. H. Nichols and W. Rostoker, Ref. 3, p. 213.

110. A. R. C. Westwood and R. G. Lye, Ref. 2, p. 3.

111. A. R. C. Westwood, C. M. Preece, and D. L. Goldheim, *Molecular Processes on Solid Surfaces*, McGraw-Hill, New York (in press).

112. A. R. C. Westwood, in *Fracture of Solids*, Interscience, New York (1963), p. 553.

113. M. H. Kamdar and A. R. C. Westwood, Ref. 3, p. 581.

114. A. R. C. Westwood, in *Strengthening Mechanisms: Metals and Ceramics*, Syracuse University Press, Syracuse, N.Y. (1966), p. 407.

115. A. R. C. Westwood and M. H. Kamdar, *Phil. Mag.* **8**, 787 (1963).

116. N. S. Stoloff and T. L. Johnston, *Acta Met.* **11**, 251 (1963).

117. A. R. C. Westwood, C. M. Preece, and M. H. Kamdar, *Trans. ASM* **60**, 763 (1967).

118. A. R. C. Westwood, C. M. Preece, and M. H. Kamdar, *Treatise on Brittle Fracture*, Academic Press, New York (in press).

119. W. H. Brattain, Ref. 4, p. 9.

120. E. A. Stern, *Phys. Rev.* **162**, 565 (1967).

121. A. Kelly, W. R. Tyson, and A. H. Cottrell, *Phil. Mag.* **15**, 567 (1967).

122. T. L. Johnston, R. G. Davies and N. S. Stoloff, *Phil. Mag.* **12**, 305 (1965); N. S. Stoloff, R. G. Davies, and T. L. Johnston, Ref. 3, p. 613.

123. M. H. Kamdar and A. R. C. Westwood, *Acta Met.* **16**, 1335 (1968).

124. M. H. Kamdar and A. R. C. Westwood, *Phil. Mag.* **15**, 641 (1967).

125. C. M. Preece and A. R. C. Westwood, *Trans. ASM* **61**, (1969).

126. N. J. Petch, *Phil. Mag.* **3**, 1089 (1958).

127. E. D. Shchukin, L. A. Kochanova, and N. V. Perstov, *Soviet Physics–Crystallography* **8**, 49 (1963).

128. R. Rosenberg and I. Cadoff, Ref. 112, p. 607.

129. F. W. J. Pargeter and M. B. Ives, *Can. J. Phys.* **45**, 1235 (1967).

130. W. Rostoker, Report No. ARF-B183-12 on Contract No. DA-11-ORD-922-3108, Armour. Res. Fdn. (Nov., 1963).

131. J. V. Rinovatore, J. D. Corrie, and J. D. Meakin, *Trans. ASM* **61**, 321 (1968); J. V. Rinnovatore and J. D. Corrie, *Scripta Met.* **2**, 467 (1968).

132. W. Klinkenberg, K. Lucke, and G. Masing, *Z. Metallk.* **44**, 362 (1953).

133. I. R. Kramer and L. J. Demer, *Prog. Metal Phys.* **9**, 133 (1961).

134. E. S. Machlin, in *Strengthening Mechanisms in Solids*, American Society for Metals, Metals Park, Ohio (1962), p. 375.

135. E. D. Shchukin, *Proc. Acad. Sci. USSR* **118**, 1105 (1958).

136. E. N. da C. Andrade, in *Properties of Metallic Surfaces*, Institute of Metals, London (1953), p. 133.

137. D. S. Kemsley, *Nature* **163**, 404 (1949).

138. S. Harper and A. H. Cottrell, *Proc. Phys. Soc. London* **B63**, 331 (1950).

139. I. R. Kramer, *Trans. AIME* **227**, 529 (1963).

140. I. R. Kramer, *Trans. AIME* **221**, 989 (1961).

141. H. A. Smith and R. M. McGill, *J. Phys. Chem.* **61**, 1025 (1957).
142. E. A. Steigerwald, *Proc. ASTM* **60**, 750 (1960).
143. H. H. Johnson and A. M. Willner, *Appl. Mater. Res.* **4**, 34 (1965).
144. H. H. Johnson and P. C. Paris, *Eng. Fract. Mech.* **1**, 3 (1968).
145. A. R. C. Westwood, Ref. 3, p. 1.
146. R. Boniszewski and G. C. Smith, *Acta Met.* **11**, 165 (1963).
147. B. A. Wilcox and G. C. Smith, *Acta Met.* **12**, 371 (1964).
148. A. H. Windle and G. C. Smith, *Met. Sci. J.* **2**, 187 (1968).
149. T. S. Liu and M. A. Steinberg, *Trans. ASM* **50**, 455 (1958).
150. A. S. Tetelman, Ref. 1, p. 446.
151. M. B. Whiteman and A. R. Troiano, *Corrosion* **21**, 53 (1965).
152. G. G. Hancock and H. H. Johnson, *Trans. AIME* **236**, 513 (1966).
153. C. Zapffe and C. Sims, *Trans. AIME* **145**, 225 (1941).
154. A. S. Tetelman and W. D. Robertson, *Acta Met.* **11**, 415 (1963).
155. N. J. Petch and P. Stables, *Nature* **169**, 842 (1952).
156. N. J. Petch, *Phil. Mag.* **1**, 331 (1956).
157. H. H. Johnson, J. G. Morlet, and A. R. Troiano, *Trans. AIME* **212**, 526 (1958).
158. E. A. Steigerwald, F. W. Schaller, and A. R. Troiano, *Trans. AIME* **218** 822 (1960).
159. E. A. Steigerwald, F. W. Schaller, and A. R. Troiano, *Trans. AIME* **215**, 1048 (1959).
160. A. R. Troiano, *Trans. ASM* **52**, 54 (1960).
161. J. W. Geus, *Surf. Sci.* **2**, 48 (1964).
162. I. R. Kramer and S. Podlaseck, *Acta Met.* **11**, 70 (1963).
163. H. Shen, S. E. Podlaseck, and I. R. Kramer, *Trans. AIME* **223**, 1933 (1965).
164. M. R. Achter, *ASTM STP-415* 181 (1967).
165. N. J. Wadsworth and J. Hutchings, *Phil. Mag.* **3**, 1154 (1958).
166. N. J. Wadsworth, *Internal Stresses and Fatigue in Metals*, Elsevier, New York (1959), p. 382.
167. K. U. Snowden, *Acta Met.* **12**, 295 (1964); *Phil. Mag.* **10**, 435 (1964).
168. M. R. Achter, G. J. Danek, Jr., and H. H. Smith, *Trans. AIME* **227**, 1296 (1963).
169. H. Shen, S. E. Podloseck, and I. R. Kramer, *Acta Met.* **14**, 341 (1966).
170. H. Shen and I. R. Kramer, in *Trans. Int. Vac. Met. Conf.* (1967), p. 263.
171. M. J. Hordon, *Acta Met.* **14**, 1173 (1966).
172. N. Thompson, N. J. Wadsworth, and N. Louat, *Phil. Mag.* **1**, 113 (1956).
173. C. Laird and G. C. Smith, *Phil. Mag.* **8**, 1945 (1963).
174. J. C. Grosskreutz and C. Q. Bowles, Ref. 3, p. 67.
175. T. Broom and A. Nicholson, *J. Inst. Metals* **89**, 183 (1960–61).
176. J. C. Grosskreutz, *Surf. Sci.* **8**, 173 (1967).
177. K. U. Snowden and J. N. Greenwood, *Trans. AIME* **212**, 626 (1958).
178. P. Shahanian and M. R. Achter, *Trans. ASM* **51**, 244 (1959); *Proc. of Joint International Conf. on Creep* (1963), pp. 7–49.
179. A. R. C. Westwood, *Materials Science Research*, Vol. 1, Plenum Press, New York (1963), p. 114.
180. R. J. Stokes, T. L. Johnston, and C. H. Li, *Phil. Mag.* **6**, 9 (1961).
181. R. J. Stokes, *Trans. AIME* **224**, 1227 (1964).
182. A. R. C. Westwood, *Phil. Mag.* **5**, 981 (1960).
183. T. Suzuki, Ref. 62, p. 215.
184. S. Mendelson, *J. Appl. Phys.* **33**, 2175, 2182 (1962).

185. A. Joffe, M. W. Kirpıtschewa, and M. A. Lewitsky, Z. Physik 22, 286 (1924).
186. A. R. C. Westwood, Ind. Eng. Chem. 56, 15 (1964).
187. W. H. Class, E. S. Machlin, and G. T. Murray, Trans. AIME 221, 769 (1961).
188. A. E. Gorum, E. R. Parker, and J. A. Pask, J. Am. Ceram. Soc. 41, 161 (1958).
189. W. Ewald and M. Polanyi, Z. Physik 28, 29 (1924).
190. A. R. C. Westwood, Adv. in Metals Research II, Microplasticity, Interscience, New York (1968), p. 365.
191. A. R. C. Westwood, D. L. Goldheim and E. N. Pugh, Disc. Faraday Soc. 38, 147 (1964); Acta Met. 13, 695 (1965); Mat. Sci. Res. 3, 553 (1966).
192. A. R. C. Westwood, D. L. Goldheim, and E. N. Pugh, Phil. Mag. 15, 105 (1967).
193. R. J. Friauf, J. Appl. Phys. Suppl. 33, 494 (1962).
194. T. B. Grimley and N. F. Mott, Disc. Faraday Soc. 1, 3 (1947).
195. I. M. Lifshitz and Ya. E. Geguzin, Soviet Physics–Solid State 7, 44 (1965).
196. J. H. Westbrook, Ref. 3, p. 247.
197. D. B. Holt, Ref. 3, p. 269.
198. J. H. Westbrook and R. E. Hanneman, Phil. Mag. 18, 73 (1968).
199. A. R. C. Westwood and D. L. Goldheim, J. Appl. Phys. 39, 3401 (1968).
200. A. R. C. Westwood, Phil. Mag. 7, 633 (1962).
201. A. R. C. Westwood, D. L. Goldheim, and R. G. Lye, Phil. Mag. 17, 951 (1968).
202. W. G. Johnston, J. Appl. Phys. 33, 2716 (1962).
203. P. L. Pratt, R. L. Harrison, and C. H. Newey, Disc. Faraday Soc. 38, 211 (1964).
204. F. F. Volkenstein, Adv. Catalysis 12, 187 (1960); Soviet Physics–Uspekhi 9, 7432 (1967).
205. C. Kuczynski and R. F. Hochman, Phys. Rev. 108, 946 (1957).
206. J. H. Westbrook and J. J. Gilman, J. Appl. Phys. 33, 2360 (1962).
207. N. Ya Goridko, P. P. Kuzmenko, and N. N. Novikov, Soviet Physics–Solid State 3, 2652 (1962).
208. E. P. Warekois, M. C. Lavine, and H. C. Gatos, J. Appl. Phys. 31, 1302 (1960).
209. M. S. Seltzer, J. Appl. Phys. 37, 4780 (1966).
210. A. R. C. Westwood and D. L. Goldheim, RIAS Annual Report, Baltimore, Maryland (1968), p. 15.
211. P. A. Rebinder, L. A. Schreiner, and K. F. Zhigach, Hardness Reducers in Rock Drilling, CSIRO Transl., Melbourne (1948).
212. F. J. P. Clarke, R. A. J. Sambell, and H. G. Tattersall, Phil. Mag. 7, 393 (1962).
213. E. N. Pugh, A. R. C. Westwood, and T. T. Hitch, Phys. Stat. Sol. 15, 291 (1966).
214. J. W. Faust, A. Sagar, and H. F. John, J. Electrochem. Soc. 109, 824 (1962).
215. R. Roscoe, Nature 133, 912 (1934); Phil. Mag. 21, 399 (1936).
216. J. C. Bilello and I. B. Cadoff, J. Metals. 14, 87 (1962).
217. R. J. Stokes, T. L. Johnston, and C. H. Li, Trans. AIME 218, 655 (1960).
218. C. S. Barrett, Acta Met. 1, 2 (1953); C. S. Barrett, R. M. Aziz, and I. Markson, Trans. AIME 197, 1655 (1953).
219. D. B. Holt, Acta Met. 10, 1021 (1962).
220. M. R. Pickus and E. R. Parker, Trans. AIME 191, 792 (1951).
221. F. R. Lipsett and R. King, Proc. Phys. Soc. B70, 608 (1957).
222. J. J. Gilman and T. A. Read, Trans. AIME 194, 875 (1952); J. J. Gilman, ASTM STP-171 3 (1955).
223. E. N. da C. Andrade and C. Henderson, Phil. Trans. Roy. Soc. A244, 177 (1951).
224. D. J. Phillips and N. Thompson, Proc. Phys. Soc. B63, 839 (1950).

225. J. Takamura, *Mem. Fec. Engng. Kyoto Univ.* **18** (3), 255 (1956).
226. J. Garstone, R. W. K. Honeycombe, and G. Greetham, *Acta Met.* **4**, 485 (1956).
227. D. H. Bradhurst and J. S. Ll. Leach, *J. Electrochem. Soc.* **113**, 1245 (1966); J. S. Ll. Leach and P. Nuefeld, *Proc. Brit. Ceram. Soc.* **6**, 49 (1966).
228. M. A. Adams, *Acta Met.* **6**, 327 (1958).
229. F. D. Rosi, *Acta Met.* **5**, 348 (1957).
230. R. M. Johnson and R. J. Block, *Acta Met.* **16**, 831 (1968).
231. A. K. Head, *Phil. Mag.* **44**, 92 (1953); *Aust. J. Phys.* **13**, 278 (1960).
232. G. H. Conners, *J. Eng. Sci.* **5**, 25 (1967).
233. *Engineering Properties of Selected Ceramic Materials*, Battelle Mem. Inst., Am. Ceram. Soc., Columbus, Ohio (1966).
234. S. F. Bubar and D. A. Vermilyea, *J. Electrochem. Soc.* **113**, 892 (1966); *J. Electrochem. Soc.* **114**, 882 (1967).
235. W. A. Jemian and C. C. Law, *Acta Met.* **15**, 143 (1967).
236. T. Evans and D. R. Schwarzenburger, *Phil. Mag.* **4**, 889 (1959).
237. D. R. Brame and T. Evans, *Phil. Mag.* **3**, 971 (1958).
238. G. E. Ruddle and H. G. F. Wilsdorf, *Appl. Phys. Letters* **12**, 271 (1968).
239. J. W. Matthews and W. A. Jesser, *Acta Met.* **15**, 595 (1967).
240. G. A. Bassett and D. W. Pashley, *J. Inst. Metals* **87**, 449 (1958–59).
241. J. H. van der Merwe, *J. Appl. Phys.* **34**, 117, 123 (1963); in *Single Crystal Films*, Pergamon Press, London (1964), p. 139.
242. W. A. Jesser and D. Kuhlmann-Wilsdorf, *Phys. Stat. Sol.* **19**, 95 (1967).
243. E. R. Thompson and K. R. Lawless, *Appl. Phys. Letters* **9**, 138 (1966).
244. U. Gradmann, *Ann. Phys. (Lpz.)* **13**, 213 (1964); **17**, 91 (1965).
245. J. W. Menter and D. W. Pashley, in *Structure and Properties of Thin Films*, John Wiley and Sons, New York (1959), p. 111.
246. R. J. Block and M. Metzger, *Phil. Mag.* **19**, 599 (1969).
247. R. W. Hoffman, in *Thin Films*, American Society for Metals, Metals Park, Ohio (1964), p. 99.
248. W. T. Brydges, *Scripta Met.* **3**, 271 (1969).
249. D. J. Duquette, *Scripta Met.* **3**, 513 (1969).
250. E. S. Machlin, *J. Appl. Phys.* **30**, 1109 (1959).

MECHANISM AND PHENOMENOLOGY OF ORGANIC INHIBITORS

Giordano Trabanelli and Vittorio Carassiti

Corrosion Study Center "A. Daccò"
Chemical Institute of the University
Ferrara, Italy

INTRODUCTION

In general, any phase constituent whose presence is not essential to the occurrence of an electrochemical process, but leads to a retardation of this process by modifying the surface state of the metallic material, will be called an inhibitor (this applies to electrolytic phases). Such modification of the surface state implies adsorption, the formation of "surface compounds," or a reaction between the metallic material and the inhibitor with separation of the corrosion products at the contact surface of the metallic material with the electrolytic conductor.[1] If we consider that corrosion is the degradation of a metallic material through the passage of its constitutive elements to a state of combination with surrounding materials, making particular reference to corrosion phenomena in electrolytic conductors, we can define corrosion inhibitors as substances that inhibit one or more of the partial processes of the corrosion process as a whole.

On the basis of these considerations, corrosion inhibition is a particular case of the inhibition of an electrochemical process.[2] For this reason, the mechanism of the action of inhibitors and, in particular, of the organic inhibitors with which this paper is concerned, has been studied by various authors on the basis of comparative measurements of corrosion inhibition and the inhibition of various electrode processes, involving, for example, the dropping mercury electrode[3–5] or electrocrystallization.[6]

MECHANISMS OF THE ACTION OF ORGANIC INHIBITORS

Adsorption

The inhibiting action exercised by organic compounds on the dissolution of metallic materials is normally attributed to interactions by adsorption between the inhibitor and the metal surface, although some authors consider that this phenomenon forms only the first stage of the inhibition process proper. The type of interaction involved may be characterized by an examination of the adsorption kinetics, the heat of adsorption, or the reversibility and specificity of the bond established.

The species that are adsorbed physically by means of electrostatic or van der Waals forces interact rapidly with the electrode but are easily removed from the surface, for example, by immersion of the metal in a solution free from any inhibitor. The process of chemisorption involving charge sharing or charge transfer, takes place more slowly and with high heat of adsorption. In this case, the adsorption is specific for certain metals and is not completely reversible.[7]

Comparison of the properties mentioned above shows that an effective inhibiting action must normally be linked with a phenomenon of chemisorption, and examples are reported in the literature which range from simple adsorption to the formation of true physical barriers. From the results of electrochemical analysis, many authors have postulated, in general terms, an action of the inhibitors based on an increase in the overvoltage of the proton discharge process forming the partial cathodic reaction of the corrosion process,[8,9] on an increase in the ohmic resistance due to the presence of an inhibitor film at the metal–solution interface,[10] or on nonspecific adsorption phenomena. These theories have been complemented and have found better definition through an analysis of the correlations between molecular structure and inhibitor characteristics. Substantial contributions to a better comprehension of the phenomenon have been made by detailed study of the electrochemical behavior of metallic materials in environments containing inhibitors, by a deeper evaluation of the potential dependence of the adsorption phenomena, and by a better knowledge of the energy parameters involved in the interaction between inhibitors and metallic materials.

Influence of Structural Parameters on Adsorption and Inhibition

The possibility of correlating structural characteristics with the inhibitor properties of organic substances is justified by the fact that the metal–

inhibitor interactions are based on chemisorption. The electron density of the organic function that can be defined as the reaction center for the establishment of the adsorption bond is then obviously important, since it is possible to assume a bond of the Lewis acid–base type,[7] generally with the inhibitor as the electron donor and the metal as the electron acceptor. The strength of this bond depends on the characteristics of both the adsorbent and the adsorbate.

Most organic inhibitors are compounds with at least one polar function, having atoms of nitrogen, sulfur, oxygen, and, in some cases, selenium and phosphorus (see Appendix). In general, the polar function is regarded as the reaction center for the establishment of the chemisorption process. In such a case the adsorption bond strength is determined by the electron density of the atom acting as the reaction center and by the polarizability of the function. The effectiveness of the function atoms with respect to the adsorption process, when the stabilities of the compounds are equal, can be taken as being in the following sequence[11-13]:

$$selenium > sulfur > nitrogen > oxygen$$

This classification has frequently been found in reviews of inhibition. An explanation may be sought in the less negative nature of the elements on the left, in which an easier polarizability can be found.

The idea of electron density acquires particular importance in aromatic or heterocyclic inhibitors whose structure may be affected by the introduction of substituents in different positions of the rings. The availability of electron pairs for the formation of chemisorption bonds can thus be altered by regular and systematic variations of the molecular structure. Thus, Hackerman,[14] investigating the inhibiting properties of pyridine and its derivatives, found that inhibition increases when the electron density at the nitrogen atom is increased in the sequence:

$$pyridine < 3\text{-picoline} < 2\text{-picoline} = 4\text{-picoline}$$

Assuming that for the compounds mentioned, as for aliphatic amines, the electron given up during the first ionization is one of the electrons of the lone pair, the ionization potential may constitute a measure of the differences in the electron density at the nitrogen atom. A more general interpretation of the importance of the electron density in chemisorption of organic substances in relation to inhibition phenomena has been attempted by Donahue,[15-17] who pointed out the changes in the corrosion current densities of iron immersed in $1N$ H_2SO_4 containing various inhibitors as a function of Hammett's σ constant or of Taft's σ^* constant. A similar study has been carried out by Grigoryev and Osipov.[18]

A rational statement of the problem of the correlations between molecular structure and inhibition has been made by Hackerman,[7,11] who developed a theory of adsorption based on numerous measurements carried out in the presence of a series of secondary aliphatic amines and cyclic imines as inhibitors in an acid medium. Hackerman concluded that the greater the percentage of π orbitals of the free electrons on the nitrogen atom the more effective is the inhibiting action. In this way, the behavior of cyclic imines with a high inhibiting efficiency (nonamethyleneimine and decamethyleneimine) as compared with the corresponding aliphatic amines finds a valid explanation, on the assumption that in the 9- and 10-atom rings the angle of the C—N—C bond increases from the 109° characteristic of sp^3 hybridization to about 120° (sp^2 hybridization), with a consequent greater availability of the nonbonding electron pair of the nitrogen atom.[19]

An analysis of the inhibiting action of aromatic amines and thiols and their compounds has been made by Riggs,[20] who interpreted the results on the basis of electron densities and obtained results in harmony with the theoretical assumptions. Other results that can be interpreted in this manner have been obtained by Trabanelli, Fiegna, and Carassiti[21] in a study of the mechanisms of action of a number of series of organic inhibitors acting in the vapor phase. Cycloaliphatic amines such as cyclohexylamine and dicyclohexylamine were tested. All these substances were shown effectively to inhibit the corrosion of iron in aggressive atmospheres. In relation to the influence of the molecular structure on the inhibiting action, it may be noted that on passing from cyclohexylamine to derivatives more highly substituted on the nitrogen atom with cycloaliphatic nuclei or aliphatic chains, the inhibiting efficiency increases markedly because of the increase in the electron density at the nitrogen due to the electron-repelling effects of the substituents.

On the other hand, analogous measurements on aromatic amines such as aniline, di- and triphenylamines, N,N-dimethyl- and N,N-diethylanilines, N,N-dimethyl- and N,N-diethyl-o-toluidines, N,N-dimethyl- and N,N-diethyl-p-toluidines, and meta-N,N derivatives have shown that the inhibiting action of these compounds is low. It is possible to imagine a conjugative effect of the electron pair of the nitrogen atom with participation in mesomerism of the aromatic ring, without excluding the relative contribution of steric hindrance effects of these substituted amines. The introduction of alkyl substituents on the nitrogen atom or in the para position of the aromatic nucleus, in contrast to the meta-derivatives, improves the inhibition efficiency in accordance with theories about the activation of various positions in aromatic nuclei by substituents.

Among other structural parameters may be mentioned the molecular area of the inhibitor projected onto the metallic surface, considering the various possibilities of arrangement of the organic ions or the molecules at the metal–solution interface. Thus, for example, the behavior of a series of pyridine derivatives has been evaluated by Hackerman[14] by comparing their inhibiting efficiencies with the corresponding values of the projected areas corresponding to the cations, to the neutral molecules oriented perpendicularly to the metallic material (regarding the nitrogen atom of the heterocycle as the center of the interaction), or to the molecules arranged parallel to the metallic surface, in which case the interaction would take place through the π-electron sextet of the heterocyclic rings. A correlation, although a slight one, was found between the values of the inhibitions and those of the circular areas relating to the arrangement parallel to the surface.

The value of the projected molecular area does not in general appear to be determinative over the whole of the metal-inhibitor chemisorption bond, as has been shown by Podobaev[22] by comparing the results of measurements of inhibition carried out on a series of derivatives of propargyl alcohol $HO—CH_2—C{\equiv}CH$, keeping the triple bond in the molecule unchanged and substituting nitrogen-containing heterocyclic nuclei in the hydroxyl group or esterifying it with fatty acids. Under such conditions the inhibiting efficiency of the derivatives of the series became smaller than that of the initial propargyl alcohol, in spite of the considerable increase in the projected molecular area. An interpretation of the phenomenon may be obtained by considering the mechanism of the action proposed by Podobaev. The author assumes that the surface action of propargyl alcohol is attributable to π conjugation between the triple bond and the C—OH bond with consequent weakening of the π bond of the acetylene group, which would facilitate the formation of a π bond with the metal. The substitutions carried out on the hydroxyl group would not give a sufficient weakening of the triple bond and this, becoming more stable, would not permit surface interaction, in spite of the high values of the molecular areas favoring the covering of the surface.

The influence exercised by the molecular weight of the additives on inhibition can be evaluated objectively by studying the behavior of a homologous series of organic substances chemically stable in the medium under examination. In such a case it appears that at the same molar concentration an increase in the length of the hydrocarbon chain of amines or imines

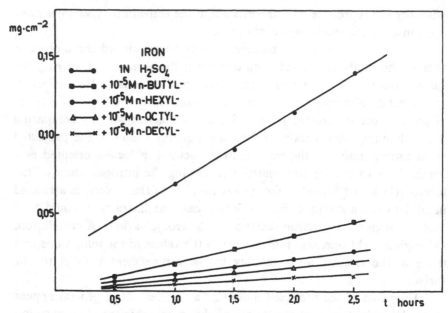

Fig. 1. Corrosion rate of iron in $1N\,H_2SO_4$ (containing 10% ethyl alcohol) in the presence of different aliphatic mercaptans (metal samples pickled in $2N\,H_2SO_4$ for 10 min).*

gives an increase in the inhibition efficiency which can be attributed to the inductive effect of the methylene groups, shown in a decrease of the ionization potential of the molecules of the free base.[23] On the other hand, on studying the inhibition efficiency of homologous mercaptans, it was shown[24] that the inhibition increases on going from n-butyl mercaptan to n-decyl mercaptan (Fig. 1).

Action of Reduction, Polymerization, or Reaction Products

The problem of the correlation between molecular weights and inhibiting characteristics is particularly apparent with substances capable of polymerization during inhibition, probably because of a cathodic reduction, and with substances suitably prepared by condensing a definite number of monomer molecules to give a repetition of similar functional groups capable of promoting better adsorption on the metal.

The first case may include acetylene compounds[22,25] which, giving rise to a bond with the metal, break the $C\equiv C$ triple bond, permitting successive

---

* All the results quoted in the figures were obtained from the Members of the Chemical Institute of the University of Ferrara.

polymerization of the materials. In the case of acetylene, according to Putilova,[25] such polymerization is proved by the change in the state of the additive with the formation of oily products and by the results of analytical and structural determinations. In a study of the inhibition efficiency of polyalkylamines,[26] it was shown that the best inhibiting characteristics, particularly for temperatures up to 90°C, are obtained with products of higher mean molecular weight and with the derivatives of monomers with low-carbon aliphatic chains.

Hackerman et al.[27,28] have compared the adsorption characteristics and the inhibition efficiencies of 4-ethylpyridine with those of polyvinyl-pyridines having different degrees of polymerization (from 4 to 30 units). The authors have shown that adsorption of the polymers takes place through several points of the molecule and that the surface area of the metal inhibited in solutions of hydrochloric acid containing polyvinylpyridine is considerably greater than the area physically covered by the adsorbed molecules.

Into this framework comes the "film theory of protective activity" formulated by Balezin et al.,[29] according to whom effective inhibiting action is explained by the formation on the metallic surface of a layer due to reaction between the metal, the inhibitor, and the ions of the corrosive medium. The films must be adherent, have a low or zero solubility, and prevent of access the solution components to the metal. According to these authors, the surface chemical compounds arising in this way are distinguished from chemisorption bonds and from physical adsorption, and it is due to their formation that there is an inhibition of the partial electrochemical reactions of the corrosion process. According to Balezin, the proposed mechanism has a final justification in that numerous organic inhibitors are capable of forming organometallic-type complexes. When the complexes are soluble we may expect an acceleration of the corrosive effect, while the formation of insoluble surface complexes guarantees optimum protection. As an example, we may cite the complexes of numerous derivatives, including the following:

with o- or p-toluidine $(C_7H_7NH_3)_4 \cdot FeCl_6 \cdot H_2O$

with quinoline $C_9H_7NH \cdot FeCl_3$

A structural factor which is not easy to evaluate is that relating to decomposition processes involving the reduction of organic inhibitors at the surface of metals,[30] which leads to the existence in solution of a mixture of the initial substance and of its reduction products, or only of the reduction products, depending on the time of the process and on the electrode potential. In the intervals of time normally adopted for the study of inhibition processes there is generally only a partial reduction of the initial additive, so that the problem changes to considering whether the reduction product or products are capable of explaining an inhibiting effect smaller than, equal to, or greater than that of the initial material. Horner[13,31] has called the inhibition attributable to the initial additive "primary," and the inhibition brought about by the reduction products "secondary."

The literature reports the results of numerous investigations designed to distinguish between the primary and secondary effects. In particular Putilova[25] investigated the mechanism of the action of acetylenic products, assuming that there is initially a reduction of the triple bond by reaction with the hydrogen evolved in the corrosion process. This phase corresponds to a slight influence on the corrosion rate which subsequently decreases appreciably because of the polymerization process which gives rise to substances that are generally slightly soluble and deposit on the surface of the metal, making it hydrophobic.[32] A similar idea has been put forward by Podobaev, who assumes the formation of polymers by acetylenic derivatives bound to the metal by an interaction of the π type following the rupture of the triple bond.[22] Proof of the secondary inhibiting action of acetylenic compounds has been obtained by Poling,[33] who recorded the infrared spectra of surface films formed on specular surfaces of iron and steel placed in solutions of hydrochloric acid inhibited with acetylenes, propargyl alcohol, or 1-ethynylcyclohexan-1-ol. These compounds react with the metallic surfaces exposed to the acid forming covering protective films. According to the author, in this particular case there cannot be a single adsorption phenomenon since the films grow until several molecular layers have been produced, as shown by their calculated thicknesses of 20 to 200 Å. The corrosion inhibition increases with increasing film thickness. Poling considers that there is initially a rapid chemisorption of the inhibitors which gives rise to metal–acetylenes surface complexes as the initial stage of the formation of the protective polymolecular film. The chemisorbed product would already be in a position to give ~90% inhibition of the corrosion process, which would facilitate the successive growth of thicker films capable of increasing the inhibition efficiency to 99.9%. The reactions leading to secondary action of the acetylenic products by the formation of a film are, according to Poling, hydrogenation of the acetylenic inhibitors adsorbed by the hydrogen evolved in the cathodic process and the catalytic hydration of the acetylenes followed by isomerization to carbonyl compounds:

$$HC\equiv CH + H_2O \xrightarrow{Fe,HCl} H_2C=CHOH \rightarrow H_3C-CHO$$

In confirmation of this hypothesis, infrared spectra have shown the presence in the film of saturated hydrocarbon chains and carbonyl, carboxyl, and hydroxyl groups. These latter, according to the author, must be attributed mainly to retention of adsorbed water and to the presence of more or less hydrated oxidized species of the metal.

A remarkable agreement of opinions on the secondary inhibiting action of dibenzyl sulfoxide can be deduced from the results obtained by various authors.[31,34-36] It has been found by various methods that the inhibiting effect can be attributed to the reduction product produced as follows:

$$
\begin{array}{c}
\text{C}_6\text{H}_5\text{—CH}_2 \\
\diagdown \\
\hspace{1em} \text{S}{=}\text{O} + \text{H}^+ \rightarrow \\
\diagup \\
\text{C}_6\text{H}_5\text{—CH}_2
\end{array}
\qquad
\begin{array}{c}
\text{C}_6\text{H}_5\text{—CH}_2 \\
\diagdown \\
\hspace{1em} \text{S}^+{-}\text{OH} \\
\diagup \\
\text{C}_6\text{H}_5\text{—CH}_2
\end{array}
$$

$$
\begin{array}{c}
\text{C}_6\text{H}_5\text{—CH}_2 \\
\diagdown \\
\hspace{1em} \text{S}^+{-}\text{OH} + \text{H}^+ + 2e \rightarrow \\
\diagup \\
\text{C}_6\text{H}_5\text{—CH}_2
\end{array}
\qquad
\begin{array}{c}
\text{C}_6\text{H}_5\text{—CH}_2 \\
\diagdown \\
\hspace{1em} \text{S} + \text{H}_2\text{O} \\
\diagup \\
\text{C}_6\text{H}_5\text{—CH}_2
\end{array}
$$

In our laboratory[36] investigations have been extended to sulfoxides with aliphatic chains or aromatic rings, the inhibiting efficiencies found being in the following sequence:

dibenzyl sulfoxide > di-*n*-butyl sulfoxide > di-*p*-tolyl sulfoxide >

diphenyl sulfoxide > tetramethylene sulfoxide > dimethyl sulfoxide

In $1N$ sulfuric acid (10% C_2H_5OH), dimethyl sulfoxide slightly stimulates the corrosion of iron (Fig. 2).

Analysis of the results obtained in the measurement of polarization shows that in this case too the inhibiting action is of the secondary type, i.e., it is due to the corresponding sulfides produced by reduction of the sulfoxides at the electrode. This secondary inhibition is greater in compounds having a higher electron density on the sulfur atom (di-*n*-butyl sulfoxide, dibenzyl sulfoxide).

The aromatic sulfoxides (diphenyl sulfoxide, di-*p*-tolyl sulfoxide) are reduced to sulfides; these are capable of becoming attached to the surface of the metal because of their low solubility, in spite of the fact that the electron density at the sulfur atom is slightly decreased by conjugation. Furthermore, one must not overlook the possibility that one of the phenyl

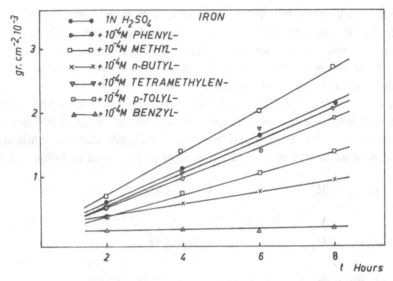

Fig. 2. Corrosion rate of iron in $1N$ H_2SO_4 in the presence of different organic sulfoxides (metal samples abraded with # 150 emery paper).

groups contributes to the adsorption via a π bond due to an arrangement parallel to the metal surface.

The low-molecular-weight dimethyl and tetramethylene sulfoxides do not inhibit the corrosion because their short hydrocarbon groups give small coverage, and because they are reduced to sulfide and have a very high vapor tension. The fact that dimethyl sulfoxide promotes corrosion may be explained by the formation of a soluble complex between the sulfide produced by reduction and metallic ions. Analogous measurements performed in the presence of different sulfides confirmed that these substances, when in the lowest concentrations tested (10^{-6}–$10^{-7}M$), have a stimulating effect on the dissolution of iron in acids.[36] The inhibiting action of sulfides at higher concentrations ($10^{-4}M$) was clearly greater than that showed by the corresponding sulfoxides at the same concentrations.

Steric Effects

The results obtained[36] with organic sulfides have underlined the difficulty of attributing the variation in inhibitor efficiency within a series of monofunctional compounds to a single molecular property. The inhibitor action of sulfides at higher concentrations is attributed to adsorption of the molecules on the metal surfaces through an unshared electron pair belonging to the sulfur atom. It would then seem obvious that an increase in the charge on the sulfur atom would increase the ability of the organic molecule to attach itself to the metal, and enhanced inhibitor efficiency would result. It is well known that alkyl groups are electron-repelling and therefore increase the electron charge on the sulfur by an inductive effect. The increase in inhibitor power on going to diethyl, to di-*n*-butyl, and to di-*n*-hexyl sulfide (Fig. 3) is explained in this way. While the electronic charge on the sulfur atom is thus an important factor, it cannot explain all the observed effects. The reduction of inhibitor power on going from di-*n*-hexyl to di-*n*-octyl and di-*n*-decyl sulfide cannot in fact be interpreted on the basis of the above hypothesis. Although admittedly the inductive effect of the alkyl groups does not increase continuously with increasing number of carbon atoms in the chain, it should reach and mantain a constant value that would endow the compounds having a longer chain with a greater inhibitor effect than was found experimentally. Reduction of solubility with increasing number of carbon atoms cannot be considered a valid cause of the reduction of the inhibitor effect of compounds having very longs chains, where in our experimental conditions[36] aqueous alcohol solutions were used with the very intention of avoiding this phenomenon as far as possible. Most probably the reduction in inhibitor effectiveness on going from six to 10 carbon atoms (Fig. 3) may be attributed to the screening action of the hydrocarbon chains, which are quite long in comparison with

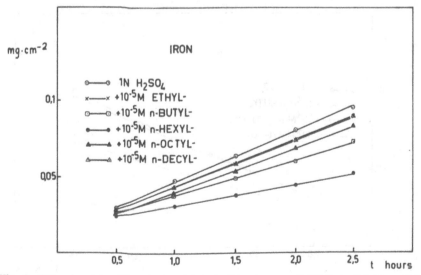

Fig. 3. Corrosion rate of iron in $1N\,H_2SO_4$ in the presence of different aliphatic sulfides (metal samples annealed, and treated in $1N\,H_2SO_4$ for 90 min).

the sulfur atom and particularly with the positions involved in the adsorption process. Working with a molecular model in which free movement of the chains was allowed, a screening action was in fact observed for a chain length of about eight carbon atoms.

The results obtained for the three butyl sulfides (di-*n*-butyl, di-*s*-butyl, and di-*t*-butyl sulfide—Fig. 4) clearly underline the importance of molecular configuration. The tertiary compound has no inhibitor action, although its sulfur atom has the highest electron charge, because the sulfur is screened by the methyl groups (as a result of the steric configuration) and is therefore unable to attach itself stably to the metal surface. In this way, the inhibitor efficiency increases on going from tertiary to secondary and primary compounds.

The effect of molecular asymmetry in increasing the inhibitor efficiency of organic sulfides is plainly apparent from a comparison of the change in inhibitor power with a changing number of carbon atoms in compounds having symmetrical and asymmetrical molecular structures.

The beneficial effect of molecular asymmetry on inhibitor action cannot easily be understood on the basis of the electron density on the sulfur atom alone (Fig. 5). It may be that the sulfur adsorption is established in relation to the chain having the smallest number of carbon atoms, while the longer chain's contribution to the inhibitor effect is its property of extensively covering the metal surface.

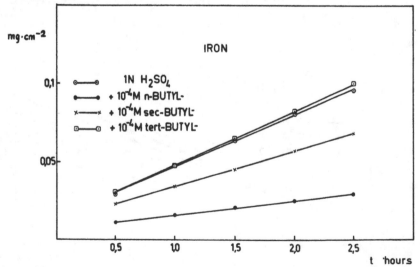

Fig. 4. Corrosion rate of iron in $1N$ H_2SO_4 in the presence of *n*-, *sec*-, and *tert*-butyl sulfides (metal samples annealed, and treated in $1N$ H_2SO_4 for 90 min).

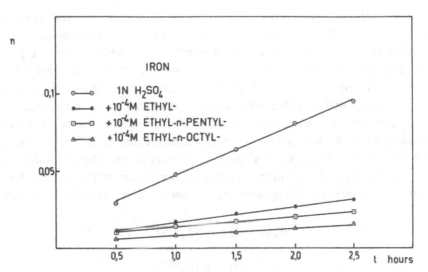

Fig. 5. Corrosion rate of iron in $1N\,H_2SO_4$ in the presence of aliphatic sulfides with asymmetric structure (metal samples annealed, and treated in $1N\,H_2SO_4$ for 90 min).

Action of the Organic Cations

The problem of the action of cations rather than molecules as inhibitors has been considered since the first investigations on the inhibiting actions of amines. Recently, Antropov[30] studied the behavior of pyridine derivatives as inhibitors of the acid corrosion of iron, coming to the conclusion, after an examination of the variations in coverage as a function of the molecular structure, that the compounds examined are adsorbed on the metal surface in the form of cations and not as neutral molecules. In this case, the nitrogen atom would be directed at the metallic surface.

Other authors[19,23,38] generally assume the existence of cations in the bulk of the acid solution and subsequent discharge with the formation of molecules at the interface with consequent chemisorption through an electron pair of the heteroatom. According to the schemes

$$R—NH_3^+ + e \rightarrow R—NH_2 + \tfrac{1}{2}H_2 \quad \text{(for aliphatic compounds)}$$

$$\diagdown\!\!\!\diagup NH^+ + e \rightarrow \diagdown\!\!\!\diagup N + \tfrac{1}{2}H_2 \qquad \text{(for heterocyclic compounds)}$$

the additives can act as transporters of protons catalyzing the electrode reaction, i.e., providing a catalytic path with a lower activation energy for the discharge of the proton as a function of the structure of the additive.[39]

Nevertheless, the opinions of the various authors do not always agree on the action mechanism of the organic cations, as is shown by an examination of the results reported by Horner[31] and Fischer.[35] According to the first author a "secondary" action, analogous to that of dibenzyl sulfoxide mentioned above, would apply to quaternary organic compounds (onium compounds), such as arsonium or phosphonium derivatives. Moreover, Fischer[37] has shown that onium derivatives such as triphenylbenzylphosphonium ion and triphenyl benzylarsonium ion inhibit corrosion in the ionic form with a "primary" action, contrary to the dibenzyl sulfoxide.

According to Jofa,[40] many organic substances are of the cationic type and develop their inhibiting activity as cations, for example the following:

$$R_3As + H^+ \rightarrow [R_3AsH]^+$$

$$R_3P + H^+ \rightarrow [R_3PH]^+$$

$$R_3N + H^+ \rightarrow [R_3NH]^+$$

$$R_2S{=}O + H^+ \rightarrow [R_2S{-}OH]^+$$

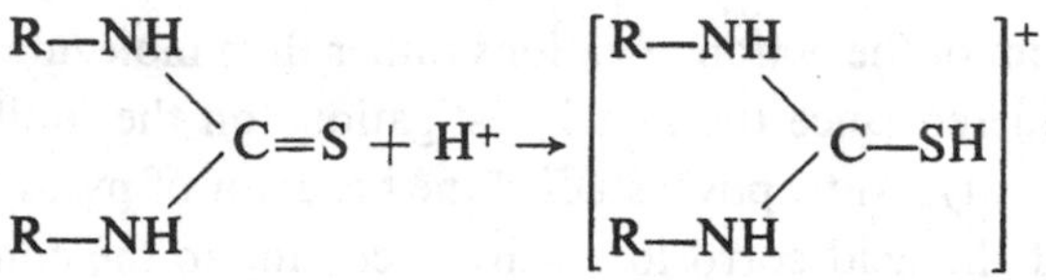

$$\begin{array}{c} R{-}NH \\ \diagdown \\ \diagup \\ R{-}NH \end{array} C{=}S + H^+ \rightarrow \left[\begin{array}{c} R{-}NH \\ \diagdown \\ \diagup \\ R{-}NH \end{array} C{-}SH \right]^+$$

Nevertheless, the adsorption of the cations from solutions of sulfuric and perchloric acids onto the surface of iron is weak, and the inhibiting effect is low.[40] In the opinion of Jofa, the substances of the nonionic type are not adsorbed on the iron surface and they have an almost nonexistent inhibitor action in acid media.

Anionic substances are assumed to be adsorbed immediately on the surface of the iron from solutions in sulfuric acid, but give rise to a stimulation of the corrosion process because of an unfavorable displacement of the adsorption potential ψ'. In solutions containing halogen anions chemisorbed on the metal surface the creation of dipoles oriented to the surface takes place, according to Jofa, with the consequent possibility of the attachment of cationic-type inhibitors to the dipole. This could explain the increase in the inhibiting action of quaternary compounds in acid solutions containing, for example, potassium iodide. Such synergistic action of inhibitors, which has also been found by other authors (Fig. 6)[41] has been interpreted[42]

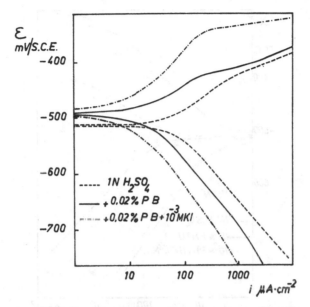

Fig. 6. Anodic and cathodic polarization curves of iron in $1N$ H_2SO_4 in the presence of polybutyrimine and poly-butyrimine + KI.

by assuming chemisorption of the halogen anion and the establishment of a bond between the nonionized inhibitor molecule and the adsorbed halogen in virtue of the covalent bond characteristics involved in this phenomenon, which cannot be considered as purely electrostatic.

On the other hand, the anion has little or no influence on adsorption and inhibition by neutral molecules, as has been shown by Cavallaro et al.[41] in the case of substituted thioureas in the presence of potassium iodide (Fig. 7). The absence of any synergistic effect that can be attributed to the halogen led these authors to conclude that in this case the action of the inhibitor is developed in the molecular form.

THE ROLE OF THE METAL IN INHIBITION

As mentioned above, the common feature of the various theories proposed for the interpretation of the action mechanism of organic inhibitors is always adsorption. It can be stated, in agreement with Frumkin,[43] that the adsorption of organic inhibitors on the surface of a metal forms an indispensable condition for their inhibiting action, which is shown only in the range of potentials at which the organic compounds are adsorbed.

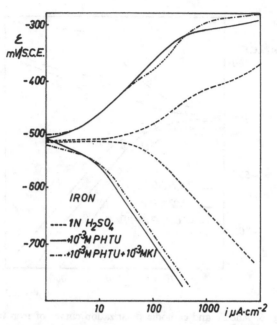

Fig. 7. Anodic and cathodic polarization curves of iron in $1N$ H_2SO_4 in the presence of phenylthiourea and phenylthiourea + KI.

Surface Charge of the Metal

Besides the structural parameters of the inhibiting substances, therefore, the importance of the magnitude and sign of the surface charge of the metal for the establishment of the adsorption bonds appears obvious. The effects exercised by organic inhibitors on the electrode reactions must be connected with the modifications induced in the structure of the electrochemical double layer because of their adsorption. This change is shown in the kinetic equations corresponding to the partial corrosion processes by the appearance of the adsorption potential ψ'.

According to Antropov,[30] in the case of a corrosion process in which the cathodic reaction is the evolution of hydrogen, ψ' can be calculated from the relation

$$\psi' = (\psi_2' - \psi_2) = \frac{b_0(b_a + b_c)}{b_0(b_a + b_c) - b_a b_c} \Delta \varepsilon_c$$

where ψ_2' and ψ_2 represent the potentials at the distance of "closest approach" to the surface of the metal in the presence and in the absence of the inhibitor, b_0 is equal to 2.3 RT/F, b_a and b_c are the slopes of Tafel's anodic and cathodic curves, and $\Delta\varepsilon_c = \varepsilon_c' - \varepsilon_c$ is the difference between the corrosion potentials in the presence and absence of the inhibitor.

The potential dependence of the adsorption can be followed by recording electrocapillary curves with dropping mercury electrodes,[30] by measuring differential capacities of the electrochemical double layer on mercury[43,44] or on solid electrodes[40,45,46] or by suppressing the polarographic maxima.[47]

By measurements of the type mentioned, it is possible to study both the adsorption of different species and the possible modification undergone by the organic inhibitors as a function of the potential of the surface on which they are adsorbed. In this way it is possible to interpret the adsorptions of organic cations or neutral molecules due to their deprotonation, the various orientations of aromatic or heterocyclic compounds on the surface (adsorption through the π electrons or through the polar functions), and the occurrence of true structural modifications of the inhibiting substances due to reduction or polymerization.

Because of the difficulties in working with solid electrodes, numerous data have been obtained with dropping mercury electrodes by recording the electrocapillary curves (Fig. 8). The maximum on the electrocapillary curve occurs at a potential corresponding to the zero-charge potential of the metal. The adsorption of surface-active inhibiting anions, molecules,

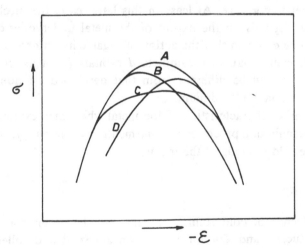

Fig. 8. Electrocapillary curves: A) solution without surface-active addition; B) the same solution + surface-active organic cation; C) the same solution + surface-active nonionized organic substance; D) the same solution + surface-active organic anion.

or cations will take place respectively on the positive branch, at the maximum, and on the negative branch of the electrocapillary curve, depressing it.

If, as said above, adsorption is considered as the first stage of inhibition, the potential assumed by the metal under consideration in a definite environment ε with respect to the value of its zero-charge potential $\varepsilon_{q=0}$ will acquire importance in order to establish whether adsorption and therefore inhibition could take place. The expression $\varepsilon - \varepsilon_{q=0} = \phi$ defines the potential of the metal on the ϕ scale of potentials, according to the idea developed mainly by Antropov,[4] and determines the charge of the metals with respect to the environment.

At equal values of ϕ for different metals, a similar behavior of a given inhibiting substance should be expected, as has been shown in some cases by Antropov and by Fischer and Seiler.[3] These latter authors studied the inhibition of various electrode reactions by β-naphthoquinoline cations on mercury, nickel, copper, and iron electrodes, coming to the conclusion that the electronic configuration of the electrode metals appears to play a secondary role.

In most cases in which the organic inhibitors act on various metallic electrodes by being adsorbed in the cationic form, the inhibiting efficiency should be the same when the potentials on the ϕ scale of the various metals are equal, while in the case where the inhibitor is adsorbed in the molecular form the chemical properties and the electronic structure of the metal assume greater importance. At least in this latter case, the mechanism of the inhibition depends on the nature of the metal undergoing corrosion; it may in fact be expected that the action of organic inhibitors chemisorbed on transition metals having incomplete d orbitals (such as iron, nickel, chromium, etc.) may be different from that developed on nontransition metals, such as zinc and cadmium.

Among other characteristics of the metal which may exhibit some influence on the inhibition processes must be mentioned the purity, the surface state, and the cold working of the metal.

Cold Working

The influence of cold rolling on inhibition has been considered by Trabanelli, Zucchi, and Zucchini[48] for iron annealed and rolled to 50% of its initial thickness. The measurements were carried out in $1N$ hydrochloric acid, either pure or in the presence of $10^{-4}M$ dibenzyl sulfoxide. Under the experimental conditions used, rolling caused an increase in the inhibiting power. These conclusions were obtained using several techniques:

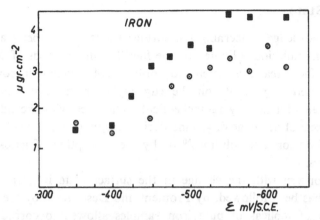

Fig. 9. Amount of inhibitor (S³⁵ dibenzyl sulfoxide) adsorbed
on iron as a function of the potential: ⊙ annealed samples;
■ cold-rolled samples.

measurements of corrosion rates, recording of polarization curves, and determinations of polarization resistance. Radiochemical determinations of
the amount of inhibitor adsorbed, carried out by using dibenzyl sulfoxide
labelled with S³⁵ (Fig. 9), led to the conclusion that the interaction of the
inhibitor is greater on electrode surfaces previously subjected to cold
rolling.

The shape of the polarization curves recorded after 4 h preimmersion
of the electrodes under conditions of free corrosion indicates the possible
formation of polylayers, even though the inhibiting power calculated from
the losses in weight does not increase with time. This behavior leads to the
belief that the efficiency of the inhibitor is basically connected with the first
layers of chemisorbed inhibitor. These results can be interpreted on the
basis of the mechanism of the action of the inhibitor already mentioned.
It has been stated that dibenzyl sulfoxide acts as a secondary inhibitor by
virtue of its reduction to the sulfide. Such reduction is shown on iron even
at the stationary value of the corrosion potential, and increases under
cathodic conditions. The presence of a higher number of dislocations acting
as points of weak hydrogen overvoltage on rolled electrodes[49] leads to a
promotion of the reduction of the inhibitor to dibenzyl sulfide, which is
responsible for the increase in inhibiting power. The considerable action
of the inhibitor on the cathodic process has been confirmed by the amount
of inhibitor adsorbed under conditions of cathodic polarization. The inhibitor is present only in small amounts on the anodically polarized electrode,
and the anodic dissolution of the metal is stimulated.

Surface State

Our knowledge concerning the surface states of the metal and their influence on inhibition is limited by the fact that the methods of determining real surface areas are difficult to apply to solid metals, in contrast to powders. Interesting results on the rugosity of metallic materials have recently been obtained by radiochemical techniques with the adsorption on iron sheets of acetic acid, glycine derivatives, and stearic acid (labelled with C^{14}) from organic solutions[50] or by the adsorption of gases at low temperatures.[51,52]

The concern with the change in the surface state in corrosion and inhibition has been followed, by Froment and Desestret,[53] by the electron microscopy of replicas of pure iron samples allowed to corrode in $1N$ sulfuric acid in the presence and absence of 2-butyne-1,4-diol. The authors showed that after a 1 h immersion in the noninhibited acid solution, the surface of the iron was affected by a considerable attack of the generalized type, and its appearance showed the crystal orientation of the grains; after 2 h of contact with the acid solution containing 30 mmole/liter of 2-butyne-1,4-diol the surface was only slightly attacked and showed no geometrical attack forms because of the leveling effect due to the inhibitor. These measurements show that the change in the state of the real surface of the electrode may lead to erroneous results if they are based only on the recording of the values of the current (e.g., determination of polarization curves). In such cases, obviously, the effects described will be still greater in the recording of the anodic polarization curve of the dissolution of the metal.

Surface Treatments

Inhibition tests on materials with different surface treatments in various environments have been carried out by several authors. By recording the polarization curves and applying the technique of linear polarization, the behavior of dibenzyl sulfoxide as inhibitor in the acid corrosion of nickel with various degrees of surface treatment obtained by treatment with abrasive papers of varying fineness has been studied.[54] It was shown that the efficiency of the inhibitor, referred to apparent surface, becomes greater, as a coarser abrasive is used. By comparing the polarization curves obtained (Figs. 10 and 11), a clear cathodic effect of the inhibitor can be observed, while the anodic effect decreases when the degree of finish of the metallic surface is increased.

Similar results have been obtained by Talbot[55] in the propargyl alcohol

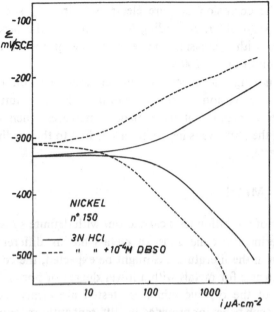

Fig. 10. Anodic and cathodic polarization curves of nickel
electrodes abraded with # 150 emery paper.

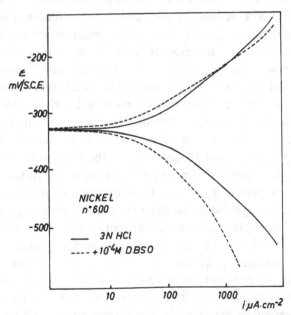

Fig. 11. Anodic and cathodic polarization curves of nickel
electrodes abraded with # 600 emery paper.

inhibition of the corrosion of pure electrolytic iron in sodium chloride solution acidified to pH 1. According to this author, the percentage inhibition increases with increasing diameter of the grains of the abrasive, tending to a limit at about 200 μ.

Gatos[56] studying the effects of surface treatment on the benzoate inhibition of the corrosion of iron in sodium chloride solution, has also shown that the adsorption of the benzoate increases when the diameter of the grains of the abrasive is increased, according to the readings observed with a profilometer.

Purity of the Metal

The results of inhibition tests carried out with definite substances under analogous experimental conditions on metals having different degrees of purity are given in the literature. As might be expected, the corrosion rates measured are greater for metals with a lower degree of purity, but on these the efficiencies of the organic inhibitors tested are clearly superior. The explanations of such behavior provided by different authors frequently do not agree, but tend essentially to justify the action of the particular types of inhibitor used.

Hoar and Holliday[57] have shown that the dissolution of pure iron in 5% sulfuric acid at 70°C is only slightly inhibited by quinoline and thiourea derivatives, in contrast to the situation under similar conditions with steel. The authors consider that the essentially inhibiting action of these substances is due mainly to blockage of the "sulfur-activated" anodic zones present in the steel but not present in pure iron. This fact is connected with the hypothesis on anodic dissolution formulated by Hoar, according to which the anodic process is clearly stimulated by sulfur-containing products derived from the dissolution of sulfur inclusions in the metal. Dissolution in nonoxidizing acids is said to give rise to the formation of hydrogen sulfide, which can be adsorbed immediately on the metal surface, bringing the surface atoms into energy states permitting easier dissolution. Since adsorption must be regarded in the dynamic sense, such energy conditions can be regarded as capable of extension almost to the entire surface of the metal. The anodic zones so created on steel are just those "sulfur-activated" zones that are blocked by quinolines and thioureas. In support of this hypothesis, Hoar reports that the anodic dissolution of pure iron is stimulated by traces of thiourea and its derivatives, since the decomposition of these sulfur compounds may lead to the formation of "sulfur-activated" anodic zones even in the absence of sulfur inclusions in the metal. Naturally, an increase in the concentration of the thioureas will mask this effect and inhibit the dissolution process.

A different view is expressed by Epelboin[58] on the basis of his study of the behavior of 2-butyne-1,4-diol as an inhibitor of Armco iron and of high-purity iron in $1N$ sulfuric acid at 25°C. According to this author, the rate of corrosion of pure iron is appreciably lower than that of Armco iron, but the action of the inhibitor is shown only in the latter case and it does not appreciably alter the corrosion of high-purity materials. Similar results have also been reported by Talbot,[55] who used Armco iron in comparison with pure electrolytic iron.

It has been shown[53,58] that 2-butyne-1,4-diol exerts its action on the cathodic evolution of hydrogen by virtue of the $C\equiv C$ triple bond, appreciably decreasing the cathode discharge current. The absence of corrosion inhibition of pure iron by 2-butyne-1,4-diol is explained by Epelboin on the assumption that in solutions of sulfuric acid in the absence of additives the discharge of hydrogen at the potential of free corrosion is already so weak that it is unaffected by the inhibitor. Epelboin provides an explanation of the observed low values of the cathodic current by assuming that on high-purity iron the cathodic zones are protected by solvated protons (H_3O^+), being screened from the action of the inhibitor. On the other hand, according to the same author, the percentage coverage of the surface of Armco iron with H_3O^+ is much smaller because Armco iron absorbs hydrogen at about a 25 times greater rate than iron prepared by zone melting. In this case the cathodic action of 2-butyne-1,4-diol is not hindered and there is an appreciable diminution in the discharge current.

Hydrogen Penetration

The already mentioned problem of the penetration of hydrogen also assumes remarkable importance in the use of organic pickling inhibitors. Bockris et al.[59] showed that naphthalene increases the rate of hydrogen permeation into iron because of the interaction between the naphthalene π bonds and the d orbitals of the metal, with a consequent decrease in the strength of the Me—H_{ads} bond. Valeronitrile, benzonitrile, and naphthonitrile, on the other hand, decrease the rate of hydrogen permeation because of the vertical adsorption of these compounds on the surface of the metal, with a consequent inhibition of proton discharge and a decrease in the coverage of the surface with hydrogen. It appears to have been also shown[60,61] that certain organic compounds containing the $\rangle C{=}S$ bond, although acting effectively as corrosion inhibitors for iron and steel, stimulate hydrogen penetration, creating the prerequisites for the embrittlement of the metal. The action of derivatives with the $\diagdown$C$=$S bond, such as the thioureas, may be interpreted by assuming that such compounds are partially reduced in the cathodic zones, liberating the hydrogen sulfide which acts as a promoter of hydrogen penetration. With other types of organic compounds, such as amines, aldehydes, acids, sulfides, and sulfoxides, which cannot decompose into hydrogen sulfide or other penetration promoters, the adsorption bonds between organic molecules and the surface atoms of the metal lead to inhibition of corrosion and simultaneously to a considerable retardation of hydrogen penetration.[62] According to Smialowski,[63] in the presence of a layer of molecules strongly adsorbed on cathodic surfaces the discharge of the hydrogen ions is not produced directly on the metal, rather at a certain distance, in contact with the adsorbed layer; the penetration of hydrogen into the metal is thus almost completely prevented.

METHODS OF STUDYING INHIBITORS

The conventional measurements adopted for the study and control of inhibition are those generally used to evaluate the corrosion phenomenon as a whole, carried out in the presence and absence of the inhibitors concerned.

Corrosion Rate Measurements

The corrosion rate can be expressed as variation in the weight per unit surface and per unit time or as penetration of the corrosive process into the metallic material in unit time. These expressions are valid if we refer to generalized and uniform corrosion processes, but they have a smaller practical significance in the case of localized attack phenomena. In the latter case, we can obtain an idea of the corrosion rate from the maximum depth of attack that can be deduced by various physical methods or by microscopic observations. The amount of metallic material dissolved can also be determined by estimating spectrophotometrically the metal that has passed into solution.

In the case of the corrosion of metallic materials in a nonoxidizing acid environment, the effects of inhibitors can be evaluated by measuring the volume of hydrogen evolved in the presence and absence of the additives using gas-volumetric procedures. This method, which enables the corrosion rates to be easily calculated, can give inaccurate results when the inhibitors

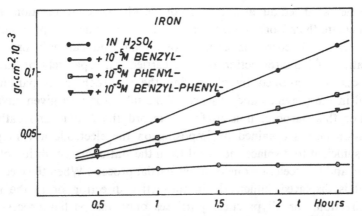

Fig. 12. Corrosion rate of iron in $1N$ H_2SO_4 in the presence of different aromatic sulfides (annealed samples, treated in $1N$ H_2SO_4 for 90 min).

under study are subjected to appreciable decomposition by reaction with the hydrogen, or where the penetration of hydrogen into the metal is not negligible in comparison with the total volume of hydrogen evolved.

The corrosion rate under conditions of both inhibition and noninhibition can be represented by the diagrams shown in Fig. 12. The effect of the inhibitor is normally expressed by the percentage inhibition (P.I.) deduced from the following formula:

$$\text{P.I.} = \frac{\text{uninhibited corr. rate} - \text{inhibited corr. rate}}{\text{uninhibited corr. rate}} \times 100$$

Electrochemical Methods

A remarkable contribution has been made to the progress of investigations in the field of inhibition by instrumental techniques. At the present time, it may be considered that the electrochemical methods form the means most widely used for the study of the behavior of inhibitors.[58,64,65]

The efficiency of organic inhibitors stable in the aggressive environment and acting by adsorption may be evaluated, for example, by the displacement of the corrosion potential in the presence of the inhibitors. According to Antropov,[30] this correlation can be deduced from the following expression:

$$\Delta \varepsilon_c = \varepsilon_c' - \varepsilon_c = \left[1 - \frac{b_a \cdot b_c}{b_0(b_a + b_c)} \right] (\psi_2' - \psi_2)$$

where the magnitudes cited have the meanings given above. In this case, measurements carried out as a function of the additive concentration enable us to calculate the adsorption isotherms. Obviously, since the phenomenon of electrochemical corrosion consists of anodic and cathodic processes, examination of inhibitor action may be carried out separately on each of the processes mentioned by comparing the electrochemical parameters recorded in the presence and absence of the inhibitor in a given environment. For this purpose, it is useful to record the anodic and cathodic polarization curves obtained by imposing on the electrode a voltage or current sufficient to displace the metal from the stationary condition of free corrosion and to accelerate one or other partial process. When the electrode potential is displaced sufficiently in the active direction or in the more noble direction, for all practical purposes only one partial process will take place at the electrode, since the rate of the other will be negligible. If the corrosion process in the absence of inhibitor takes place according to the following partial reactions

$$\text{Me} \rightarrow \text{Me}_{\text{aq}}^{z+} + ze$$

$$\text{H}_3\text{O}^+ + e \rightarrow \tfrac{1}{2}\text{H}_2 + \text{H}_2\text{O}$$

the rate of which is determined by a slow kinetic step, it is possible to find the current densities relating to the partial processes i_a and i_c. Thus, from Frumkin, Antropov[30] got the equations which also account for the adsorption potential:

$$i_a = k_a \exp\left(2.3 \; \frac{\varepsilon_c - \psi_2}{b_a}\right)$$

$$i_{\text{H}_2} = k_c[\text{H}^+] \exp\left\{2.3 \left(\frac{\psi_2 - \varepsilon_c}{b_c} - \frac{\psi_2}{b_0}\right)\right\}$$

where k_a and k_c represent the rate constants. In the case in which the cathodic process takes place according to the equation

$$\text{O}_2 + 4\text{H}^+ + 4e \rightarrow 2\text{H}_2\text{O}$$

the rate of which is controlled by mass transfer, we have

$$i_{\text{O}_2} = k_{c'} \, [\text{O}_2]$$

In the preceding cases, the slopes of the Tafel curves calculated from the voltage–current curves enable us to deduce the step controlling the partial electrochemical process under study. It is possible to distinguish between activation polarization, resistance polarization, and concentration

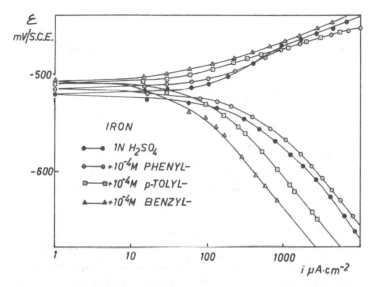

Fig. 13. Anodic and cathodic polarization curves of iron in $1N$ H_2SO_4 in the presence of different organic sulfoxides. Potential sweep rate, 2 V/h.

polarization. Displacements of the anodic and cathodic polarization curves toward smaller values of the current will give a measure of the action of the inhibitor on the individual partial corrosion processes. Possible stimulations will be characterized by displacements of the curves with the additives to values of the current greater than for the blank tests. The corrosion currents can be measured by extrapolating the anodic and cathodic curves to the value of the corrosion potential. The polarization curves can be recorded by intermittent methods or by imposing a linear variation of the variable chosen with time. The rate of variation of the variable chosen leads to the establishment of a stationary or nonstationary condition at the electrode in the recording.

The influence of the rate of imposition of the potential has been shown by Greene[66] in recording anodic polarization curves for stainless steels and by Trabanelli, Zucchi, and Gullini in the case of the inhibition of the corrosion of Armco iron in acid solutions by sulfoxides (Figs. 13 and 14).

Polarization techniques have been applied to the study of inhibition in a wide variety of environments. Rozenfeld,[67] Agarwala and Tripathi,[68] and Trabanelli, Fiegna, and Carassiti[21] have used these methods to study the action of inhibitors in the vapor phase on electrodes covered with thin films of electrolytes (Fig. 15), or immersed in neutral solutions. Amines, amine salts, thiourea, and thiourea derivatives were examined.

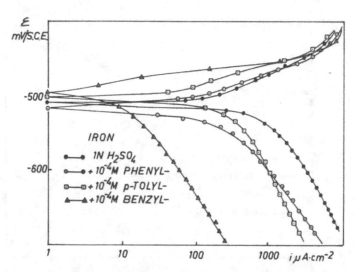

Fig. 14. Anodic and cathodic polarization curves of iron in $1N$ H_2SO_4 in the presence of different organic sulfoxides. Potential sweep rate, 0.2 V/h.

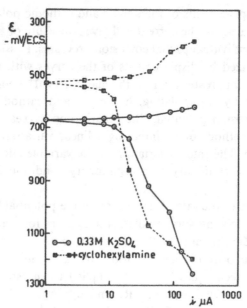

Fig. 15. Galvanostatic anodic and cathodic polarization curves of iron covered with a thin film of electrolyte ($\sim$100 μ) in the absence and in the presence of an organic amine inhibitor.

Among the investigations on the inhibition by organic substances of metal corrosion in salt solutions carried out with the aid of polarization methods, we should mention those of Sathianandhan[69] and Smialowski.[70] Inhibition in acid solutions has in many cases been investigated by analyzing the results of polarization measurements. Among the numerous investigations are those of Fischer,[8] Hackerman,[19] Kaesche,[71,72] Cavallaro *et al.*,[73] Smialowski,[12] Hoar,[39] Jofa,[74] Talbot,[55] Grubitsch,[75] and others. ·

If the influence of large surface changes of the electrode in the course of anodic polarization or a considerable influence of hydrogen penetration in the course of cathodic polarization is feared, polarization can be limited to the linear section of the polarization curve near the stationary potential of the metal. The technique of linear polarization enables us to determine the ratio existing between the overvoltage ($\Delta\varepsilon = \varepsilon - \varepsilon_c$) and the current applied (ΔI) for overvoltages of 10 mV at a maximum. The slope of the line obtained has the dimensions of resistance and is connected with the corrosion rate by Stern and Geary's[64] relationship:

$$\frac{\Delta\varepsilon}{\Delta I} = R = \frac{b_a \cdot b_c}{2.3 \cdot I_{corr} \cdot (b_a + b_c)}$$

where R is the "polarization resistance," I_{corr} is the corrosion current, and b_a and b_c are the Tafel constants for the anodic and cathodic processes. This technique has been applied by Kaesche[76] to the study of the behavior of phenylthiourea on iron in acidified sodium perchlorate solutions to define the variations in the inhibition with the concentration of the additive. Froment and Desestret[53] found the "polarization conductance" ($K_s = 1/\Delta\varepsilon$) in the case of the inhibition of iron in sulfuric acid by 2-butyne-1,4-diol, subsequently deducing the values of I_{corr} by means of an empirical relation developed by Menenoh and Engell ($I_{corr} = K_s \cdot 40$) for corrosion inhibitors in sulfuric acid.

By determining polarization resistances, Trabanelli, Zucchi, and Zucchini[54] have confirmed the action of dibenzyl sulfoxide on the dissolution of nickel subjected to various surface treatments in $3N$ hydrochloric acid (Fig. 16). The same authors[48] made use of this method to follow the variations in the inhibiting action of dibenzyl sulfoxide with time on iron electrodes in $1N$ sulfuric acid.

The technique of pulse polarization, using very short times and oscillographic recording, provides information on transient electrode phenomena and has been found to be particularly useful in studies of inhibition, particularly for the purpose of defining the primary or secondary inhibiting

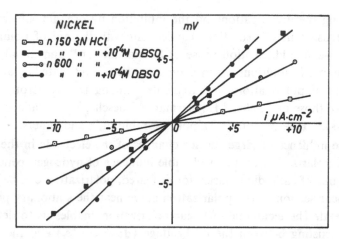

Fig. 16. Polarization resistances on nickel electrodes (prepared with different emery papers) in the absence and in the presence of dibenzyl sulfoxide.

action of organic substances in cases in which such action cannot be clearly characterized by polarization curves. Lorenz and Fischer[77] have developed a galvanostatic pulse technique characterized by reversal of the current, which permits adsorption of the inhibitor on the electrode surface under definite potential conditions and only for a limited time. Using this method, the authors analyzed the inhibitor behavior of quaternary "onium" compounds, the rate of reduction of which was relatively small as compared with the times involved in the measurement. From a comparison of the shapes of the potential/time oscillograms at constant current, Lorenz and Fischer were able to show that quaternary phosphonium and arsonium compounds exert a primary inhibiting action by being adsorbed on the electrode as "onium" salts, while dibenzyl sulfoxide acts only as a secondary inhibitor through previous reduction to dibenzyl sulfide.

A rapid polarization method has been developed by Okamoto[78] for the study of the inhibition due to amines and thioureas.

Other electrochemical techniques applicable to investigations on inhibition phenomena are those described in connection with the determination of the potential-dependence of the adsorption of organic compounds. Thus, recording of electrocapillary curves enabled Antropov to show that the effectiveness of physical inhibitors of the acid corrosion of metals is directly connected with the reduction of surface tension at the mercury–solution interface at a potential corresponding to the corrosion potential of the metal on the ϕ scale of potentials. Previously, Indelli and Pancaldi[47] studied the depression in the polarographic maxima of the reduction waves of copper and nickel ions at the dropping mercury electrode due to aromatic amines, isothiocyanates, and thiourea derivatives in relation to their adsorption characteristics (Fig. 17). The authors recognized that the adsorption of amines is highly dependent on the potential, in contrast to what was found for thioureas and isothiocyanates. Measurements of the same type were carried out by Montel with thioglycolic acid.[79] Gatos,[80] has shown that the effect of amines in depressing the negative branch of the electrocapillary curve on mercury increases with increasing molecular weight to the same degree as the inhibiting efficiency of these compounds in acid corrosion.

Interesting results on correlations between adsorption and the inhibition of acid corrosion have been obtained with these techniques by Ostrowski[81] for amines and sulfoxides, by Meakins[82] for quaternary ammonium cations, by Jeannin[83] for thioglycolic acid and propargyl alcohol, and by Bockris[84] for aromatic amines. De[95] has studied the action of benzoate, salicylate, and cinnamate on the dropping mercury electrode, these being anions that inhibit the corrosion of metals in a neutral medium. The ions investigated by De were all found to be surface active, in contrast to ions having no inhibiting characteristics. Determinations of the differential capacity of the

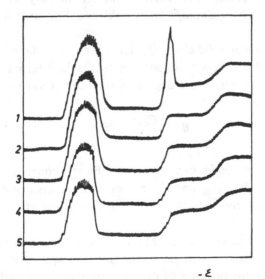

Fig. 17. Depression in the polarographic maxima of Cu^{2+} (left) and Ni^{2+} by o-toluidine: 1) without o-toluidine; 2–5) increasing concentration of the additive.

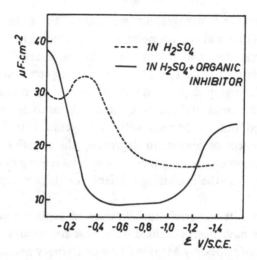

Fig. 18. Differential capacity curves of the electrical double layer on dropping mercury electrode.

double layer were carried out both on mercury electrodes and on solid electrodes in order to determine from the capacity lowering, the range of potentials corresponding to the adsorption of the organic inhibitors (Fig. 18).

Among the investigations carried out on mercury we may mention those of Frumkin,[43] Epelboin,[58] Ostrowski and Fischer,[44] Meakins,[82] and Antropov.[4]

Measurements on solid electrodes have been carried out by Hackerman and Murakawa,[46] Murakawa, Nagaura, and Ohashi,[45] Jeannin,[83] and Jofa.[40] This last author calculated the degree of surface coverage θ with inhibitor from the relation

$$\theta = \frac{C_0 - C_{inh}}{C_0 - C_\infty}$$

where C_0 is the differential capacity at a definite frequency in the absence of the inhibitor, while C_{inh} and C_∞ represent the corresponding capacities found at a definite concentration of the inhibitor and at the limiting concentration.

In a series of investigations, Machu[10] developed a method of determination of the variations in resistance and in capacity at the electrode brought about by the presence of organic additives in the solution, basing interpretations of the mechanism of the action of the substances tested (including substituted thioureas) on the increase in resistance and the decrease in capacity.

Epelboin[58] carried out impedance measurements as a function of frequency on iron and nickel electrodes at definite potentials in sulfuric acid solutions in the presence and absence of 2-butyne-1,4-diol to obtain information on various steps of the electrochemical reactions.

Radiochemical Methods

Radiochemical methods with labeled inhibitors have made it possible to extend our knowledge of inhibition phenomena, permitting trace detection of substances adsorbed even under conditions of extreme dilution, and have provided an objective method of evaluating the amount of inhibitor adsorbed on the electrode (Fig. 19).[86]

The advantages of radiochemical measurements become particularly obvious when it is shown to be possible to carry out adsorption determinations without having to remove the metal sample from the conditions of thermodynamic equilibrium achieved in the environment under study. In addition, the development of techniques for measuring the amounts of substance adsorbed at various values of the potential imposed on an electrode has enabled problems of inhibition to be resolved.

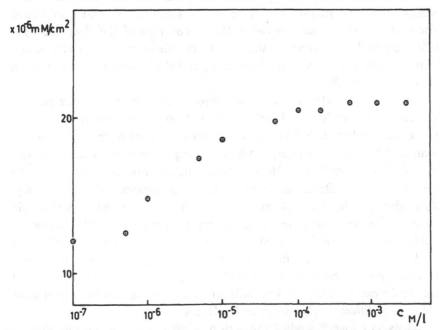

Fig. 19. Adsorption curve of n-dodecylamine (tritium-labeled) on Nickel in $1N$ HCl as a function of the inhibitor concentration; temperature, 25°C.

The simplest techniques consist of bringing about the adsorption of the compound under study on metallic material from various media, subsequently extracting the metallic sample, and subjecting it, after washing, to a count determination to measure its activity. Determinations of activity on metal samples can be carried out by keeping the samples in the medium under study, using cells in which the counter tube is placed behind the metal on which the adsorption is taking place, or measuring the decrease in the concentration of the labeled additive in the solution as a result of adsorption. In an accurate experimental technique developed by Bockris,[87] a special cell with ribbon electrodes is used which permits working under definite conditions of applied potential. The electrodes leaving the cell are not displaced from the equilibrium conditions achieved since they are subjected to counting between thin sheets of Mylar, enabling the solution to remain in contact with the metal. Using this technique, Bockris studied the adsorption of decylamine[87] and naphthalene[88] on nickel, copper, iron, lead, and platinum electrodes, while Gileadi[89] obtained the isotherms for the adsorption of benzene on platinum electrodes from solutions of sulfuric acid.

Radiochemical measurements with counting on a sample after extraction have been carried out by Gatos[56] in a study of the behavior of C^{14}-labeled benzoate anions as iron corrosion inhibitors in NaCl solutions. The author came to the conclusion that a coverage of the electrode by a 10% monomolecular layer is sufficient to inhibit corrosion. In very dilute benzoate solutions there is no inhibition, and the benzoate is adsorbed on the corrosion products.

Bordeaux and Hackerman[90] used similar methods to study the inhibiting action of stearic acid labeled with C^{14} on the corrosion of iron in an aqueous solution. Schwabe,[91] using accurate experimental techniques, compared the action of compounds inhibiting atmospheric corrosion with that of pickling inhibitors. He was thus able to draw conclusions on the adsorption from different environments, and at various pH's, of C^{14} dicyclohexylamine, dicyclohexylamine nitrite, and S^{35} dibenzyl sulfoxide. He came to the conclusion that a necessary requirement for the action of pickling inhibitors of the type studied (dibenzyl sulfoxide and its reduction product, dibenzyl sulfide) in an acid medium is the formation of at least a monolayer on the surface, while dicyclohexylamine nitrite is adsorbed only to the extent of 10% of a monolayer, in agreement with the conclusions deduced in different circumstances by Gatos.

Ross and Jones[92] studied the action of S^{35} thiourea as an inhibitor of the corrosion of steel in $1N\ H_2SO_4$ by adsorption measurements, and found

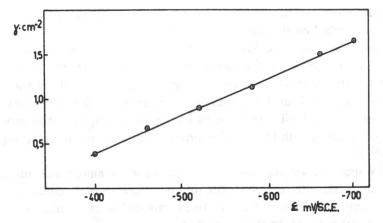

Fig. 20. Amount of adsorbed inhibitor as a function of the potential on iron electrodes. Test solution: $1N$ HCl containing $10^{-4}M$ S^{35} dibenzyl sulfoxide; test time: 30 min; corrosion potential: ~ -500 mV(SCE).

that the adsorption of thiourea on specimens in free corrosion takes place in a uniform manner and reaches a limiting value within 10 min. On carrying out activity measurements on samples kept in inhibited solutions at different potential values, the authors were able to show that the adsorption varies with the applied potential and is greater under anodic conditions.

The potential dependence of the inhibitor adsorption has been shown on several occasions by Carassiti et al.[48,54,93] in studies of the action of dibenzyl sulfoxide in the inhibition of the acid corrosion of iron, nickel, and stainless steels. In all these cases the inhibitor mentioned, which tends to form polylayers, is preferentially adsorbed under cathodic conditions of the electrode and the adsorption increases almost linearly when the potential is changed in the direction of cathodic values (Fig. 20).

IR and UV Spectroscopic Methods

IR and UV spectroscopic analyses performed directly on the adsorbed products or on substances desorbed from metallic electrodes have been found to be particularly advantageous for the interpretation of inhibition phenomena. IR techniques applied to substances adsorbed on finely subdivided metals are connected with studies of heterogeneous catalysis[33,94–96] and enable conclusions to be drawn on the interesting functions of the adsorption bonds and on the arrangement of the inhibitor molecules on the surface of the metal. In the case of corrosion inhibition with dibenzyl sulfoxide, Schwabe[34] was able to show by means of this technique that the product adsorbed on the electrode was dibenzyl sulfide.

Rozenfeld[97] examined films of varnishes treated with various corrosion inhibitors, such as guanidine chromate, by IR spectroscopy. From the changes in the IR spectra before and after the aging of the film under the action of light and heat, the author came to the conclusion that guanidine chromate exerts its inhibiting action by strongly retarding the anodic process and by interacting with the complex products formed during the drying of the varnish.

UV spectroscopy has been used to determine the amounts of inhibitor adsorbed on the electrode by evaluating the decrease in its concentration in the solution under conditions of free corrosion[9] or at definite values of the potential imposed on the electrode.[98]

Hackerman et al.[27] studied the adsorption of 4-ethylpyridine, 4-ethylpiperidine, and poly-4-vinylpyridine from solutions in cyclohexane or aqueous solutions of HCl onto reduced iron powder. Working at various concentrations, the authors obtained the adsorption isotherms and drew conclusions concerning the characteristics of the adsorption of the compounds mentioned. The authors assumed that the substances investigated interact at the surface by being adsorbed chemically.

In various investigations on the inhibition of the acid corrosion of iron, nickel, and stainless steels by means of dibenzyl sulfoxide, Carassiti et al.[48,54,93] applied UV spectroscopic analysis to the products desorbed from the electrodes. The existence of bands of dibenzyl sulfide, besides those of dibenzyl sulfoxide, was demonstrated, in confirmation of the secondary inhibiting effect of dibenzyl sulfoxide.

A particularly interesting technique was developed by Suetaka to determine directly the amount of inhibitor adsorbed by reflection spectra recorded on metallic electrodes. He studied the adsorption and reflection spectra of quinoline adsorbed on nickel surfaces,[99] coming to the conclusion that the adsorption takes place through the lone electron pair of the nitrogen atom. Similar measurements[100] with thiourea have shown that the intensity of the adsorption band decreases considerably, demonstrating interaction of the compound with the surface through the thiocarbonyl group and the possible formation of surface complexes.

The results obtained by the determination of IR spectra permit also the discrimination of adsorption phenomena on the metal from those on the surface layers and oxide.

Poling and Eischens[96] have determined the adsorption spectra of butyl nitrite adsorbed on iron–iron oxide, coming to the conclusion that during chemisorption the butyl nitrite decomposes to give chemisorbed hydrocarbon and nitrogen oxides, which are also adsorbed. The results obtained by these authors indicate that the inhibiting characteristics shown by butyl nitrite must be attributed both to the oxidizing action of the nitrogen oxides and to the pronounced chemisorption of neutral or cationic inhibiting species on the surface of the oxide.

Mass Spectrometry and NMR Methods

Mass spectrometry has also been found useful for the interpretation of the action of organic corrosion inhibitors. Using this method, Yu Yao[101] carried out an interesting study on the adsorption processes of substances from the vapor phase on metals. The author examined a series of amines, interpreting their action on the surface of iron as due to a process of non-dissociative chemisorption both on the oxide film and directly on the metal. He assumed that the chemisorbed inhibitor makes the oxide barrier more effective, decreasing the electric field gradient across the oxide and consequently slowing down the transport of metal ions through the oxide.

Mass spectrometry has also been used by Van Peteghem and Vanderkelen[102] to discern the causes of a severe corrosive attack manifesting itself on a lead coil immersed in a sulfuric pickling bath containing an inhibitor. From the spectra of the metal before and after the corrosive attack, the authors concluded that the main cause of the phenomenon was to be imputed to the particular pickling inhibitor used, which prevented attack of the acid on ferrous materials but greatly stimulated the corrosion of lead, being adsorbed on its surface.

We have already pointed out the importance of the electron density on the atoms of the polar functions in the determination of inhibitors adsorption characteristics in describing the various mechanisms proposed for the interpretation of the action of corrosion inhibitors. One method of investigation that can provide data capable of direct correlation with the values of the electron density is that of nuclear magnetic resonance. Cox, Every, and Riggs[103] obtained high-resolution NMR spectra of aniline, and of a series of substituted anilines. It was noted that the inhibition efficiency of these compounds varies with the nature of the substituent and with its position in the aromatic ring. Riggs found no significant relationship between the inhibiting properties and solubility, molecular weight, dipole moment, or surface area of the metal covered by the inhibitor, but did find a linear relationship between the proton resonance frequency of the amine hydrogens and the inhibiting characteristics of these compounds.

Other Methods

The adsorption phenomena of inhibitors on metal surfaces induce variations in the surface state of the metal itself, as shown by observations under the electron microscope by Froment and Desestret[53] of the acid corrosion of iron and its inhibition. A versatile technique that enables these phenomena to be evaluated is ellipsometry, which is essentially based on the sensitivity with which polarized light reflects the variations in the surface states of the metal. In this way the variations that surface films bring about in the polarization state of reflected light are measured. This technique has been applied to the study of organic inhibitors by Maddox[104] and Dettorre,[105] supplementing other methods.

The corrosion products formed on the surface of iron immersed in sodium chloride solutions inhibited with sodium benzoate, salicylate, acetate, or tartrate were examined by Sathianandhan et al.[69] by means of thermogravimetric analysis. In this way the authors were able to define the stability conditions of the complexes formed by the organic ions with the ferric ions as a function of temperature.

Where the formation of volatile products takes place, such as in the liberation of H_2S from acid solutions inhibited with decomposable sulfur-containing compounds, or where it is desired to analyze inhibitors in the vapor phase, gas chromatography can be used.

Where the surface-active characteristics of the additives are very pronounced, one method that enables the amount of inhibitor adsorbed on a metal surface to be determined consists of evaluating the decrease in the surface tension σ of the inhibitor solution. To avoid adsorption on materials other than the metallic sample under study (such as parts of the reaction cells), Le Boucher and Lefebvre[106] used metal samples in the form of beakers which were subjected to the attack of an acid solution containing tetradecylamine. Once the relationship between $\Delta\sigma$ and the concentration of the inhibitor had been found, they were able to find the adsorption characteristics of the amine under consideration.

Determination of Inhibitor Behavior vs Hydrogen Penetration

The methods outlined above always lead to an evaluation of the behavior of inhibitors with regard to their efficiency in decreasing the dissolution rate of the metal, which does not take into account the effects produced by the inhibitor on the metal. We wish to refer, in particular, to the previously mentioned possibility, that the inhibitor stimulates hydrogen penetration into the metal. In fact, an inhibitor can be considered completely effective if it simultaneously inhibits dissolution of the metal and hydrogen penetration into the metal itself. For this reason, it appears desirable to complete the description of the methods of investigation on inhibitors by referring also to the techniques of evaluating hydrogen penetration. For this purpose, Cavallaro et al.[61] calculated the ratio between the volume of hydrogen evolved and the hydrogen equivalent to the amount of metal dissolved, determined colorimetrically. In this way they were able to show that substituted thioureas, although acting as inhibitors of the dissolution of iron in an acid medium, strongly stimulate hydrogen penetration.

Other indirect methods[63] consist of the determination of the brittleness of metal samples previously subjected to acid attack in inhibited solutions or charged cathodically with hydrogen in acid solutions containing the inhibitors under study. Brittleness data can be deduced from variations in the breaking load, from measurements of the bending angles necessary for fracture of wires, from the number of double flexures before fracture, or from measuring the maximum number of rotations (torsions) of a metal wire before fracture. Comparative measurements carried out on metal specimens kept in an acid medium in the presence and absence of inhibitors will enable the percentage embrittlement produced by the penetration of hydrogen to be calculated, and consequently will offer an evaluation of the characteristics of the compounds under examination with respect to the phenomenon of hydrogen penetration.

Denoting by r_0 fracture data in the absence of adsorbed hydrogen and by r_H that after acid corrosion or cathodic charging with hydrogen, the percentage embrittlement E is found to be

$$E = \frac{r_0 - r_H}{r_0} \cdot 100$$

A very accurate experimental technique, consisting of determining the permeation rate of electrolytic hydrogen through metal membranes has been developed by Devanathan and Stachurski.[107] Bockris et al.[59] applied this method to an investigation of the permeation rate of hydrogen through Armco-iron membranes in $0.1N$ H_2SO_4 containing various additives, such as naphthalene, valeronitrile, benzonitrile, and naphthonitrile. Antropov et al.[108] studied the effect of numerous inhibitors on the corrosion of iron and on the diffusion of hydrogen through the metal. They showed that the presence of pyridine derivatives in the solution practically eliminates the diffusion of hydrogen through iron membranes.

ORGANIC INHIBITORS IN VARIOUS AGGRESSIVE ENVIRONMENTS

The mechanisms of organic inhibition and the methods of characterizing them experimentally can be usefully demonstrated by a short review of the use of inhibitors in various aggressive environments.

Atmospheric Corrosion Inhibitors

Atmospheric corrosion of metals—the mechanism of which can be considered as of the electrolytic type, taking place in a film of electrolyte present on the surface of the metals—can be prevented or inhibited by introducing a corrosion inhibitor into the environment. The use of inhibiting materials in the vapor phase is encountered in many industrial fields, particularly for the preservation and storage of finished articles or for the treatment of machines or machine parts having shutdown periods. They have also found use in the protection of military materials and electrical and electronic apparatus.

Atmospheric corrosion inhibitors have the advantage, as compared with other methods of protection, of preserving without regard to article shape, dimensions, and type of surface treatment. Furthermore, as stated above, the mechanism of their action may be based on adsorption phenomena with the formation of a film sometimes less than monomolecular, which does not involve appreciable contamination of the metal surface. The methods of applying such inhibitors are essentially as follows:

> Use of packaging papers impregnated with the inhibitor to wrap the metallic material to be protected.
>
> Spraying of the metal surfaces with solutions of the inhibitor which are generally dissolved in volatile organic solvents.
>
> Inhibitor saturation of a closed environment, which is achieved by placing in the environment a certain amount of inhibitor in liquid or solid form.

Numerous problems arise in connection with a correct application of inhibitors in the vapor phase, relating both to the composition and aggressivity of the corrosive atmosphere, and to the nature of the metals to be protected.

The aggressivity of an atmosphere with a relative humidity greater than a 65% critical value can be increased by various components, such as CO_2, H_2S, or SO_2, found under both urban and industrial conditions. Considerable attention has been devoted[109] to the effects of SO_2, which has been shown to be a vigorous depolarizing agent of the corrosion process and therefore capable of stimulating the rate of metal attack and of interfering with the action of inhibitors.

The nature of the metal to be protected can determine the choice of inhibitor, or a mixture of special inhibitors in those cases when dealing with machines or machine parts of both ferrous and nonferrous materials. An inhibitor effective for one type of metallic material can stimulate attack on a different material. Thus, amines and amine salts, which inhibit the atmospheric corrosion of iron and steel, can stimulate attack on nonferrous materials. For example, amine nitrites accelerate the attack of copper, brass, bronze, cadmium, lead, magnesium, and zinc. It has been recorded[110] that the stimulating action decreases with increasing molecular weight of the amines, for which reason dicyclohexylamine nitrite can be used, in some cases, in the presence of zinc, lead, and cadmium, unlike diisopropylamine nitrite.

Cotton[111] has prevented the staining of copper-based materials by the use of benzotriazole, $C_6H_5N_3$. Excellent results have been obtained under drastic conditions, such as exposure to atmospheres containing salts, SO_2, and H_2S, by wrapping the material in paper impregnated with inhibitors or placing it in an atmosphere containing inhibitor vapors. Inhibitor action is said to exhibit itself fundamentally at weak points or at defects in the cuprous oxide film normally present on the surface of copper. Cotton[111] considers it possible that various metal-inhibitor surface compounds are formed, in which the copper is bound by substitution of the NH-group hydrogen of the benzotriazole and by the formation of coordinate links by the lone pair of electrons of one of the nitrogens.

The corrosion of brass has also been inhibited by cyclohexylamine chromate[112] and by thiourea derivatives, particularly phenylthiourea.[68] Results obtained on tin, aluminum, zinc, and nickel[112] have been reported.

Working on numerous materials, such as Al, Ni, Zn, Mg, Sn, special bronzes, steel, and cast iron, Levin[113] demonstrated the remarkable efficiency of dicyclohexylamine chromate and of oil-soluble salts of dicyclohexylamine in preventing the corrosion of the materials mentioned, even in cases of galvanic coupling. The oil-soluble salts of cyclohexylamine were prepared by treating the amine with equimolecular amounts of fatty acids containing a definite number of carbons in the chain. The most interesting results were obtained with acids having a moderate number of carbon atoms in the chain—C_{11} and C_{18}. Naturally, in these cases the inhibitors were brought into contact with the metal surface by spraying on inhibitor–organic solvent solutions. Levin assumes that the inhibiting action of these substances must be connected with the formation of a hydrophobic layer acting as a surface barrier. Other results with various nonferrous metals have been reported by Baker,[110] Agarwala and Tripathi,[114] and Rozenfeld.[67] The latter, after having studied the mechanism of atmospheric corrosion and the electrochemical phenomena on thin films of electrolytes,[109] tested the inhibiting action of a series of benzoates and nitrobenzoates of cycloaliphatic amines and cyclic imines.[67] The efficiency of the inhibiting action of the nitrobenzoates of nitrogenous bases is interpreted by the author on the assumption of an electrochemical reduction of the inhibitor with consequent depolarization of the cathodic process. The reduction of NO_2 is facilitated by its introduction into an aromatic ring. As a consequence of the reduction, the potential becomes more noble, without, however, passivation conditions being reached. If, simultaneously, the organic cation of the amine obtained by hydrolysis is adsorbed on the surface of the metal, covering an appreciable fraction of the surface, the current density on the noncovered areas can easily exceed the corresponding critical anodic density, putting the metal in a state of passivation. It would therefore be the remarkable acceleration of the cathodic process combined with the effect of surface coverage by the cation to which the action of the nitrobenzoates should be ascribed.

Investigations relating in particular to inhibitors for the protection of ferrous materials are also the most numerous owing to the wide use of these materials in comparison with nonferrous materials. Various amines[21,110] and salts of cyclic amines, among which the best known is dicyclohexylamine nitrite[115–117] have been tested.

Mestreit and Guenard[118] suggest the use of fatty acid salts of polyamines with long carbon chains. They also discuss the effects of double bonds in the chain, of *cis–trans* isomerism, and of the chain length on inhibition. Since this concerns high-molecular derivatives with a low vapor pressure, the treatment of the metallic surface must be carried out in organic solvents.

Accurate methods for the study of inhibitor actions in the vapor phase have been given by Mindovich[119] and Volrabova.[120]

Inhibitors in the Steam Zone of Industrial Installations

The corrosion of metals due to high-temperature steam in the superheaters of industrial boilers can be reduced by the use of volatile inhibitors. The corrosion in such cases is due mainly to oxygen and carbon dioxide; the presence of oxygen can be reduced by treatment with "oxygen scavengers," while carbon dioxide can be neutralized with volatile amines,[121] such as morpholine (C_4H_9NO) and cyclohexylamine ($C_6H_{11}NH_2$), which are less aggressive than ammonia on copper-based components. Thus, the corrosion in the vapor zone of the casing and tubing of oil wells and crude-oil storage tanks has been combated by the addition of volatile inhibitors to the liquid phase. Among the substances used are methylpropylamine ($H_3C—NH—CH_2—CH_2—CH_3$), diethylamine ($(C_2H_5)_2NH$), and triethyl-amine ($(C_2H_5)_3N$).

Mixtures of urea and urease have also been used successfully. The enzyme hydrolyzes the urea to NH_3, which prevents corrosion in the vapor zone.[122]

Volatile amines of the types mentioned above have also found application in treatment of the vapor zone of storage tanks of the finished products in the petroleum industry.

Inhibitors in Aqueous Solutions

The inhibition of the corrosion of metals in contact with aqueous solutions at a pH not far from neutrality has in many cases been achieved by the use of inorganic substances. The sodium salts of organic acids, such as the benzoate,[56,123] cinnamate, salicylate, phthalate, and tartrate,[69] have been suggested as alternatives to these compounds, particularly with ferrous metals. The action of these substances appears to be due to the adsorption of the organic anion on the metal surface[56] or to the formation of complexes with the metal ions, which in some cases can be found in the corrosion products.[69]

Benzotriazole has been found to be very effective in preventing the corrosion of copper and its alloys[111] and of cadmium[124] in aqueous media.

The dissolution of aluminum over a wide pH range has been inhibited by Machu[123] by the use of hexamethylenetetramine, gelatin, tannic acid, thiourea, and di-o-tolylthiourea. He also showed that sodium benzoate acts effectively on zinc and aluminum, as well as on iron.

Balezin[125] has proposed the use of ethanolamine salts, especially ethanolamine phosphate, for the protection of ferrous and nonferrous materials. The possibility of unwanted corrosion acceleration by organic inhibitors on different metallic materials must always be borne in mind. For example, ethanolamine benzoate and amines in general inhibit the corrosion of ferrous materials and at the same time promote attack on nonferrous metals such as copper and nickel.

Various types of organic inhibitors have found use in numerous industries to prevent corrosion in boilers and water-cooling systems, and in the primary and secondary extraction of oil. In preboilers, corrosion is generally controlled by adjustment of the pH with neutralizing volatile amines, such as cyclohexylamine and morpholine, which ensure a prevention of attack even in the boilers. The use of volatile amines with a neutralizing action to inhibit corrosion by CO_2 in the vapor zone in connection with steam-condensing and recycling systems has already been mentioned.

A different method for preventing corrosion in steam-condensing systems is based on the use of film-forming substances, whose action is due to the formation of a hydrophobic film on the surface in contact with the condensate. This film acts as a barrier between the metal and the condensate.[126] Examples of these substances are octadecylamine and its salts, such as the acetate. Tests have also been carried out with mixtures of octadecylamine and cyclohexylamine with a nonionic surfactant, which, as compared with the salt mentioned, eliminates attacks that can be attributed to the acetate anion found at the points of injection of the inhibitor.[127] The use of the reaction products of fatty acids and polyamine[128] and of imidazolines or pyrimidines with side chains containing a large number of carbon atoms has also been suggested. The compounds concerned are in general those with at least one polar function and a hydrocarbon chain. Comparative measurements carried out on amines with equal numbers of carbon atoms in the chain and with double bonds or branched chains have shown that the greatest inhibiting power is obtained with the straight-chain saturated derivatives. Primary amines were found to be more effective than tertiary. The mechanism of their action has been interpreted by assuming the formation of a compact monomolecular layer by adsorption of the compounds on the metal surface through the polar function, with the hydrocarbon chain oriented toward the aqueous phase. The covering and water-repellent characteristics of the film ensure protection against corrosion. The greatest coverage density can be obtained with saturated long-chain compounds rather than with those having branched chains because of their lateral orientations. Again, because of their spatial arrangement, the chains of tertiary amines give a smaller coverage and therefore a lower degree of inhibition. Some authors[129] have stated that film-forming substances are dangerous inhibitors because of the possibility that localized attacks on the metal may develop in the zones not covered by the inhibitor when this is added in concentrations lower than the critical amount necessary to ensure complete inhibition. In view of the particular mechanism of their action, these substances are particularly advantageous in systems with a high concentration of CO_2, where the use of substances with a neutralizing action is not convenient. The inhibitor concentration necessary to obtain inhibition is independent of the content of CO_2 and O_2 and slows attacks on the metal due to these gases.

Important corrosion problems are encountered in industrial cooling systems, heat exchangers, piping, and finally, cooling towers. In these cases, too, corrosion inhibitors, particularly inorganic inhibitors, can contribute to resolution of the problems if, of course, for the reasons mentioned above, account is taken of the nature of the metal to be protected. Mention must be made here of the possibility of using organic inhibitors in these installations.

The corrosion of ferrous materials has been inhibited with tannins and alkali-metal tannates, sodium sulfoglucosates, and acylation products of polyamines. The use of the products of the reaction of CrO_3 with organic substances has also been suggested.[121] The inhibition in this last case can be considered as due to the passivating action of the chromate and to the simultaneous formation of a protective film of adsorbed inhibitor.

Besides the benzotriazole already mentioned, the use of 2-mercapto-benzothiazole has been suggested for the inhibition of copper corrosion (Fig. 21). The action consists of the formation of an adherent and protective film. The binding characteristics of 2-mercaptobenzothiazole can be interpreted by substituting the hydrogen atom of the sulfydryl group by copper cations and by means of bonds coordinated with the electron pairs of the other sulfur atom or of the nitrogen atom.

The use of oil-soluble sulfonates,[130] of amides, and of tannin derivatives in aluminum cooling towers has been suggested.

A remarkable application of organic inhibitors is found in the primary and secondary production of crude oil. The corrosive attack of the electro-chemical type shown by metallic materials must be connected with the aqueous phase that accompanies the crude oil rather than with the hydro-carbon phase. The corrosion problems are very diverse, both because of the variation in the water–oil ratio, which may range from 1% water to 1% oil, and because of the absence of H_2S (sweet wells), or because of its presence (sour wells), in the primary production of the crude oil. In secondary production (water flooding), in which water under pressure is introduced into some wells in order to force the raw material to flow out from a central well, the problems of corrosion are due to the injection water which, being recycle material, generally possesses highly corrosive properties. The problems of inhibition in this phase of exploitation have been treated broadly by Cavallaro,[131] Negreev,[132] Gonik,[133] and in particular by Bregman.[121]

In this account we shall confine ourselves to mentioning certain types

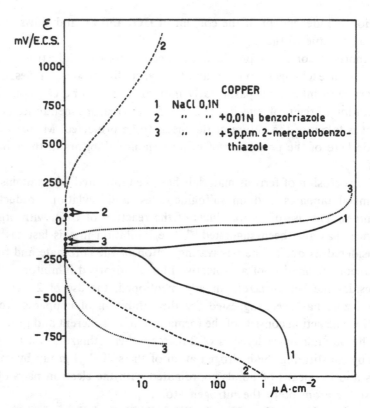

Fig. 21. Anodic and cathodic polarization curves of copper electrodes in 0.1N NaCl in the absence and in the presence of 5 ppm of 2-mercaptobenzothiazole or 0.01N benzotriazole.

of substances that are widely used. Some quaternary nitrogen compounds combine with their corrosion-inhibiting action a pronounced bactericidal effect. Among ammonium salts of the type $[(R_1R_2):N:(R_3R_4)]^+X^-$, we may mention those in which $R_1 = R_2 = CH_3$ and R_3 and R_4 consist of hydrocarbon chains containing from 12 to 18 carbon atoms. Other examples are dodecylbenzyldimethylammonium chloride and dioctadecyldimethylammonium chloride. Examples with two nitrogen functions exist, such as the decamethylene-*bis*-(dimethylhexadecylammonium bromide) proposed by Lytle. Effective inhibiting action is also shown by numerous derivatives of imidazoline such as 1-hydroxyethyl-2-octadecylimidazoline and 1-ethylamino-2-octadecylimidazoline. Quaternary derivatives such as 1-(2-hydroxyethyl)-1-benzyl-2-tridecyl-2-imidazolinium chloride and salts of imidazoline with salicylic acid, sebacic acid, and sulfonic acids have also been studied.

Among nitrogen compounds with a film-forming action, the most numerous group is that of the derivatives of aliphatic fatty acids. Under this term are included salts of oleic and naphthenic acids, amides of fatty acids doubly substituted on the nitrogen atom, and monoamines and diamines with aliphatic chains.

Certain substances with nitrogen and sulfur in the molecule, such as the thioimidazolines and thiourea derivatives with the nitrogen atoms bound to aliphatic chains containing amide functions, are also effective inhibitors.

Compounds containing sulfur used with very good results in corrosion inhibition in the extraction of crude oil are sulfonates having the general formula $(C_nH_{(2n-10)}SO_3)_x$ A, where A may be a metal or an aliphatic amine.

Inhibitors in Acid Solutions

The use of organic inhibitors in acid solutions is very common, particularly in view of the high rate of corrosion shown by metallic materials in such media. The mechanism of the action of organic inhibitors has been given above, but before referring to some practical applications of these substances it appears suitable to mention the influence that certain parameters, such as temperature, pressure, and rate of stirring, may exert on the efficiency of the inhibitors.

An increase in the temperature leads to different effects according to the type of inhibitor. Interesting results have been reported by Balezin,[134] Anoshchenko,[135] Radovici,[136] and Riggs.[137] Balezin has shown that an increase in pressure of up to 300 atm leads to a decrease in the corrosion rate of the metal and to a reduction in the efficiency of organic inhibitors.

The dependence of the inhibition efficiency on the concentration of the additive is generally shown in an increase of inhibiting power when the concentration is raised. Nevertheless, cases have been recorded in which decomposable substances such as the thioureas cause stimulation at relatively low concentrations,[76] generating depolarizing agents under conditions of inadequate coverage of the electrode. At higher concentrations, inhibition is obtained because of the favorable change in the ratio of decomposed materials to inhibiting materials, and because of the increase in surface coverage. When the additives under consideration are capable of forming soluble complexes with the ions of the metal, a stimulation effect may appear at higher concentrations. Data are also reported in the literature on the use of mixtures of organic inhibitors with inorganic substances. Anoshchenko[135] has used mixtures of thiourea and aluminum sulfate to inhibit the dissolution of iron in mixtures of nitric, hydrochloric, and sulfuric acids. Experiments carried out at different additive ratios may lead to antagonistic, additive, or synergistic effects. In the last mentioned case the use of the mixtures will be particularly suitable. Results in this direction have been reported by Balezin[125] for mixtures of organic substances (acetaldehyde with 2,4-dimethylpyridine and formaldehyde with furfuralimine). Furthermore, the effects of anions, such as I^- and SH^-, in the promotion of a pronounced inhibiting action by organic cations in acid solution are well known.[41,42,74]

In some industrial applications of corrosion inhibitors in an acid medium, substances with a surface-active effect are also added to the solutions to increase the wettability of the metal surface, together with foaming or antifoaming agents, according to the type of treatment.

An increase in the flow rate of the acid solution generally leads to an increase in the dissolution rate of the metal and to a decrease in the inhibitor efficiency. This appears to be shown by measurements carried out by Ross[92] on iron in solutions of H_2SO_4 in the presence of thiourea.

Balezin[134] has shown that in sulfuric acid solutions with flow rates of 2 m/sec, the dissolution rate of metal increases by a factor of 5–10, while hexamethylenetetramine, which is known to have an inhibiting action in static solutions, behaves as a corrosion stimulant under the flow conditions mentioned above. In confirmation of these effects, the inhibition produced by a mixture of hexamethylenetetramine and potassium iodide in flowing conditions is reduced by a factor of three as compared with a static medium.

Among the processes, on an industrial level, in which organic inhibitors have found application in an acid medium are, for example, the pickling of ferrous materials, descaling of stainless steels, cleaning of industrial apparatus, and decontamination of nuclear reactors. The pickling operation for removing rolling mill scale or rust formed in various ways on iron and steel during their manufacture or on finished products are always carried out in the presence of organic inhibitors to avoid severe direct attack on the metal. These processes use dibenzyl sulfoxide, acetylenic alcohols, α- and β-naphthylamines, thiourea derivatives, pyridine and quinoline bases, hexamethylenetetramine, and products of the condensation of aldehydes with ammonia and amines.

The use of corrosion inhibitors has also been studied in descaling operations of austenitic stainless steels carried out in nitric acid–hydrofluoric acid or ferric sulfate–hydrofluoric acid solutions or in an electrolytic bath of sulfuric acid in which the iron functions as the anode. Thorpe[138] studied the effect of thiourea, propylamine, and octylamine, coming, however, to the conclusion that the advantages of using these additives are rather modest.

A process basically similar to pickling and scaling from the operative point of view is that of the decontamination of nuclear reactors. Contamination takes place mainly as a result of corrosion of the primary system and of fuel-element failures. Decontamination consist of the removal of the surface film from metals contaminated with active atoms, such as Cr^{51}, Co^{60}, Mn^{56}, and Cu^{64} depending on the type of metallic materials used. Ayres[139] used aggressive solutions consisting of oxalic acid, either in the pure state or in mixtures with K_2SO_4, sulfamic acid, phosphoric acid, sodium bisulfate, and ammonium citrate. Among the inhibitors tested by that author there were various amines (triethyl-, tripropyl-, tributyl-, and diphenylamines) and numerous derivatives of thiourea (phenyl-, 1,3-diphenyl-, 1,1-diphenyl-, 1,3-di-o-tolyl-, 1,3-di-p-tolyl-, allyl-, and acetylthioureas).

Dissolution and removal of encrustations in heat exchangers, evaporators, and boilers is generally performed with solutions of hydrochloric acid in the presence of organic inhibitors of the type already mentioned in connection with their use in acid media.

In oil-well operations it is sometime necessary to remove calcareous products. This makes it necessary to carry out acidification with very concentrated acid solutions (for example, 15% HCl) which, if carried out in the absence of inhibitors, would lead to rapid destruction of the metallic parts. Here, very satisfactory results have been obtained by Hugel[140] using organic inhibitors such as mercaptans, aldehydes, glycol xanthates, and glycol dibenzylxanthates. In general, these are substances of very low solubility added to the acid solutions in concentrations that can be called critical, since the inhibition decreases for lower and higher concentrations. They have been called "insoluble inhibitors," and act by giving rise to macroscopic films of products from surface reactions with metals.

Remarkable corrosion problems are also encountered in the distillation and refining of petroleum products. Among the main causes of attack are the hydrochloric acid produced by hydrolysis of the chlorides initially present in the brines, certain lower aliphatic acids, and, in particular, hydrogen sulfide. Besides neutralizing inhibitors, imidazolines, straight-chain aliphatic amines with a high number of carbon atoms (C_{18}), polyamines and their derivatives, and amides of fatty acids are used.

Nottes[141] has proposed the use of acetylene derivatives as inhibitors in petroleum refineries. The efficiency of 3-diethylamino-1-propyne, 3-dimethylaminobutyne, 3-benzylaminobutyne, 3-isopropylaminobutyne, 5-di-

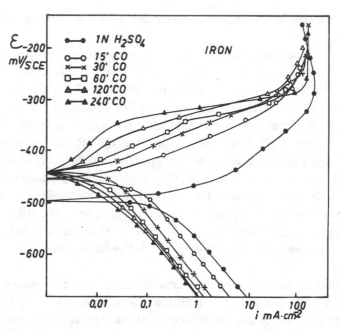

Fig. 22. Anodic and cathodic polarization curves on iron in $1N$ H$_2$SO$_4$ in the presence of carbon monoxide (15, 30, 60, 120, and 240 min are the bubbling times of CO in the solution at a constant flow rate of 800 ml/h).

ethylamino-4-pentyn-2-ol, and other compounds with a triple bond between carbon atoms has been studied by him in solutions of hydrochloric, acetic, and formic acids. This type of substances combines all the best inhibiting characteristics of the acetylenic derivatives with those due to the presence of nitrogen functions in the molecule, while the presence of hydroxyl groups affects their solubilities.

For uses of a more general nature, it may be interesting to mention some types of inhibitors proposed for various metals and alloys in acid media. Here the presentation is carried out with the structural characteristics of the various inhibitors taken into account. The bulk of the results that can be found in the literature refer to measurements on ferrous materials. A simple substance capable of acting as an effective inhibitor by adsorption on iron is carbon monoxide. It will retard the evolution of hydrogen and inhibit anodic dissolution of the metal up to moderately high anodic overvoltages, above which the polarization curves show a zone of impolarizability attributable to desorption[72] or to participation in the process of anodic dissolution of CO (Fig. 22).[142,143]

Among hydrocarbons, the most effective are undoubtedly the acetylenic hydrocarbons.[32,144] The effects of the introduction of alcohol groups[33] or other functions[22] into their molecules have been studied.

Other compounds that have been tested are acetylene, propargyl alcohol,[55,83] 2-butyne-1,4-diol,[53] ethynylcyclohexanol,[33,145] and more complex molecules.[22,144] In view of the proposed action mechanism of the substances mentioned, it appears obvious that the inhibiting action will be reduced when the saturation of the bonds between the carbon atoms is increased (Fig. 23).

A class of compounds that has been the object of numerous investigations is the nitrogen compounds, including aliphatic[39] and aromatic[71] amines, cyclic imines,[19] products of the reaction of aldehydes with ammonia and amines,[146] azoles,[147] pyridine derivatives,[57,38,148,14,30,28] quinoline bases,[57] and naphthoquinoline bases.[2,3] To these compounds may be added quaternary ammonium salts,[40] such as tetrabutylammonium halides. The "onium" compounds of arsenic, phosphorus, and sulfur have also found application as inhibitors in acid media.[31,35]

Organic derivatives of phosphorus with selenium or sulfur in the molecule have been studied by Smialowska.[12] The results of the electrochemical and gas-volumetric measurements showed adsorptions of the inhibitors

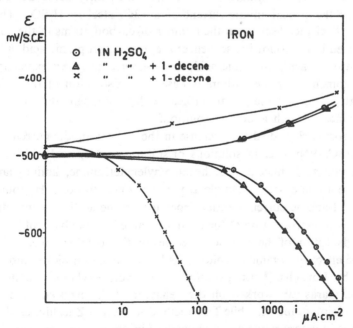

Fig. 23. Anodic and cathodic polarization curves of iron electrodes in $1N\ H_2SO_4$ in the presence of n-decene or n-decyne.

following the Freundlich isotherm. The inhibitor efficiency of compounds containing the same aliphatic substituent was higher in the presence of selenium than in the presence of sulfur in the molecule.

Organic substances containing atoms of both nitrogen and sulfur in the molecule have also found application in pickling treatments. Among these are thiourea and its derivatives substituted on the nitrogen atom with aliphatic chains or aromatic rings.[10,60,73,76,139] Karagounis[149] has proposed the adoption of a polymeric product with a high barrier effect derived from the condensation of melamine with thionyl chloride.

Numerous organic substances with sulfur atoms in the molecule have been reported as inhibitors of the acid corrosion of iron and steel. Luttringhaus and Grubitsch[35] have studied the inhibiting action of the trithiones, and Hugel[140] tested the effectiveness of a homologous series of mercaptans, while other authors[31,34-36] devoted their attention particularly to sulfoxides, especially the dibenzyl compound.

The inhibition of the acid corrosion of stainless steels has been studied by Carassiti[150] using decylamine, quinoline, phenylthiourea, and dibenzyl sulfoxide.[93] The passivating effect of organic inhibitors on steel in acid solutions has also been studied.

Smialowski and Smialowska[70] showed that fatty acids decrease the density of the critical anodic current of a 13Cr steel in H_2SO_4. The passivating effect increases when the number of carbon atoms in the additives is increased in the following sequence: acetic acid, propionic acid, n-butyric acid, n-valeric acid, n-caproic acid, and n-capric acid. An explanation of the phenomenon by the authors is based on adsorption of the anions of these fatty acids on the metal surface, with subsequent reduction of the active zones through anodic dissolution.

Grigoryev[151] studied the decrease in the critical anodic current density for an 18Cr–9Ni steel in solutions of sulfuric acid up to 14.3M, working with acrylic acid, maleic acid, hexamethylenetetramine, aminoguanidine, aminophenol, and dicyandiamide at various concentrations. He found that the critical anodic current density depended on the additive concentration, and gave an interpretation of the results obtained on the basis of the theory of the adsorption of the organic molecules on the metal surface.

The effect of organic inhibitors has been studied on nickel, monel, and copper by Foroulis,[152] using n-butylamine, cyclohexylamine, aniline, and pyridine. Carassiti[23] worked on the behavior of dibenzyl sulfoxide in the acid corrosion of nickel, while Trabanelli, Zucchi, and Zucchini used carbon monoxide in comparative measurements of inhibition on iron and nickel in solutions of hydrochloric and sulfuric acids (Fig. 24).[143]

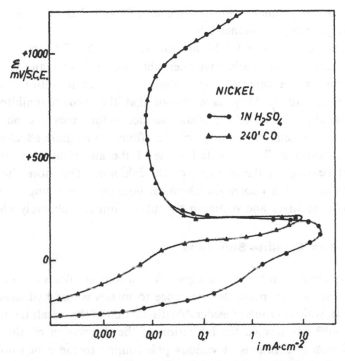

Fig. 24. Anodic polarization curves of nickel in $1N$ H_2SO_4 in the absence and in the presence of carbon monoxide (240 min CO bubbling time in solution with a constant flow rate of 800 ml/h).

The dissolution of copper in acid solutions has been inhibited by sulfur-containing organic compounds, benzyl thiocyanate, and furfural. Furfural has been found by Desai[153] to be an excellent inhibitor for copper in hydrochloric acid solutions. This inhibitor affords protection to copper at all concentrations of HCl, the efficiency being 90–100%. The amounts of furfural required to bring about efficient inhibition in dilute acidic solutions is greater than that required at higher concentrations of HCl solutions. According to Desai, the resin-forming tendency of furfural appears to be responsible for the inhibiting power.

Furfural has also shown very good properties in inhibiting corrosion of brass, aluminum, and aluminum alloys.[154] Its action must probably be connected with the tendency to form resinous products. Among the substances used to inhibit the acid dissolution of aluminum we may mention acridine, quinoline, dextrin, and thiourea,[155] n-butyl-, di-n-butyl-, and tri-n-butylamines, phenol, pyrocatechol, resorcinol, and hydroquinone.[156]

The corrosion of Al–Zn alloys has been prevented by means of acridine, quinine, strychnine, and nicotinic acid,[157] and that of Al–Mg alloys by the butylamines mentioned above.

Interesting results in inhibiting corrosion of Zn, Cd, and Pb using di-o-tolylthiourea and gelatin have been obtained by Machu and Gouda,[158] by determining the capacities and ohmic resistance in sulfuric acid at various temperatures. They have shown that the studied inhibitors are first chemically bonded to the metal surface before they are physically adsorbed to it. Chemical bonding has been found to be more effective than physical adsorption. The potential value of the adsorbing metal has an important bearing on the bonding of the inhibitor. The more the metal is electropositive, the poorer the chemical bonding. According to Machu and Gouda for silver and platinum the adsorption is exclusively physical.

Inhibitors in Alkaline Solutions

Information on inhibition in aqueous solutions of alkalis can be found in the literature with particular reference to metals whose hydroxides are amphoteric, which permits considerable dissolution of the metals themselves.

Machu[123] has studied the inhibition of the dissolution of aluminum, zinc, and iron in solutions at various pH, coming to the conclusion that the behavior of the inhibitors is strictly related to the solubility of the surface oxides or hydroxides. The inhibitors, he states, widen the stability region of the layers of oxide and hydroxide of amphoteric metals such as aluminum and zinc in weakly alkaline solution. The action of the inhibitors is explained by assuming that they have an effect in repairing the pores in oxide and hydroxide films. Obviously, the inhibitors can no longer exercise any effective action in a clearly alkaline medium in which the oxides and hydroxides of aluminum and zinc are soluble.

In the experimental work carried out by Machu[123] a quaternary ammonium derivative, gelatin, and thiourea inhibited (though only weakly) the dissolution of aluminum at about pH 13, while various other substances stimulated attack on the metal. Balezin[125] showed that saponin also inhibits the dissolution of aluminum in alkaline solutions.

Inhibiting corrosion of zinc in alkali is currently applied to the attack of zinc anodes in batteries with an alkaline electrolyte. Among the organic substances proposed as inhibitors we find p-dicyclohexylbenzene,[159] triphenylchloromethane, 1-chloro-3-fluorobenzene, terephthalic acid, biphenyl-4-carboxylic acid, benzyl-t-butanol, and biphenyl-4-carbonitrile. Fairly satisfactory results have been obtained with n-dodecanol, dicyclohexylamine, N,N-dimethylcyclohexylamine, and N,N-diethyl-p-toluidine.

Desai[160] has studied the effect of 30 organic substances on the corrosion of copper in NaOH solutions up to $1N$. Among the most effective inhibitors are resorcinol, phloroglucinol, pyrogallol, tannin, m- and p-aminophenols, furfuraldehyde, 8-hydroxyquinoline, and quinalizarin. In many cases the action of the additives is explained by assuming the formation of a surface reaction product between the inhibitor, the metal, and the corrosive medium.

The formation of metal-inhibitor surface compounds has also been proposed by Levin in inhibiting corrosion of iron in alkali by the addition of tannin.

Ayres[139] used phenylthiourea in an alkaline solution of potassium permanganate as an inhibitor in the decontamination of nuclear reactors.

Inhibitors in a Nonaqueous Environment

Corrosion phenomena in nonaqueous environments are found in installations of the modern chemical and petrochemical industries. These corrosive attacks are frequently combated by the use of inhibitors, the choice in some cases made difficult by our inadequate knowledge of the mechanism by which the attack takes place.

Among nonaqueous media capable of promoting corrosive attacks are hydrocarbon phases, particularly in relation to the storage and special treatment of petroleum products, halogen derivatives of hydrocarbons, alcohols, phenols, etc. Inhibiting corrosion due to crude-oil or petroleum products[121] is achieved with organic substances such as arylsarcosines and their salts with amines, and with salts of aliphatic amines and diamines with fatty acids, including olein.

The inorganic and organic salts of dinonylnaphthalenesulfonic acid act as inhibitors in gasolines. Similar properties are exhibited by oleic acid esters with higher glycols.

Other efficient rust inhibitors are amine salts of nonyl-phenoxyacetic acid. Salts of aliphatic amines and of imidazolines with long-chain carboxylic acids have also achieved remarkable use.[121]

Treatments with inhibiting and antioxidant substances are also carried out on lubricating oils. The compounds used are organic phosphates or phosphites, salts of alkyldithiophosphoric acids, derivatives of amines and amides, or alkaline-earth metal salts of long-chain fatty acids.

Putilova has shown that the addition of sulfonated stearic acid, β-naphthol, α- or β-naphthylamine, or diphenylamine inhibits the corrosion of carbon steels and low-alloy steels in paraffinic oils. Marked corrosive attacks are found even on particularly highly valued metals in the course of the preparation and use of halogen derivatives of the hydrocarbons.[161] Such phenomena have been interpreted by assuming the formation of halohydric acids by hydrolysis of the organic compounds, promoted by the presence of even minute traces of water.

The use of various organic substances as inhibitors has been proposed. A classification is difficult both because of the variety of such compounds and because of their different action on different metallic materials. Aliphatic amines, diphenylamine, acetone condensation products, formamide, and olefin epoxides have been used.

SUMMARY

The ideas developed in synthetic form in the first part of this review offer a picture of our present knowledge of the action mechanisms of organic inhibitors. This picture must be considered as highly incomplete, because it is limited both in time and in the number of researchers whose ideas have been outlined, and some of whose results have been mentioned.

It appeared interesting to give some data on the role of the metal in inhibition processes and on methods permitting a study of inhibition phenomena.

After an examination of the various aggressive environments, some typical inhibitors for the various environments were mentioned. This has been done not for the purpose of compiling a general list of inhibiting substances but with the idea of providing reference material for persons interested in this problem from the point of view of study and application.

A more complete picture of the problem of inhibition can be obtained by considering also the use of inorganic substances, which have not been referred to in the present account.

APPENDIX

Since the detailed structures and composition of many of the organic compounds used in organic inhibition are unfamiliar to a broad range of metallurgists, materials scientists, engineers, electrochemists, and physical chemists, it was considered appropriate to include the structural formulas of the various organic compounds described in this article. On the following pages, reference numbers are given in parentheses following the names.

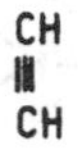

CH
∥
CH

Acetylene (25)

CH
∥∥
C
|
CH₂OH

Propargyl alcohol (22) (55) (144)

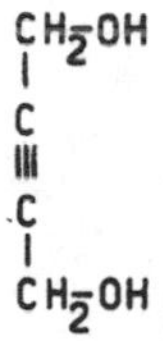

CH₂OH
|
C
∥∥
C
|
CH₂OH

2-Butyne-1,4-diol (53) (83)

C≡CH
C—OH
H₂C   CH₂
H₂C   CH₂
    C
    H₂

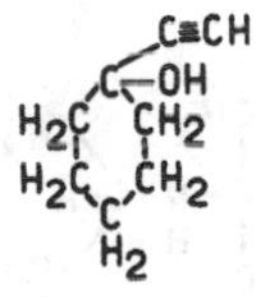

1-Ethynylcyclohexane-1-ol (145) (33)

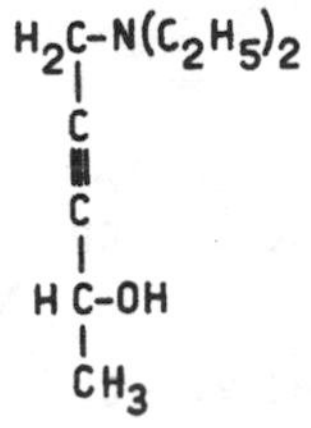

H₂C-N(C₂H₅)₂
|
C
∥∥
C
|
H C-OH
|
CH₃

5-Diethylamino-4-pentyne-2-ol (141)

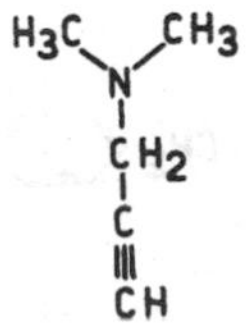

H₃C   CH₃
    N
    |
    CH₂
    |
    C
    ∥∥
    CH

3-Dimethylamino-1-propyne (141)

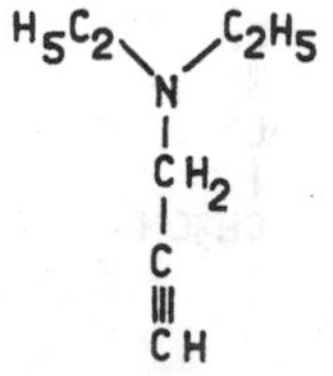

3-Diethylamino-1-propyne (141)

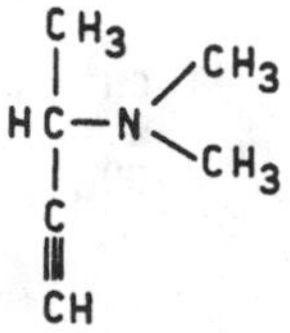

3-Dimethylamino-1-butyne (141)

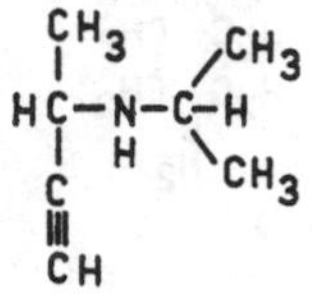

3-Isopropylamino-1-butyne (141)

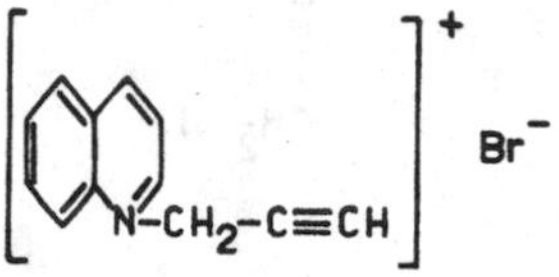

Propargyl quinolinium bromide (22)

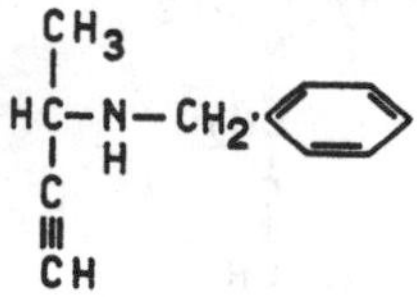

3-Benzylamino-1-butyne (141)

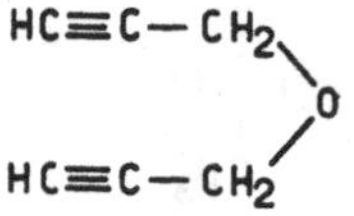

Dipropargyl ether (144)

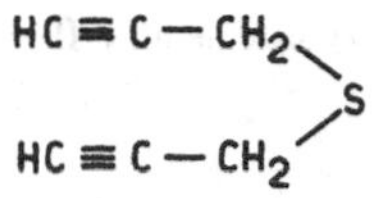

Dipropargyl thioether (144)

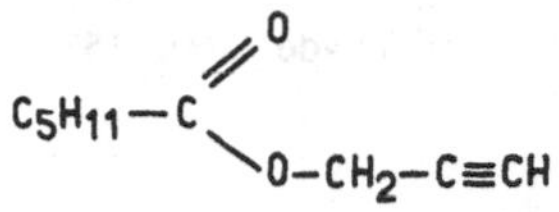

Propargyl caproate (22)

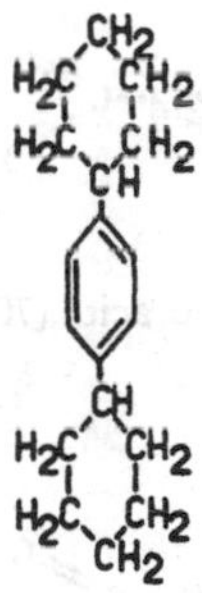

p-Dicyclohexylbenzene (159)

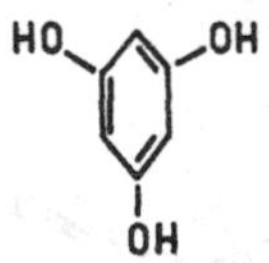

Phloroglucinol (160)

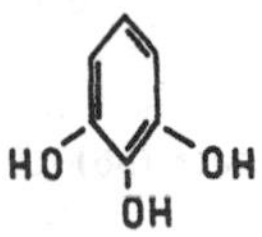

Pyrogallol (160)

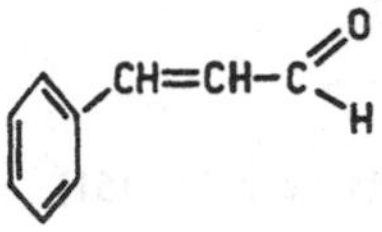

Cinnamic aldehyde (140)

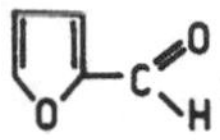

Furfuraldehyde (160) (153)

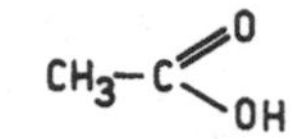

Acetic acid (70)

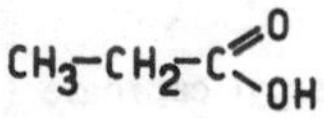

Propionic acid (70)

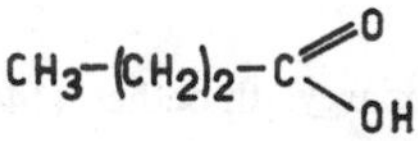

n-Butyric acid (70)

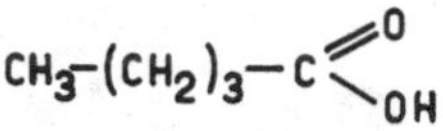

n-Valeric acid (70)

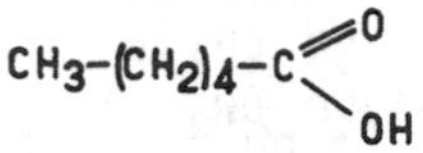

n-Caproic acid (70)

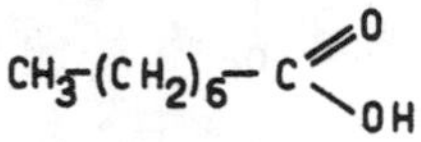

n-Caprylic acid (70)

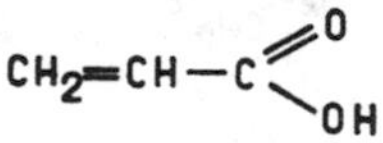

Acrylic acid (151)

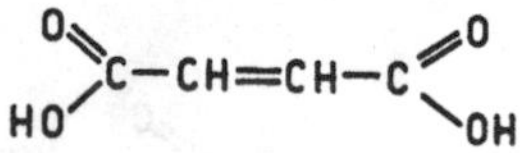

Maleic acid (151)

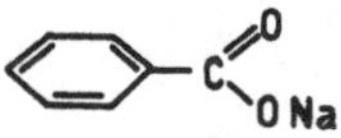

Sodium benzoate (56) (69) (123)

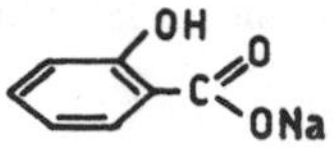

Sodium salicylate (85) (69)

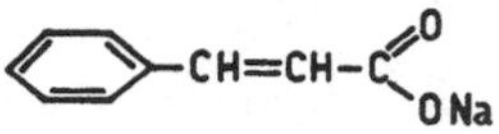

Sodium cinnamate (85)

$$CH_3-NH_2$$

Methylamine (71)

$$CH_3-CH_2-NH_2$$

Ethylamine (71)

$$CH_3-CH_2-CH_2-NH_2$$

n-Propylamine (71)

$$CH_3-(CH_2)_3-NH_2$$

n-Butylamine (152) (156)

$$CH_3-(CH_2)_7-NH_2$$

n-Octylamine (138) (39)

$$CH_3-(CH_2)_9-NH_2$$

n-Decylamine (87)

$$CH_3-(CH_2)_{13}-NH_2$$

n-Tetradecylamine (106)

$$CH_3-(CH_2)_{17}-NH_2$$

r-Octadecylamine (126)

$$CH_3(CH_2)_2 \begin{matrix} CH_3 \\ \end{matrix} NH$$

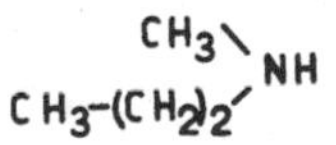

Methylpropylamine (121)

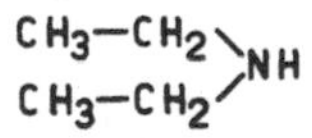

$$\begin{matrix} CH_3-CH_2 \\ CH_3-CH_2 \end{matrix} NH$$

Diethylamine (121)

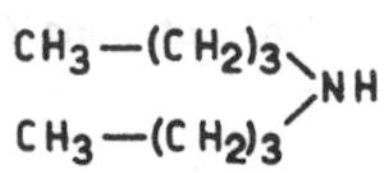

$$\begin{matrix} CH_3-(CH_2)_3 \\ CH_3-(CH_2)_3 \end{matrix} NH$$

Dibutylamine (156)

$$\begin{matrix} CH_3-(CH_2)_4 \\ CH_3-(CH_2)_4 \end{matrix} NH$$

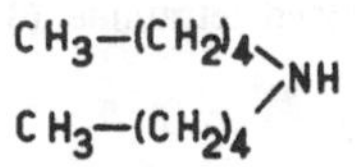

Diamylamine (19)

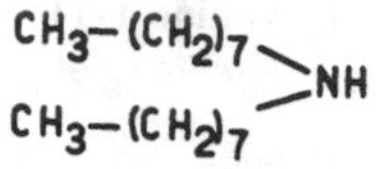

$$CH_3-(CH_2)_7$$
$$CH_3-(CH_2)_7$$ NH

Di-*n*-octylamine (39)

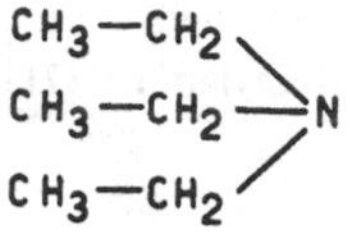

$$CH_3-CH_2$$
$$CH_3-CH_2$$ N
$$CH_3-CH_2$$

Triethylamine (121)

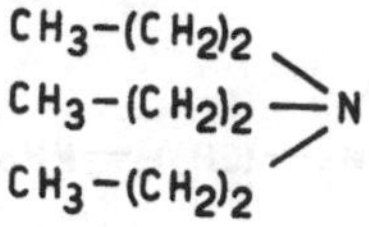

$$CH_3-(CH_2)_2$$
$$CH_3-(CH_2)_2$$ N
$$CH_3-(CH_2)_2$$

Tripropylamine (139)

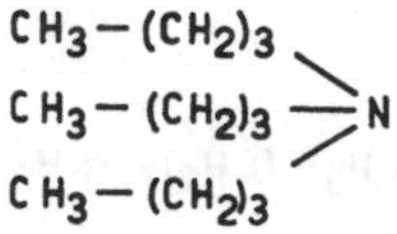

$$CH_3-(CH_2)_3$$
$$CH_3-(CH_2)_3$$ N
$$CH_3-(CH_2)_3$$

Tributylamine (139) (156)

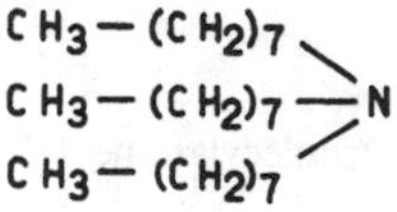

$$CH_3-(CH_2)_7$$
$$CH_3-(CH_2)_7$$ N
$$CH_3-(CH_2)_7$$

Tri-*n*-octylamine (39)

$$H_2N-(CH_2)_2$$
$$H_2N-(CH_2)_2$$ NH

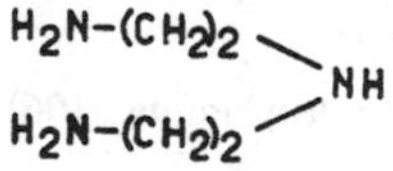

Diethylenetriamine (39)

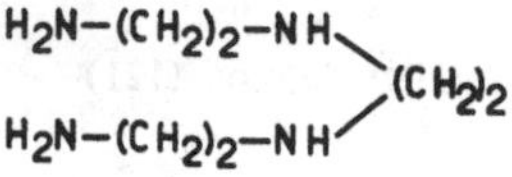

$$H_2N-(CH_2)_2-NH$$
$$H_2N-(CH_2)_2-NH$$ (CH_2)_2

Triethylenetetramine (39)

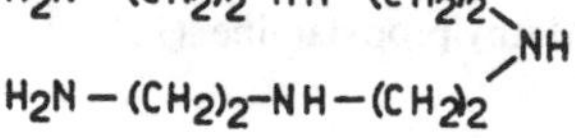

$$H_2N-(CH_2)_2-NH-(CH_2)_2$$
$$H_2N-(CH_2)_2-NH-(CH_2)_2$$ NH

Tetraethylenepentamine (39)

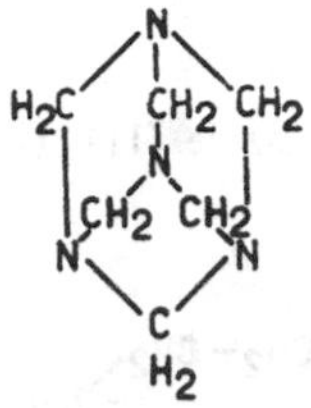

Hexamethylenetetramine
(123) (134) (151)

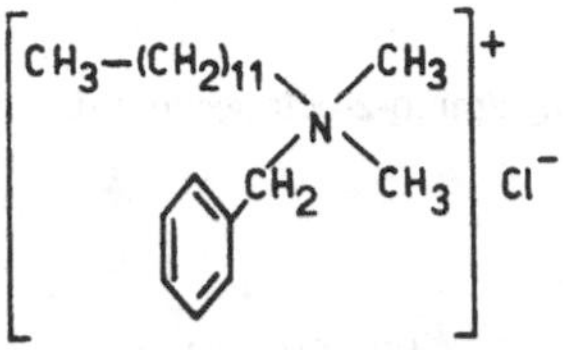

Dodecylbenzyldimethylammonium
chloride (121)

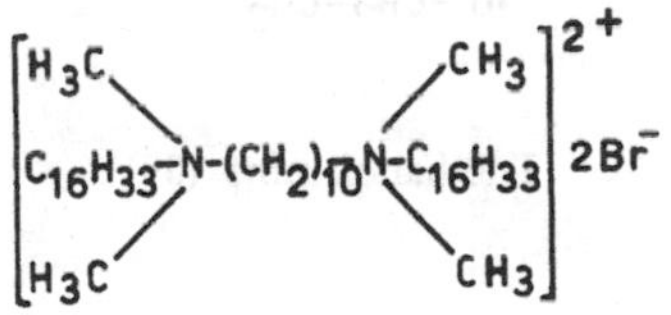

Decamethylene *bis*-dimethyl hexadecyl
ammonium bromide (121)

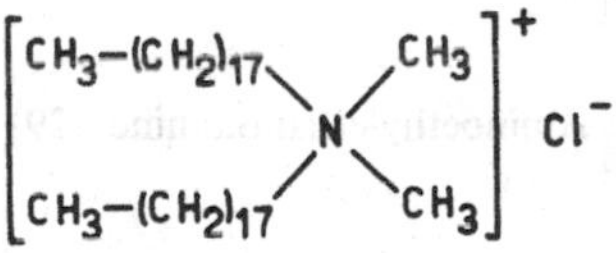

Dioctadecyldimethylammonium
chloride (121)

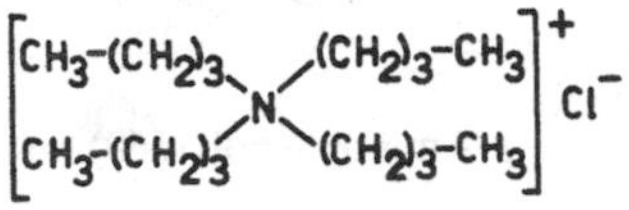

Tetrabutylammonium chloride (40)

Aniline (20) (71) (152)

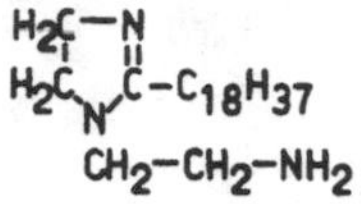

1-Ethylamino-2-octadecylimidazoline (121)

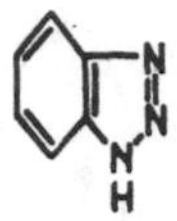

Benzotriazole (111) (124)

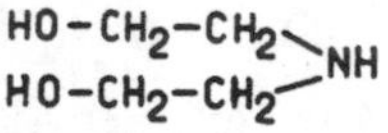

Diethanolamine (39)

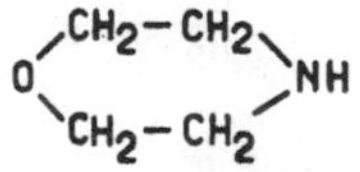

Morpholine (121) (39)

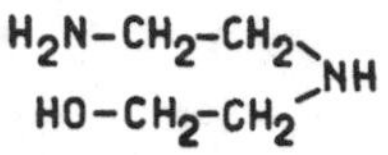

Aminoethylethanolamine (39)

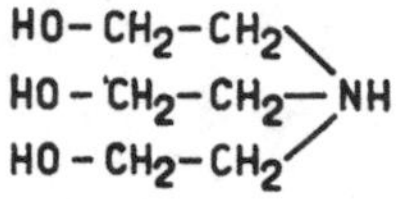

Triethanolamine (39)

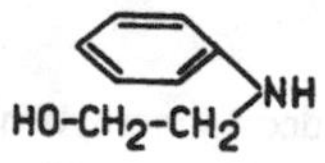

Phenylethanolamine (39)

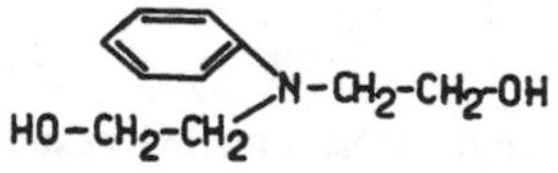

Phenyldiethanolamine (39)

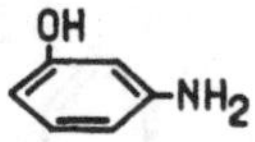

m-Aminophenol (160)

p-Aminophenol (160)

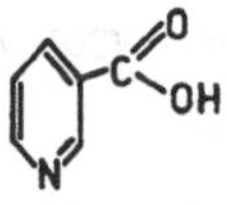

Nicotinic acid (157)

8-Hydroxyquinoline (160)

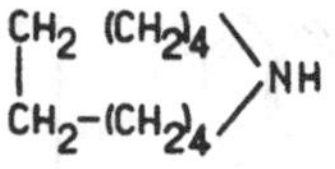

Decamethyleneimine (19)

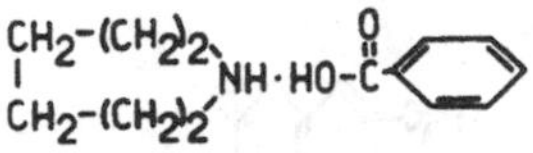

Hexamethyleneiminebenzoate (67)

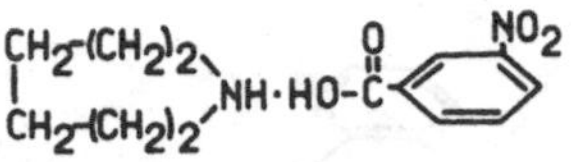

Hexamethyleneimine-*m*-nitrobenzoate (67)

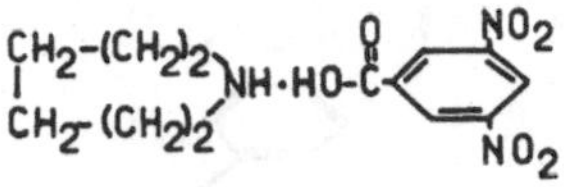

Hexamethyleneimine-3,5-dinitrobenzoate (67)

Pyridine (14) (152)

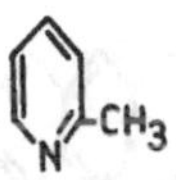

2-Picoline (14)

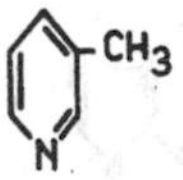

3-Picoline (14)

4-Picoline (14)

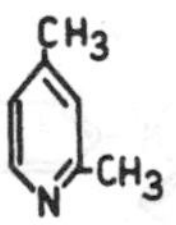

2,4-Lutidine (125)

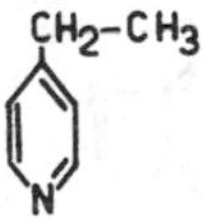

4-Ethylpyridine (27) (28)

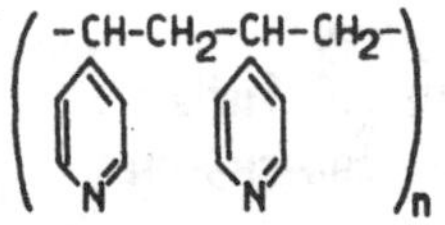

Poly-(4-vinylpyridine) (27) (28)

Quinoline (29) (150)

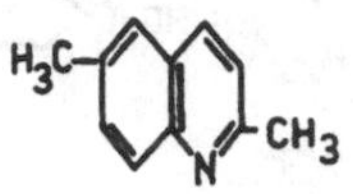

2,6-Dimethylquinoline (57)

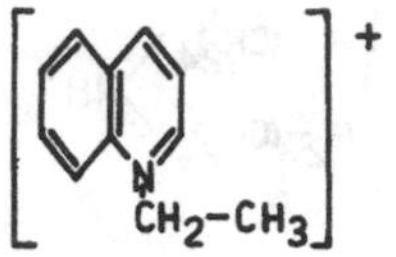

Ethylquinolinium ion (57)

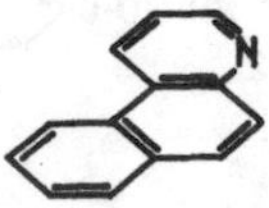

β-Naphthoquinoline
(5,6-Benzoquinoline) (2) (3)

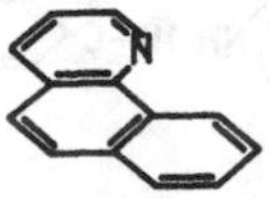

α-Naphthoquinoline
(7,8-Benzoquinoline) (137) (2) (3)

Acridine (157)

Imidazole (147)

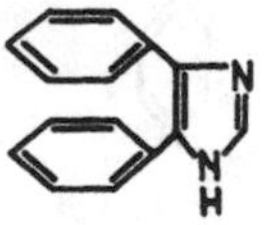

4,5-Diphenylimidazole (147)

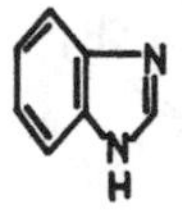

Benzimidazole (147)

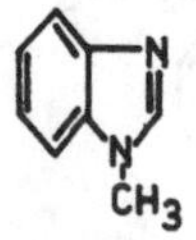

1-Methylbenzimidazole (147)

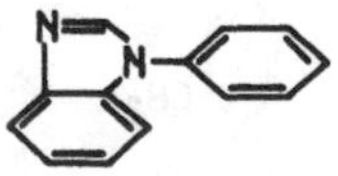

1-Phenylbenzimidazole (147)

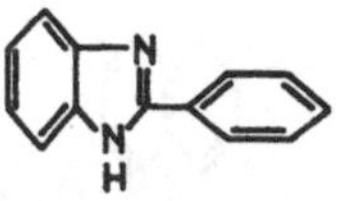

2-Phenylbenzimidazole (147)

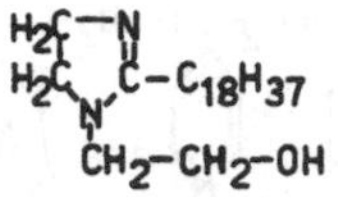

1-Hydroxyethyl-2-octadecylimidazoline (121)

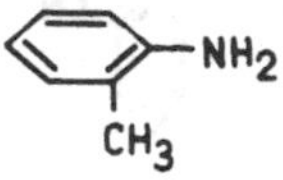

o-Toluidine (20) (29) (47) (71)

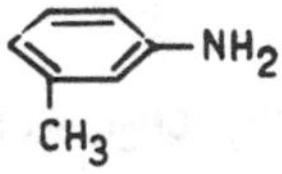

m-Toluidine (20) (71)

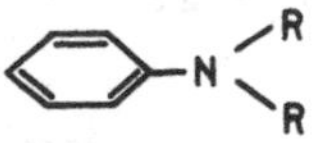

N,N-Dialkylaniline (21) (71)
R = methyl or ethyl

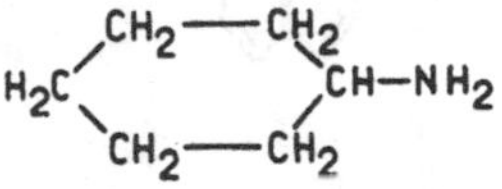

Cyclohexylamine (21) (121) (152)

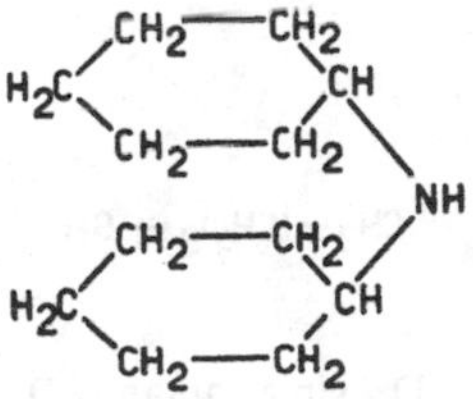

Dicyclohexylamine (91)

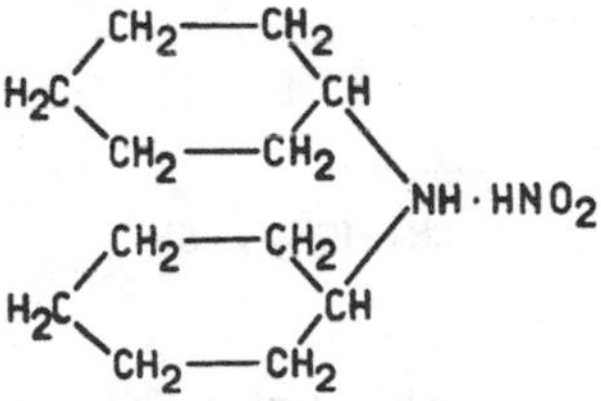

Dicyclohexylamine nitrite (91) (110)
(67) (115) (117)

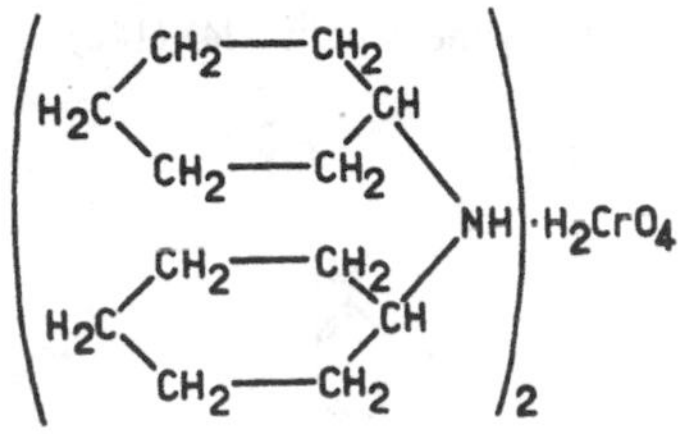

Dicyclohexylamine chromate (113)

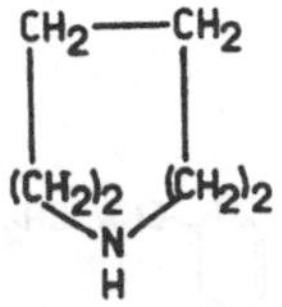

Hexamethyleneimine (67)

$CH_3-(CH_2)_4-CN$

Valeronitrile (59)

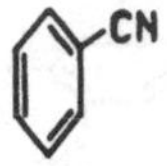

Benzonitrile (59)

Naphtonitrile (59)

$CH_3-(CH_2)_3-O-N=O$

Butylnitrite (96)

$CH_3-(CH_2)_3-SH$

Butylmercaptan (24)

$CH_3-(CH_2)_5-SH$

Hexylmercaptan (24)

$CH_3-(CH_2)_7-SH$

Octylmercaptan (24)

$CH_3-(CH_2)_9-SH$

Decylmercaptan (24)

$CH_3-(CH_2)_{11}-SH$

Laurylmercaptan (140)

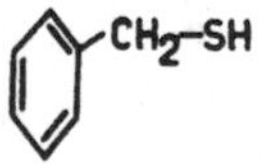

Benzylmercaptan (24) (140)

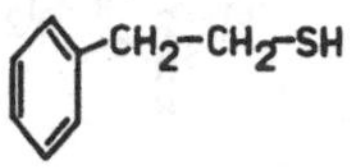

Phenylethylmercaptan (24)

Thiophenol (20)

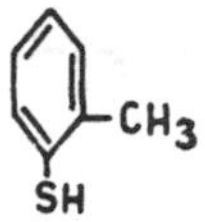

o-Thiocresol (20)

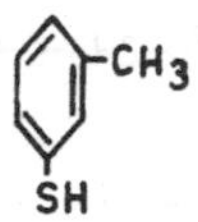

m-Thiocresol (20)

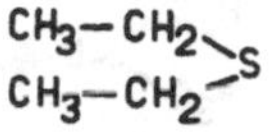

Diethylsulfide (36)

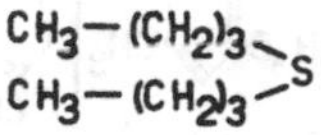

Dibutylsulfide (36)

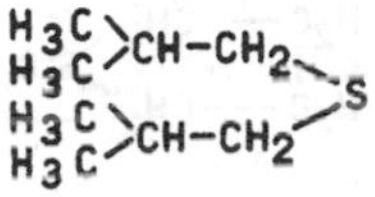

Di-*sec*-butylsulfide (36)

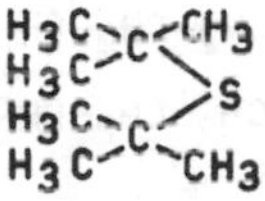

Di-*ter*-butylsulfide (36)

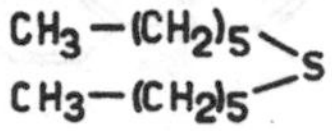

Dihexylsulfide (36)

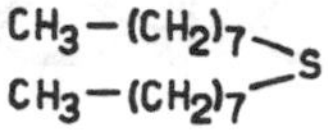

Dioctylsulfide (36)

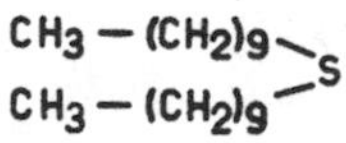

Didecylsulfide (36)

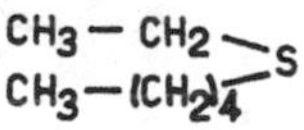

Ethyl-*n*-pentyl sulfide (36)

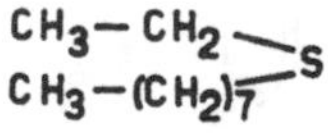

Ethyl-*n*-octyl sulfide (36)

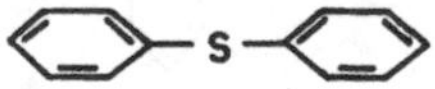

Diphenylsulfide (36)

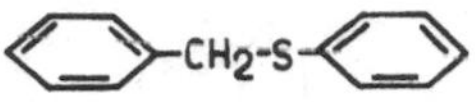

Phenylbenzyl sulfide (36)

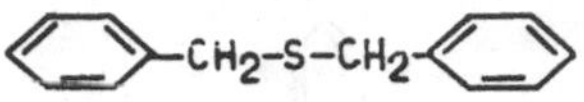

Dibenzylsulfide (36)

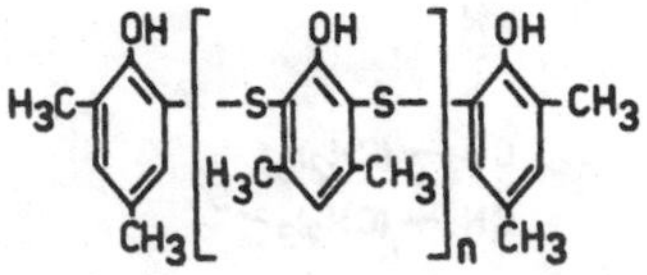

Xylenol polysulfide (136)

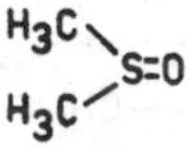

Dimethylsulfoxide (36)

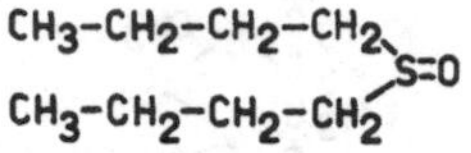

Di-*n*-butylsulfoxide (36)

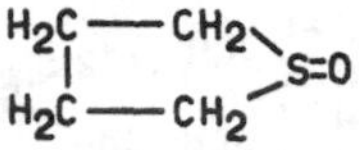

Tetramethylensulfoxide (36)

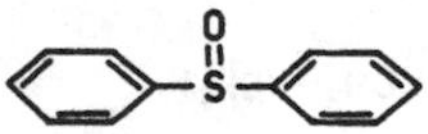

Diphenylsulfoxide (36)

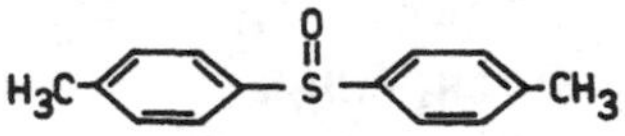

Di-*p*-tolylsulfoxide (36)

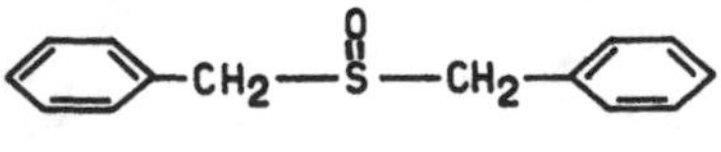

Dibenzylsulfoxide (31) (34) (35) (36) (150)

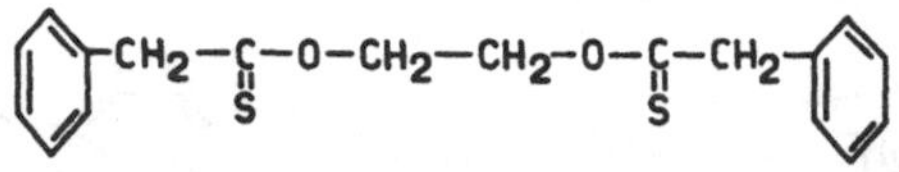

Ethyleneglycol *bis*-dibenzylxanthate (140)

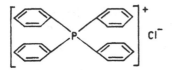

Tetraphenylphosphonium chloride (31)

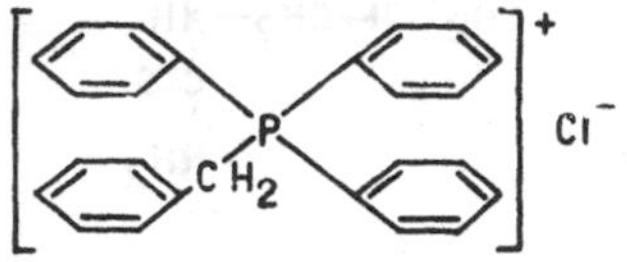

Triphenylbenzylphosphonium chloride
(31) (37)

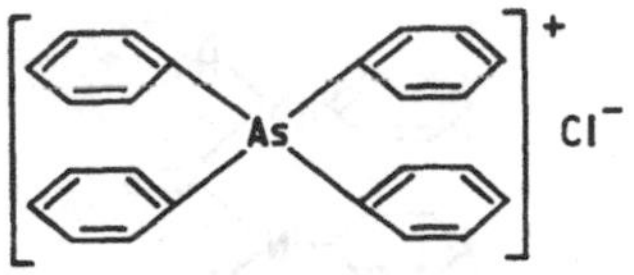

Tetraphenylarsonium chloride (31)

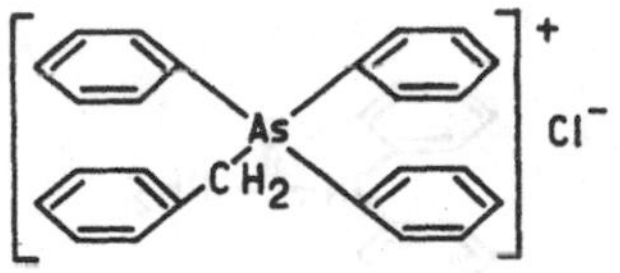

Triphenylbenzylarsonium chloride (31)

CH_3-CH_2-O
CH_3-CH_2-O——PS
CH_3-CH_2-O

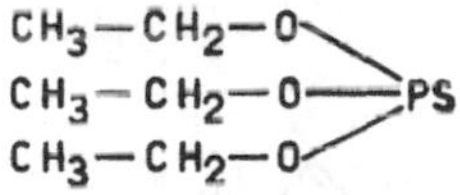

Triethylthiophosphate (12)

CH_3-CH_2-O
CH_3-CH_2-O——PS
HO

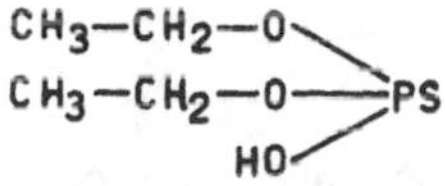

Diethylthiophosphate (12)

$CH_3-(CH_2)_3-O$
$CH_3-(CH_2)_3-O$——PS
HO

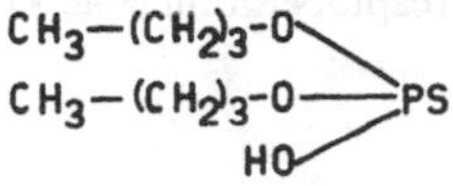

Dibutylthiophosphate (12)

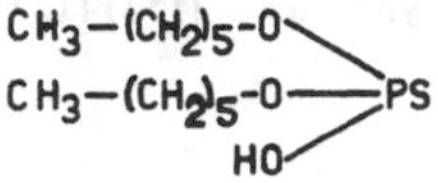

$CH_3-(CH_2)_5-O$
$CH_3-(CH_2)_5-O$——PS
HO

Dihexylthiophosphate (12)

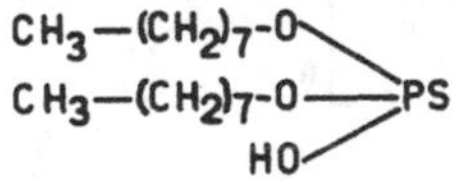

$CH_3-(CH_2)_7-O$
$CH_3-(CH_2)_7-O$——PS
HO

Dioctylthiophosphate (12)

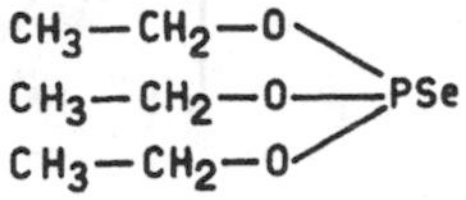

CH_3-CH_2-O
CH_3-CH_2-O——PSe
CH_3-CH_2-O

Triethylselenophosphate (12)

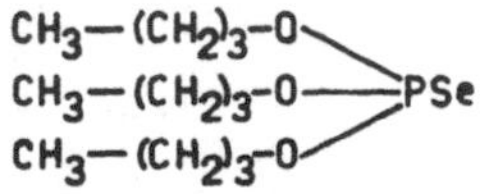

$CH_3-(CH_2)_3-O$
$CH_3-(CH_2)_3-O$——PSe
$CH_3-(CH_2)_3-O$

Tributylselenophosphate (12)

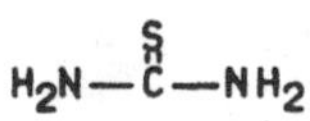

$$H_2N-\overset{\overset{S}{\parallel}}{C}-NH_2$$

Thiourea (57) (92) (135) (155)

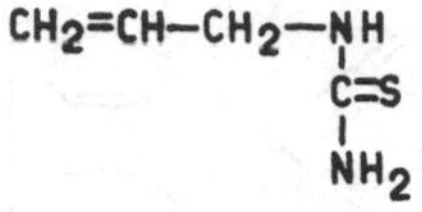

Allylthiourea (68) (139)

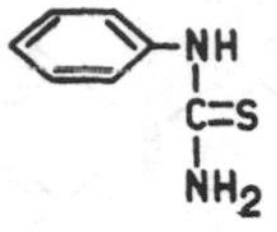

Phenylthiourea (68) (139) (150)

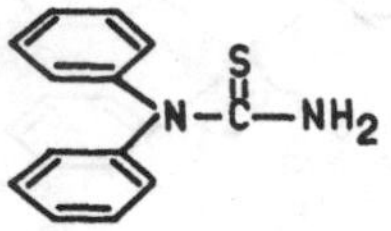

1,1-Diphenylthiourea (139)

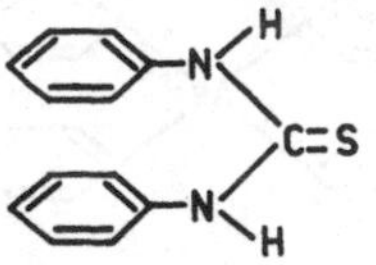

1,3-Diphenylthiourea (68) (139)

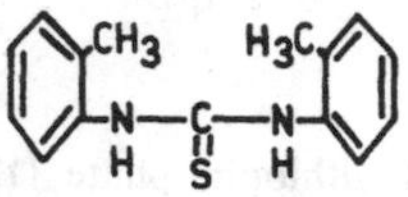

Di-o-tolylthiourea (153) (123) (139)

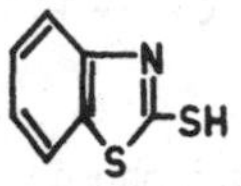

2-Mercaptobenzothiazole (121)

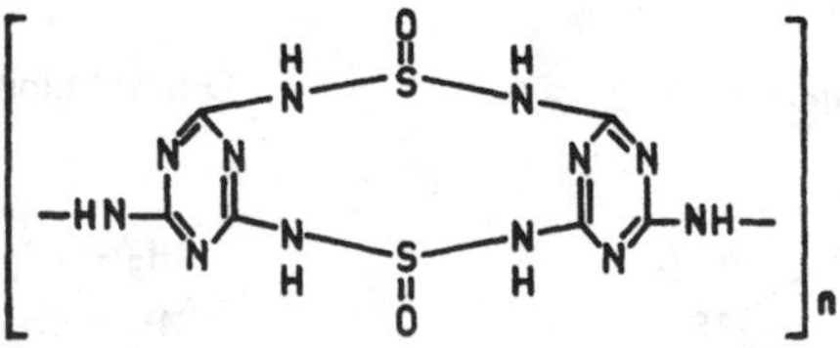

Triazine high polymer (149)

REFERENCES

1. R. Piontelli, Elementi di Teoria della Corrosione a Umido dei Materiali Metallici, Longanesi & Co., Milano (1961), p. 47.

2. H. Fischer and W. Seiler, Potentialabhängige Inhibitionsmechanismen in verschiedenen Systemen Metall/Elektrolyt, Comptes Rendus du 2ème Symposium Européen sur les Inhibiteurs de Corrosion, Annali Univ. Ferrara, N. S., Sez. V, Suppl. n. 4 (1966), pp. 19–58.

3. H. Fischer and W. Seiler, Potentialabhänginge Inhibitionsmechanismen in verschiedenen Systemen, verglichen mit der potentialabhängigen Sorption an Quecksilber, Corrosion Science 6, 159 (1966).

4. L. I. Antropov, Inhibitors of Metallic Corrosion and the ϕ-scale of Potentials, First International Congress on Metallic Corrosion, Butterworths, London (1962), pp. 147–155.

5. L. I. Antropov, Some Contributions to the Theory of Corrosion Inhibition by Organic Compounds, Comptes Rendus du 2ème Symposium Européen sur les Inhibiteurs de Corrosion, Annali Univ. Ferrara, N. S., Sez. V, Suppl. n. 4 (1966), pp. 878–881.

6. M. Froment, C. Georgoulis, and R. Wiart, Sur le rôle des inhibiteurs organiques dans la formation de revêtements de nickel brillant par voie électrolytique, 8ᵉ Série, Tome IV, n. 48 (1963), pp. 1–12.

7. N. Hackerman and R. M. Hurd, Corrosion Inhibition and Molecular Structure, First International Congress on Metallic Corrosion, Butterworths, London (1962), pp. 166–172.

8. H. Fischer, Sorptions- und Elektrolytfilminhibitoren der Korrosion, Comptes Rendus du Symposium Européen sur les Inhibiteurs de Corrosion, Annali Univ. Ferrara, N. S., Sez. V, Suppl. n. 3 (1961), pp. 1–71.

9. L. Cavallaro, L. Felloni, and G. Trabanelli, On the Inhibition of the Dissolution of Metals in Acid Media, Comptes Rendus du Symposium Européen sur les Inhibiteurs de Corrosion, Annali Univ. Ferrara, N. S., Sez. V, Suppl. n. 3 (1961), pp. 111–135.

10. W. Machu, Über die Bedeutung von Filmwiderständen, Polarisationen und Kapazitäten für den Reaktionsmechanismus von Inhibitoren, Comptes Rendus Symposium Européen sur les Inhibiteurs de Corrosion, Annali Univ. Ferrara, N. S., Sez. V, Suppl. n. 3 (1961), pp. 183–203.

11. N. Hackerman, An Adsorption Theory of Corrosion Inhibition by Organic Compounds, Comptes Rendus Symposium Européen sur les Inhibiteurs de Corrosion, Annali Univ. Ferrara, N. S., Sez. V, Suppl. n. 3 (1961), pp. 99–109.

12. Z. Szklarska-Smialowska and B. Dus, Effect of Some Organic Phosphorus Compounds on the Corrosion of Low Carbon Steels in Hydrochloric Acid Solutions, Corrosion 23, 130 (1967).

13. L. Horner and F. Röttger, Über Inhibitoren der Korrosion, I., Korrosion 16, Verlag Chemie-GMBH-Weinheim/Bergstr. (1963), pp. 57–70.

14. R. C. Ayers, Jr. and N. Hackerman, Corrosion Inhibition in HCl Using Methyl Pyridines, J. Electrochem. Soc. 110, 507 (1963).

15. F. M. Donahue and K. Nobe, Theory of Organic Corrosion Inhibitors. Adsorption and Linear Free Energy Relationships, J. Electrochem. Soc. 112, 886 (1965).

16. F. M. Donahue, A. Akiyama, and K. Nobe, Theory of Organic Corrosion Inhibitors, II Electrochemical Characteristics of Iron in Acid Solutions Containing Ring-Substituted Anilines, *J. Electrochem. Soc.* **114**, 1006 (1967).

17. F. M. Donahue and K. Nobe, Theory of Organic Corrosion Inhibitors. III LFER Correlation of Inhibition of Armco Iron by Ring-Substituted Anilines, *J. Electrochem. Soc.* **114**, 1012 (1967).

18. V. P. Grigoryev and O. A. Osipov, Correlation between Structure and Inhibition Effect of Some Organic Compounds, Extended Abstracts of Papers, Third International Congress on Metallic Corrosion, Moscow (May, 1966), pp. 114–115.

19. N. Hackerman, Recent Advances in Understanding of Organic Inhibitors, *Corrosion* **18**, 332t (1962).

20. O. L. Riggs, Jr. and R. L. Every, Study of Organic Inhibitors for Hydrochloric Acid Attack on Iron, *Corrosion* **18**, 262t (1962).

21. G. Trabanelli, A. Fiegna, and V. Carassiti, Relations entre la structure et l'efficacité d'inhibiteurs en phase vapeur, La Tribune du CEBEDEAU, n. 288 (1967), pp. 1–7.

22. N. I. Podobaev, A. G. Voskresenskii, and G. F. Semikolenov, A Study of Some Acetylenic Compounds as Inhibitors of Acid Corrosion, *Protection of Metals* **3** (1), 88 (1967).

23. G. Trabanelli, F. Zucchi, and V. Carassiti, Inhibition of the Corrosion of Nickel and its Alloys in HCl by Amines (in press).

24. G. Trabanelli, F. Zucchi, G. Gullini, and V. Carassiti, Inhibierung der Korrosion von Eisen in Säuren mit Hilfe von organischen Schwefelverbindungen, *Werk. u. Korr.*, **20**, 1012 (1969).

25. J. N. Putilova, The Chemical and Physical Changes of Inhibitors during the Corrosion Process, Comptes Rendus du 2ème Symposium Européen sur les Inhibiteurs de Corrosion, Annali Univ. Ferrara, N. S., Sez. V, Suppl. n. 4 (1966), pp. 139–151.

26. G. Trabanelli, F. Zucchi, F. Pulidori, and G. Gilli, Sulle relazioni esistenti fra struttura di polialchilimmine e loro potere inibitore della corrosione acida di materiali ferrosi, *Annali Accad. Sci. Ferrara* **40** (1962–63).

27. R. R. Annand, R. M. Hurd, and N. Hackerman, Adsorption of Monomeric and Polymeric Amino Corrosion Inhibitors on Steel, *J. Electrochem. Soc.* **112**, 138 (1965).

28. R. R. Annand, R. M. Hurd, and N. Hackerman, Inhibition of Acid Corrosion by Soluble Monomer and Polymer Amines Containing Identical Functional Groups, *J. Electrochem. Soc.* **112**, 144 (1965).

29. J. N. Putilova, S. A. Balezin, and V. P. Barannik, *Metallic Corrosion Inhibitors*, Pergamon Press, London (1960), p. 55.

30. L. I. Antropov, A Correlation between Kinetics of Corrosion and the Mechanism of Inhibition by Organic Compounds, *Corrosion Science* **7**, 607 (1967).

31. H. Ertel and L. Horner, Die Inhibition der Eisenauflösung in Säuren mit Onium-Salzen (Sekundär-Inhibition), Comptes Rendus du 2ème Symposium Européen sur les Inhibiteurs de Corrosion, Annali Univ. Ferrara, N. S., Sez. V, Suppl. n. 4 (1966), pp. 71–80.

32. J. N. Putilova, Organic Compounds with Multibonds between Carbon or Nitrogen Atoms as Inhibitors of Acid Corrosion, Extended Abstracts of Papers, Third International Congress on Metallic Corrosion, Moscow (May, 1966), pp. 99–101.

33. G. W. Poling, Infrared Studies of Protective Films Formed by Acetylenic Corrosion Inhibitors, *J. Electrochem. Soc.* **114**, 1209 (1967).

34. K. Schwabe and W. Leonhardt, Über die Wirkungsweise von Dicyclohexylammoniumnitrit und Dibenzylsulfoxyd als Korrosionsinhibitoren, *Chemie-Ingenieur-Technik* **38** (1), 59 (1966).

35. J. W. Lorenz and H. Fischer, Reaktionshemmung im System Eisen/Säure durch adsorbierte organische Oniumionen, Comptes Rendus du 2ème Symposium Européen sur les Inhibiteurs de Corrosion, Annali Univ. Ferrara, N. S., Sez. V, Suppl. n. 4 (1966), pp. 81–92.

36. G. Trabanelli, F. Zucchi, G. L. Zucchini, and V. Carassiti, Sulle relazioni fra struttura ed azione inibitrice, *Electrochimica Metallorum* **2** (4), 463 (1967); G. Trabanelli, F. Zucchi, G. Gullini, V. Carassiti, Correlation of the Structure and the Inhibitive Action of Some Sulphoxides *Br. Corr. J.*, **4**, 212–215 (1969).

37. J. W. Lorenz and H. Fischer, Electrochemical Measurements for the Distinction between Primary and Secondary Inhibition, Extended Abstracts of Papers, Third International Congress on Metallic Corrosion, Moscow (May, 1966), pp. 115–116.

38. L. Felloni and A. Cozzi, On the Effect of some Pyridine Derivatives on the Acid Corrosion of Iron, Comptes Rendus du 2ème Symposium Européen sur les Inhibiteurs de Corrosion, Annali Univ. Ferrara, N. S., Sez. V, Suppl. n. 4 (1966), pp. 253–275.

39. T. P. Hoar and R. P. Khera, The Inhibition by Formaldehyde, Pyridine, Polyamines, Alkanolamines and Octylamines of the Acid Dissolution of Mild Steel, Comptes Rendus du Symposium Européen sur les Inhibiteurs de Corrosion, Annali Univ. Ferrara, N. S., Sez. V, Suppl. n. 3 (1961), pp. 73–97.

40. Z. A. Jofa, Adsorption of Inhibitors of the Acid Corrosion of Iron and the Mechanism of their Action, Comptes Rendus du 2ème Symposium Européen sur les Inhibiteurs de Corrosion, Annali Univ. Ferrara, N. S., Sez. V, Suppl. n. 4 (1966), pp. 93–108.

41. L. Cavallaro, L. Felloni, G. Trabanelli, and F. Pulidori, The Anodic Dissolution of Iron and the Behaviour of Some corrosion Inhibitors Investigated by the Potentiodynamic Method, *Electr. Acta* **9**, 485 (1964).

42. N. Hackerman and E. S. Snavely, Jr., Effects of Anions on Corrosion Inhibition by Organic Compounds, *J. Electrochem. Soc.* **113**, 677 (1966).

43. A. Frumkin, Adsorption of Organic Substances at the Metal/Electrolyte Solution Interface and its Influence on Electrochemical Processes, Preprinted from the Proceedings of the Second International Congress of Surface Activity, London (1957), pp. EV 578–586.

44. Z. Ostrowski and H. Fischer, Potentialabhängige Sorption von konkurrierenden Dipolen und Ionen an Quecksilber, Korrosion 16, Verlag Chemie-GMBH-Weinheim/Bergstr. (1963), pp. 253–266.

45. T. Murakava, S. Nagaura, and K. Ohashi, Study on Corrosion Inhibitors by Impedance Measurements (1), *J. Electrochem. Soc. Japan* **29** (2), 96 (1961).

46. T. Murakava and N. Hackerman, The Double Layer Capacity at the Interface between Iron and Acid Solutions With and Without Organic Materials, *Corrosion Science* **4**, 387 (1964).

47. A. Indelli and G. Pancaldi, Sull'inibizione della dissoluzione dei metalli negli acidi. Nota II. Azione di alcuni inibitori sui massimi polarografici, *Gazz. Chim. Ital.* **83**, 555 (1953).

48. F. Zucchi, G. L. Zucchini, G. Trabanelli, and L. Baldi, Corrosione acida del ferro e inibizione. Influenza della lavorazione a freddo, *Electrochimica Metallorum* 1 (4), 400 (1966).

49. Z. A. Foroulis and H. H. Uhlig, Effect of Cold Work on Corrosion of Iron and Steel in Hydrochloric Acid, *J. Electrochem. Soc.* 111, 522 (1964).

50. R. J. Adams, H. L. Weisbecker, and W. J. McDonald, The Effect of Abrasion on the Specific Surface Area of Metals and Glass, *J. Electrochem. Soc.* 111, 774 (1964).

51. K. Volenik, L. Vlasakova, E. Volrabova, and O. Lashtokova, Use of Krypton Adsorption for Measuring the Surface of Industrial Metals, *Protection of Metals* 1, 504 (1965).

52. S. Thibault, J. Amouroux, and J. Talbot, Détermination de la surface réelle d'échantillons métalliques massifs, Paper Presented at 2ᵉ Colloquie International CEFRACOR–IRSID, Paris, décembre, 1967.

53. M. Froment and A. Desestret, Mécanisme de la Corrosion du Fer en Milieu Acide par le Butyne-2 diol-1-4, Comptes Rendus du 2ème Symposium Européen sur les Inhibiteurs de Corrosion, Annali Univ. Ferrara, N. S., Sez. V, Suppl. n. 4 (1966), pp. 223–238.

54. G. Trabanelli, F. Zucchi, and G. L. Zucchini, Influence de l'état de surface sur l'action inhibitrice du dibenzylsulfoxyde dans la dissolution du nickel en HCl, *Corrosion et Anticorrosion* 14, 375 (1966).

55. J. Amouroux, S. Jeannin, and J. Talbot, Influence de l'état de surface sur la vitesse de corrosion du fer et sur l'action inhibitrice de l'alcool propargylique, Comptes Rendus du 2ème Symposium Européen sur les Inhibiteurs de Corrosion, Annali Univ. Ferrara, N. S., Sez. V, Suppl. n. 4 (1966), pp. 239–251.

56. H. C. Gatos, Benzoates and Cinnamates as Corrosion Inhibitors, Comptes Rendus du Symposium Européen sur les Inhibiteurs de Corrosion, Annali Univ. Ferrara, N. S., Sez. V, Suppl. n. 3 (1961), pp. 257–272.

57. T. P. Hoar and R. D. Holliday, The Inhibition by Quinolines and Thioureas of the Acid Dissolution of Mild Steel, *J. Appl. Chem.* 3, 502 (1953).

58. I. Epelboin, Application de Méthodes électrochimiques, en Regime non Stationnaire, à l'Etude de l'Inhibition de la Corrosion, Comptes Rendus du 2ème Symposium Européen sur les Inhibiteurs de Corrosion, Annali Univ. Ferrara, N. S., Sez. V, Suppl. n. 4 (1966), pp. 663–677.

59. J. O'M. Bockris, J. McBreen, and L. Nanis, The Hydrogen Evolution Kinetics and Hydrogen Entry into α-Iron, *J. Electrochem. Soc.* 112, 1025 (1965).

60. G. Andersson, U. Trägårdh, and G. Wranglén, Influence of Organic Inhibitors and Iodide Ions on Corrosion and Hydrogen Embrittlement in Pickling and Corrosion of Carbon Steel, *Current Corrosion Research in Scandinavia*, Almqvist and Wiksell, Stockholm (1965), pp. 11–19.

61. L. Cavallaro, G. P. Bolognesi, and L. Felloni, Neue Untersuchungen über die Adsorption von Wasserstoff bei der Auflösung von Eisen, *Werk. u. Korr.* 10, 81 (1959).

62. I. L. Rozenfeld, D. M. Kramarenko, and E. N. Lantseva, Electrolytic Hydrogenation of Steel. III. Hydrogenation Inhibitors, *Protection of Metals* 3, 136 (1967).

63. M. Smialowski, Inhibition et Pénétration de l'Hydrogène dans les Aciers, Comptes Rendus du 2ème Symposium Européen sur les Inhibiteurs de Corrosion, Annali Univ. Ferrara, N. S., Sez. V, Suppl. n. 4 (1966), pp. 203–222.

64. M. Stern and A. L. Geary, Electrochemical Polarisation. I. Theoretical Analysis of the Shape of Polarisation Curves, *J. Electrochem. Soc.* 104, 56 (1957).

65. M. Pourbaix, Ein Vergleich zwischen den Ergebnissen der elektrochemischen Korrosionsprüfmethoden und dem Verhalten der Werkstoffe in der Praxis, *Werk. u. Korr.* **15**, 821 (1964).
66. N. D. Greene and R. B. Leonard, Comparison of Potentiostatic Anodic Polarisation Methods, *Electr. Acta* **9**, 45 (1964).
67. I. L. Rozenfeld, V. P. Persiantseva, and P. B. Terentiev, Mechanism of Metal Protection by Volatile Inhibitors, 2nd International Congress on Metallic Corrosion, National Association of Corrosion Engineers, Houston (1966), pp. 651–664.
68. V. S. Agarwala and K. C. Tripathi, Thioharnstoffderivate als Dampfphaseninhibitoren der Korrosion, *Werk. u. Korr.* **18**, 15 (1967).
69. B. Sathianandhan, K. Balakrishnan, and K. S. Rajagopalan, Effect of Sodium Salts of Some Organic Acids on the Corrosion of Mild Steel in Sodium Chloride Solution, Comptes Rendus du 2ème Symposium Européen sur les Inhibiteurs de Corrosion, Annali Univ. Ferrara, N. S., Sez. V, Suppl. n. 4 (1966), pp. 649–659.
70. M. Smialowski and Z. Szklarska-Smialowska, Effect of Fatty Acids on the Passivation of Steel, Comptes Rendus du 2ème Symposium Européen sur les Inhibiteurs de Corrosion, Annali Univ. Ferrara, N. S., Sez. V, Suppl. n. 4 (1966), pp. 389–400.
71. H. Kaesche and N. Hackerman, Corrosion Inhibition by Organic Amines, *J. Electrochem. Soc.* **105**, 191 (1958).
72. H. Kaesche, Untersuchungen über die Korrosionshemmung durch Adsorptionsinhibitoren, Comptes Rendus du Symposium Européen sur les Inhibiteurs de Corrosion, Annali Univ. Ferrara, N. S., Sez. V, Suppl. n. 3 (1961), pp. 137–160.
73. L. Cavallaro, L. Felloni, G. Trabanelli, and F. Pulidori, Potentiodynamic Measurements of Polarisation Curves on Armco Iron in Acid Medium in the Presence of Thiourea Derivatives, *Electr. Acta* **8**, 521 (1963).
74. Z. A. Jofa, V. V. Batrakov, and K. Ngok Ba, Influence of Anions Adsorption on the Effect of Inhibitors on the Corrosion of Iron and Cobalts in Acids, *Protection of Metals* **1**, 44 (1965).
75. H. Grubitsch and F. Hilbert, Die Inhibitorwirkung von Phenyl-Trithioniummethosulfat auf die Korrosion des Eisens in Salzsäure, Comptes Rendus du 2ème Symposium Européen sur les Inhibiteurs de Corrosion, Annali Univ. Ferrara, N. S., Sez. V, Suppl. n. 4 (1966), pp. 109–137.
76. H. Kaesche, Das Elektroden Verhalten von Eisen in Perclorsäure Lösungen von Phenylthioharnstoff, *Z. für Elektrochem.* **63**, 492 (1959).
77. W. J. Lorenz and H. Fischer, Reaktionshemmung im System Eisen/Säure durch adsorbierte organische Oniumionen, *Ber. Bunsenges. Physik. Chem.* **69**, 689 (1965).
78. G. Okamoto, M. Nagayama, and N. Sato, Application of the Rapid Method for the Measurement of Polarisation Characteristics of Iron in Acid Solutions, Proceedings of the 8th Meeting of the C.I.T.C.E., Butterworths, London (1958), pp. 72–89.
79. M. Cappellaere, S. Jeannin, and G. Montel, Sur le Mode d'Action de Certains Inhibiteurs Organiques dans la Corrosion du Fer, Comptes Rendus du Symposium Européen sur les Inhibiteurs de Corrosion, Annali Univ. Ferrara, N. S., Sez. V, Suppl. n. 3 (1961), pp. 359–389.
80. H. C. Gatos, Electrocapillary Action of Corrosion Inhibitors, *Nature* **181**, 1060 (1958).
81. Z. Ostrowski, Adsorptionsvermessungen einiger organischer Inhibitoren, Comptes Rendus du Symposium Européen sur les Inhibiteurs de Corrosion, Annali Univ. Ferrara, N. S., Sez. V, Suppl. n. 3 (1961), pp. 239–254.

82. R. J. Meakins, Quaternary Ammonium Inhibitors of Acid Corrosion, and their Adsorption at Metal Surfaces, *Australasian Engineering* **11** (2), 5 (1967).
83. S. Jeannin, Application de la Mesure de la Capacité de double couche à l'Etude de l'Inhibition de la Corrosion Acide du Fer, Comptes Rendus du 2ème Symposium Européen sur les Inhibiteurs de Corrosion, Annali Univ. Ferrara, N. S., Sez. V, Suppl. n. 4 (1966), pp. 59–69.
84. E. Blomgren and J. O'M. Bockris, The Adsorption of Aromatic Amines at the Interface: Mercury–Aqueous Acid Solutions, *J. Phys. Chem.* **63**, 1475 (1959).
85. C. P. De, Theory of Inhibition of Corrosion of Iron in Acid and Neutral Media Based on the Electrocapillary Behaviour of Ions, *Nature* **180**, 803 (1957).
86. P. Lacombe, Les Applications des Isotopes Radioactifs à l'Etude des Inhibiteurs de Corrosion, Comptes Rendus du 2ème Symposium Européen sur les Inhibiteurs de Corrosion, Annali Univ. Ferrara, N. S., Sez. V, Suppl. n. 4 (1966), pp. 517–541.
87. J. O'M. Bockris and D. A. J. Swinkels, Adsorption of *n*-Decylamine on Solid Metal Electrodes, *J. Electrochem. Soc.* **111**, 736 (1964).
88. J. O'M. Bockris, M. Green, and D. A. J. Swinkels, Adsorption of Naphtalene on Solid Metal Electrodes, *J. Electrochem. Soc.* **111**, 743 (1964).
89. E. Gileadi, Radiotracer Studies of Electrosorption of Neutral Molecules, Comptes Rendus du 2ème Symposium Européen sur les Inhibiteurs de Corrosion, Annali Univ. Ferrara, N. S., Sez. V, Suppl. n. 4 (1966), pp. 543–558.
90. J. Bordeaux and N. Hackerman, Adsorption from Solution of Stearic Acid on Iron; Effect on Electrode Potential, *J. Phys. Chem.* **61**, 1323 (1957).
91. K. Schwabe, Adsorptionsuntersuchungen mit markierten Inhibitoren, DECHEMA Monographien, 45, Verlag Chemie-GMBH-Weinheim/Bergstr. (1962), pp. 273–283.
92. T. K. Ross and D. H. Jones, Thiourea as an Inhibitor of the Acid Corrosion of Mild Steel, Comptes Rendus du Symposium Européen sur les Inhibiteurs de Corrosion, Annali Univ. Ferrara, N. S., Sez. V, Suppl. n. 3 (1961), pp. 163–182.
93. V. Carassiti, L. Baldi, G. Trabanelli, F. Zucchi, and G. L. Zucchini, Inhibition by Dibenzylsulfoxide of the Active Dissolution of 316 Stainless Steel, Extended Abstracts of Papers, Third International Congress on Metallic Corrosion, Moscow (May, 1966), pp. 106–107.
94. A. Bertoluzza, G. B. Bonino, G. Fabbri, and V. Lorenzelli, Quelques résultats récents sur les spectres infrarouges des molécules adsorbées, *J. Chimie Phys.* **63**, 395 (1965).
95. G. Karagounis and H. Gellert, Makromolekularer Bau und molekulare Zwischenräume als entscheidende Faktoren bei der Wirkung von Korrosionsinhibitoren, Korrosion 16, Verlag Chemie-GMBH-Weinheim/Bergstr. (1963), pp. 219–225.
96. G. W. Poling and R. P. Eischens, Infrared Study of Adsorbed Corrosion Inhibitors: Butyl Nitrite and Nitric Oxide on Iron–Iron Oxide, *J. Electrochem. Soc.* **113**, 218 (1966).
97. I. L. Rozenfeld, F. I. Rubinstein, V. P. Persiantseva, and S. V. Yakubovich, The Modification of Polymer Coatings by Inhibitors, Comptes Rendus du 2ème Symposium Européen sur les Inhibiteurs de Corrosion, Annali Univ. Ferrara, N. S., Sez. V, Suppl. n. 4 (1966), pp. 751–764.
98. B. E. Conway and R. G. Barradas, Spectrophotometer Determination of Adsorption of Organic Molecules at Solid Metals, Transactions of the Symposium on Electrode Processes, John Wiley and Sons, New York (1961), pp. 299–306.

99. W. Suëtaka, Adsorption Spectra of Quinoline Adsorbed on Ni Sheets from Aqueous Solution, *Bull. Chem. Soc. Japan* **38**, 148 (1965).

100. W. Suëtaka, Adsorption Spectra of Thiourea and KJ on Metallic Nickel, *Bull. Chem. Soc. Japan* **37**, 1121 (1964).

101. Y. F. Yu Yao, Chemisorption of Amines and Its Effect on Subsequent Oxidation of Iron Surfaces, *J. Phys. Chem.* **68**, 101 (1964).

102. A. P. Van Peteghem and G. Vanderkelen, Application de la Spectrométrie de Masse à l'Etude de la Corrosion, *Corrosion et Anticorrosion* **11**, 41 (1963).

103. P. F. Cox, R. L. Every, and O. L. Riggs, Jr., Study of Aromatic Amine Inhibitors by Nuclear Magnetic Resonance, *Corrosion* **20**, 299t (1964).

104. J. Maddox, Jr. and R. H. Graves, Ellipsometric Studies of the Adsorption–Desorption Characteristics of Organic Films, *Corrosion Abs. NACE* **5** (5), 362 (1966).

105. J. F. Dettorre and D. A. Vaugham, Observation of Surface Film Behaviour during Inhibition Treatments by Electronic Ellipsometry, *Corrosion Abs. NACE* **6** (2), 107 (1967).

106. Y. Lefebvre and B. Le Boucher, Etude Directe de l'Adsorption d'une Amine lors de la Corrosion Acide de l'Acier Doux, Preprint du 3ᵉ Congrès de la Fédération Européenne de la Corrosion, Bruxelles, Juin, 1963.

107. M. A. V. Devanathan and Z. Stachurski, The Mechanism of Hydrogen Evolution on Iron in Acid Solution by Determination of Permeation Rates, *J. Electrochem. Soc.* **111**, 619 (1964).

108. L. I. Antropov, M. A. Gerasimenko, Yu. S. Gerasimenko, and Yu. A. Savgira, Hydrogen Diffusion through Iron Membranes Corroding in Acidic Solutions, Extended Abstracts of Papers, Third International Congress on Metallic Corrosion, Moscow (May, 1966), pp. 97–98.

109. I. L. Rozenfeld, Atmospheric Corrosion of Metals, Some Questions of Theory, First International Congress on Metallic Corrosion, Butterworths, London (1962), pp. 243–253.

110. H. R. Baker, Volatile Rust Inhibitors, *Ind. Eng. Chem.* **46**, 2592 (1954).

111. J. B. Cotton and I. R. Scholes, Benzotriazole and Related Compounds as Corrosion Inhibitors for Copper, *Brit. Corros. J.* **2**, 1 (1967).

112. S. A. Gintzberg and A. V. Shreider, Chromates of Amines and Esters of Chromic Acid as Inhibitors of Corrosion, *Chem. Abs.* **54**, 20791f (1960).

113. S. Z. Levin, S. A. Gintzberg, I. S. Dinner, and V. N. Kuchinsky, Synthesis and the Protective Action of Some Inhibitors on the Basis of Cyclohexylamine and Dicyclohexylamine, Comptes Rendus du 2ème Symposium Européen sur les Inhibiteurs de Corrosion, Annali Univ. Ferrara, N. S., Sez. V, Suppl. n. 4 (1966), pp. 765–776.

114. V. S. Agarwala and K. C. Tripathi, Some Vapor Phase Inhibitors Experiments, *Materials Protection* **5** (12), 26 (1966).

115. K. Schwabe, Zur Theorie der Inhibitorwirkung organischer Verbindungen, *Z. für Phys. Chem. (Leipzig)* **226**, 1 (1964).

116. L. Cavallaro and G. Mantovani, Neue Untersuchungen auf den Gebiete der Dampfphasen-Inhibitoren, *Werk. u. Korr.* **10**, 422 (1959).

117. J. Nemcovà, Les Inhibiteurs Volatils de Corrosion à Base d'Urée ou d'Urotropine, Comptes Rendus du 2ème Symposium Européen sur les Inhibiteurs de Corrosion, Annali Univ. Ferrara, N. S., Sez. V, Suppl. n. 4 (1966), pp. 791–812.

118. J. Mestreit and J. Guenard, Les Sels de Diamines a Longue Chaîne dans l'inhibition de la Corrosion Atmosphérique de l'acier, Comptes Rendus du Symposium Européen sur les Inhibiteurs de Corrosion, Annali Univ. Ferrara, N. S., Sez. V, Suppl. n. 3 (1961), pp. 597–617.

119. E. Mindovich, The Investigation of Mechanisms of Metal Protection by Several Volatile Inhibitors, Extended Abstracts of Papers, Third International Congress on Metallic Corrosion, Moscow (May, 1966), p. 107.

120. E. Volrabova, New Investigation Methods on Mechanisms of Metal Protection by Volatile inhibitors, Extended Abstracts of Papers, Third International Congress on Metallic Corrosion, Moscow (May, 1966), pp. 124–125.

121. J. J. Bregman, Corrosion Inhibitors, The MacMillan Co., New York (1963).

122. M. L. Lytle, Corrosion Inhibitors for Ferrous Metals, U. S. 2,721,175, Patented October 18, 1955.

123. W. Machu, Über den Einfluss des pH-Wertes auf die Inhibition von Eisen, Aluminium und Zink durch anorganische und organische Inhibitoren, Comptes Rendus du 2ème Symposium Européen sur les Inhibiteurs de Corrosion, Annali Univ. Ferrara, N. S., Sez. V, Suppl. n. 4 (1966), pp. 153–178.

124. D. M. Brasher, J. G. Beynon, A. D. Mercer, and J. E. Rhoades-Brown, The Role of the Metal in Relation to Inhibition of Corrosion, Comptes Rendus du 2ème Symposium Européen sur les Inhibiteurs de Corrosion, Annali Univ. Ferrara, N. S., Sez. V, Suppl. n. 4 (1966), pp. 559–568.

125. J. N. Putilova, S. A. Balezin and V. P. Barannik, Metallic Corrosion Inhibitors, Pergamon Press, London (1960).

126. M. F. Olbrecht, Cause and Cure of Corrosion in Steam-Condensate Cycles, 2nd International Congress on Metallic Corrosion, National Association of Corrosion Engineers, Houston (1966), pp. 624–645.

127. W. L. Denman, Corrosion Inhibitors, U. S. 2,882,171, Patented April 14, 1969.

128. J. W. Ryznar and W. H. Kirkpatrick, Inhibition of Corrosion in Steam-Condensate Lines, U. S. 2,771,417, Patented November 20, 1956.

129. R. C. Ulmer and J. W. Wood, Inhibitors for Eliminating Corrosion in Steam and Condensate Lines, Ind. Eng. Chem. 44, 1761 (1952).

130. R. Simonoff, Sulfonates as Rust Preventives, Petrochemical Ind. 1, 35 (1958).

131. L. Cavallaro, L'emploi des inhibiteurs de corrosion dans l'industrie du pétrole, Corrosion et Anticorrosion 7, 417 (1959).

132. V. Ph. Negreev, I. A. Mamedov, and D. M. Abramov, Corrosion Inhibitors in Liquid Hydrocarbons–Water Solution System, Extended Abstracts of Papers, Third International Congress on Metallic Corrosion, Moscow (May, 1966), pp. 108–110.

133. A. A. Gonik and S. A. Balezin, Some Aspects on the Protective Action of Surface-Active Compounds Inhibiting Hydrogen Sulfide Corrosion in the System of Electrolyte–Hydrocarbon, Extended Abstracts of Papers, Third International Congress on Metallic Corrosion, Moscow (May, 1966), pp. 110–111.

134. S. A. Balezin, Principles of Corrosion Inhibition, Comptes Rendus du 2ème Symposium Européen sur les Inhibiteurs de Corrosion, Annali Univ. Ferrara, N. S., Sez. V, Suppl. n. 4 (1966), pp. 277–283.

135. I. P. Anoshchenko, On the Effect of Some Inhibitors on Corrosion of Iron in Mixtures of Hydrochloric, Sulfuric and Nitric Acids, Comptes Rendus du 2ème Symposium Européen sur les Inhibiteurs de Corrosion, Annali Univ. Ferrara, N. S., Sez. V, Suppl. n. 4 (1966), pp. 179–200.

136. O. Radovici, Corrosion of Iron in Acid Solutions in Presence of an Organic Inhibitor, Comptes Rendus du 2èrhe Symposium Européen sur les Inhibiteurs de Corrosion, Annali Univ. Ferrara, N. S., Sez. V, Suppl. n. 4 (1966), pp. 449–456.

137. O. L. Riggs, Jr., and R. M. Hurd, Temperature Coefficient of Corrosion Inhibition, *Corrosion* **23**, 252 (1967).

138. A. Thorpe and L. Fairman, Descaling of Austenitic Stainless Steels in Aqueous Solutions, *J. Iron and Steel Inst.* **203**, 922 (1965).

139. J. A. Ayres, Corrosion Aspects of Reactor Decontamination, *Corrosion of Reactor Materials*, *I*, International Atomic Energy Agency, Wien (1962), pp. 199–240.

140. G. Hugel, Les Inhibiteurs de Corrosion. Etude du Mécanisme de leur Activité, Comptes Rendus du Symposium Européen sur les Inhibiteurs de Corrosion, Annali Univ. Ferrara, N. S., Sez. V, Suppl. n. 3 (1961), pp. 229–238.

141. E. G. Nottes, Zur Chemie einiger Korrosionsinhibitoren für Erdöldestillationsanlagen, Comptes Rendus du 2ème Symposium Européen sur les Inhibiteurs de Corrosion, Annali Univ. Ferrara, N. S., Sez. V., Suppl. n. 4 (1966), pp. 507–514.

142. K. E. Heusler and G. H. Cartledge, The Influence of Iodide Ions and Carbon Monoxide on the Anodic Dissolution of Active Iron, *J. Electrochem. Soc.* **108**, 732 (1961).

143. G. Trabanelli, F. Zucchi, and G. L. Zucchini, Comportamento elettrochimico di metalli in presenza di ossido di carbonio, *Atti Accad. Sci. Ferrara* **44** (1967); Influence du mono-oxyde de carbone sur les processus de dissolution des métaux, *Corrosion—Traitement—Protection—Finition*, **16**, 335–345 (1968).

144. S. A. Balezin, N. I. Podobaev, A. G. Voskresenski, and V. V. Vasiliev, On the Mechanism Underlying the Inhibiting Effect of Acetylene Compounds during the Corrosion of Steel in HCl, Extended Abstracts of Papers, Third International Congress on Metallic Corrosion, Moscow (May, 1966), pp. 103–105.

145. E. J. Duwell, J. W. Todd, and H. C. Butzke, The Mechanism of Corrosion Inhibition of Steel by Ethynylcyclohexanol in Acid Solution, *Corrosion Science* **4**, 435 (1964).

146. L. Felloni, G. Gilli, and G. Trabanelli, Analisi potenziodinamica dell'effetto di alcune polialcanimmine su di un elettrodo di ferro in ambiente acido, *La Metallurgia Ital.* **56**, 219 (1964).

147. V. P. Grigoryev and V. V. Kuznetsov, Inhibiting Effects of Certain Azoles on Acid Corrosion of Iron, *Protection of Metals* **3**, 141 (1967).

148. N. Shyamala, N. Subramanyan, and K. S. Rajagopalan, Effect of Some Acid Inhibitors on the Hydrogen Evolution Reaction on Iron, Comptes Rendus du 2ème Symposium Européen sur les Inhibiteurs de Corrosion, Annali Univ. Ferrara, N. S., Sez. V, Suppl. n. 4 (1966), pp. 345–359.

149. G. Karagounis and D. Besse, Über das Verhalten einiger Hochpolymere als Korrosionsinhibitoren, Comptes Rendus du Symposium Européen sur les Inhibiteurs de Corrosion, Annali Univ. Ferrara, N. S., Sez. V, Suppl. n. 3 (1961), pp. 205–213.

150. V. Carassiti, G. Trabanelli, and F. Zucchi, Organic Inhibitors of the Active Dissolution of Stainless Steels, Comptes Rendus du 2ème Symposium Européen sur les Inhibiteurs de Corrosion, Annali Univ. Ferrara, N. S., Sez. V, Suppl. n. 4 (1966), pp. 417–448.

151. V. P. Grigoryev, Decrease in the Passivation Current of Steel 1Kh18 N9T in the Presence of Some Organic Compounds, *Protection of Metals* **1**, 245 (1965).

152. Z. A. Foroulis, Effect of Amines on Polarisation and Corrosion of Iron, Copper and Nickel in Acids, Comptes Rendus du 2ème Symposium Européen sur les Inhibiteurs de Corrosion, Annali Univ. Ferrara, N. S., Sez. V, Suppl. n. 4 (1966), .pp. 285–324.

153. M. N. Desai and S. S. Rana, Inhibition of the Corrosion of Copper in Hydrochloric Acid Solutions, Comptes Rendus du 2ème Symposium Européen sur les Inhibiteurs de Corrosion, Annali Univ. Ferrara, N. S., Sez. V, Suppl. n. 4 (1966), pp. 849–862.

154. M. N. Desai and S. M. Desai, Inhibition of the Corrosion of Aluminium-2S and Aluminium-3S in Hydrochloric Acid Solutions, Comptes Rendus du 2ème Symposium Européen sur les Inhibiteurs de Corrosion, Annali Univ. Ferrara, N. S., Sez. V, Suppl. n. 4 (1966), pp. 863–877.

155. T. L. Rama Char and K. G. Sheth, Inhibition of Corrosion of Aluminium in Hydrochloric Acid Solutions, 2nd International Congress on Metallic Corrosion, National Association Corrosion Engineers, Houston (1966), pp. 584–589.

156. G. De Angelis, Corrosion Inhibitors for Aluminium. 3. Action of Phenol, Pyrocathecol, Resorcinol and Hydroquinone on the Corrosion of 99,0% Aluminium Immersed in 1 N Hydrochloric Acid, Comptes Rendus du Symposium Européen sur les Inhibiteurs de Corrosion, Annali Univ. Ferrara, N. S., Sez. V, Suppl. n. 3 (1961), pp. 435–462.

157. K. G. Sheth and T. L. Rama Char, Corrosion Inhibition of Aluminium–Zinc Alloy in Hydrochloric Acid Solutions, *Corrosion* 18, 218t (1962).

158. W. Machu and V. Gouda, Über die Inhibierung der Säurekorrosion verschiedener elektronegativer und elektropositiver Metalle durch organische Inhibitoren, *Werk. u. Korr.* 13, 745 (1962).

159. H. F. Schaefer, Corrosion Inhibitors, U. S. 3,281,276, Patented October 25, 1966.

160. M. N. Desai and S. S. Rana, Inhibition of the Corrosion of Copper in Sodium Hydroxide Solutions, Comptes Rendus du Symposium Européen sur les Inhibiteurs de Corrosion, Annali Univ. Ferrara, N. S., Sez. V, Suppl. n. 4 (1966), pp. 609–647.

161. E. Rabald, Korrosion von Metallen durch intermediäre Verbindungen, DECHEMA Monographien, 45, Verlag Chemie-GMBH-Weinheim/Bergstr. (1962), pp. 105–112.

ANODIC OXIDATION OF ALUMINUM

Sakae Tajima

Tokyo Toritsu Daigaku
Tokyo City University, Japan

INTRODUCTION

It was toward the end of the previous century that the commercial production techniques of aluminum were established, owing their success to
the electrolytic reduction process. This development made it possible to
use aluminum commercially and to exploit its many desirable properties.
As a result aluminum now occupies a position of importance equal to iron
and copper.

Aluminum is lighter and has a better corrosion resistance than iron
and steel but, because of its low melting point, has an inherently lower
mechanical strength. Attempts to overcome various deficiencies by alloying
were particularly successful: e.g., the remarkable increase in mechanical
strength by addition of Cu; improvement of castability by addition of Si;
improvement of machinability and corrosion resistance by addition of Mg;
and enhancement of weldability by addition of Mg and Zn. However, the
metal could not have enjoyed the "aluminum age" by alloying only; the
problem of strengthening the surface still remains. It was fortunate for
aluminum that a variety of desirable surface conditions could be obtained.
The remarkable increase in surface hardness, abrasion and corrosion resistance, and dye-absorption properties by anodic treatment has expanded
greatly the application of aluminum. The possibility of obtaining a bright
surface by chemical and electropolishing has overcome the difficulty inherent
in polishing soft metal. As a result aluminum has been widely used for
reflectors, decorative products, and, recently, for spacecraft (because of its
thermal control properties). The improvement in corrosion resistance and

paint adhesion produced by chemical conversion treatments has also contributed greatly to its popularity.

Since there already exist several excellent monographs and technical reviews on the anodic treatment of aluminum and scientific reviews on barrier oxides,[1-22] special attention will be paid in this review to the anodizing processes of current importance, to the modern theory of anodic oxidation with a brief mention of barrier films, and to the problems relating to the corrosion of anodized aluminum.

SHORT HISTORY

In 1857, H. Buff,[22] in the course of research to find an improved primary cell, found that when Al is coupled with Pt in dilute sulfuric acid, Al becomes anodic and a large current flows, immediately reduced to a very small current. Wheatstone found a similar phenomenon in 1854 and published his observations in *Philosophical Magazine* that year. He attributed this behavior to an Si layer formed on the Al surface, since Al at that time contained appreciable amounts of Si. In the same year, Wöhler and Buff[23] found that current is not reduced for aluminum anodes in chloride solutions. Ducretet[24] attributed the diminution of current to the formation of Al_2O_3; he also found that if this anode were made the cathode, a large amount of current would flow immediately, and he suggested that this behavior could be utilized in an electric circuit. Norden[27] suggested the idea that an Al-anode/film/electrolyte system could be regarded as an electrolytic capacitor. In 1891, Hutin and Leblanc proposed the idea that the aluminum cell could be used as an AC rectifier. In 1897, Pollak[25] and Graetz[26] independently published papers on an electrolytic capacitor. Pollak used alkaline or neutral phosphates, dibasic acids, oxyacids, aldehydes, etc., as electrolytes. With these electrolytes, the cell serves as a rectifier if one electrode is Al and as capacitor if both electrodes are Al. He also stated that the large capacity is due to a very thin anodic oxide film produced in neutral or alkaline solutions, and that the capacity can be increased if the anode surface area were enlarged. Wilson pointed out that in sulfuric acid and alum solutions, the Al anode loses its high polarizability at higher temperature, and Liebenow suggested that this could be prevented by the use of ammonium bicarbonate. In 1900, AEG in Germany patented the use of weakly acidic organic Mg or Al salts for electrolytic capacitors and rectifiers. Siemens and Halske AG recommended the use of similar electrolytes.

The work of Norden[27] in 1899 is still valid. He succeeded in separating anodic oxide films from the aluminum substrate by making Al alternatively anodic and cathodic in dilute sulfuric acid. He further confirmed that the anodic oxide film could not be reduced during cathodic polarization by measuring hydrogen gas evolution. He found that the anodic film produced in dilute sulfuric acid (sp. wt. $= 1.050$) contains 13.2% of SO_3.

This classical research was performed to study and utilize the anodic behavior of Al in "solution"; the processes we are now interested in making use of involve anodic films in the "dried state," with improved corrosion and abrasion resistance and with decorative attraction. Chubb in 1907 and Chubb and Skinner[28] in 1914 attempted to obtain an electrically insulating film on Al by anodic oxidation in sodium phosphate or silicate solutions.

In about 1917, at the Rikagaku Kenkyujo (Institute of Physical and Chemical Research, Tokyo), Kujirai and Ueki were attempting to obtain insulating films on Al wire by electrolysis; they found that, by the use of oxalic acid, a dense film could be produced. This process was not practically used on electric wire, but it was found that it could be used for kitchen utensils. Various kinds of electrolytes were found to be effective in producing an anodized film, but oxalic acid gave excellent results. Two patents[29-33] were granted in 1924 (application, Dec. 20 and 28, 1923) claiming anodizing Al and its alloys in acids, bases, and salts, and especially in oxalic acid by applying ac or dc current. Also in 1924, Bengough and Stuart[34,35] patented the anodizing of duralumin sheet in chromic acid to improve corrosion resistance of aircraft components (application, Aug. 2, 1923) and showed that the film could be dyed. In 1928, a small pilot plant was installed at the Institute of Physical and Chemical Research in Tokyo to produce anodized articles commercially. Although the films thus produced were dense and could be dyed if desired, the corrosion resistance was not always satisfactory because they were porous and permitted the passage of electric current.

In order to obtain a dense and corrosion resistant film, Setoh (the successor of Kujirai's electric engineering Laboratory) and Miyata, by chance, invented the steam-sealing method in 1928.[36] In this process, the anodized part is rinsed and kept in an autoclave at 55–70 psi of dried steam for about 20–30 min. The film becomes somewhat yellowish and more transparent. The electrical insulating capability increases by 50%, and, above all, there is a significant increase in corrosion resistance. This invention was an epoch making one in the Japanese anodizing industry, and the process was named the "Alumite" process. The term is still widely used in Japan for aluminum anodizing regardless of the process employed. Considering that Röhrig[37] reported in 1931 and a British patent[38] was granted in 1934 for hot-water sealing, steam sealing may be said to be quite original. In 1929, the World Engineering Congress was held in Tokyo, and Setoh and Miyata[36] presented a paper reporting the status of aluminum anodizing with steam sealing at the time. An anodized and sealed set square was forwarded to the participants. Meanwhile, in England, Gower and O'Brien[39,40] recommended about 10% H_2SO_4, which later became very popular in the USA and UK.

In Germany, VAW (Vereinigte Aluminum Werke) installed anodizing equipment in 1924 at Lautawerk, and in 1927 they set up a cooperative group of "Eloxalpool" with Siemens und Halske AG, Langbein–Pfanhauser-werke AG, and Schering–Kahlbaum AG. Later, in 1931, a Swiss group, "Vernet," joined the pool. The electrolyte used was oxalic acid modified by the addition of small quantities of salts such as chromates or phosphates.[41] In 1931, anodizing equipment was first demonstrated at the Leipzig Fair. The details were reviewed recently by Ginsberg, Neunzig, and Sautter[42] and by Wetzki.[43]

Güntherschulze and his school[1] conducted extensive research mainly on the mechanism of formation of barrier-type anodic films in electrolyte without solvent action (the thickness is usually below 1 μ), the phenomena of their electrical breakdown, and their luminescence during formation. They found that Ohm's law does not hold in such films, but the anodic current is an exponential function of the field strength across the film.

$$I_1 = A_1 \exp(B_1 F)$$

where A_1 contains the Boltzmann factor for the zero-field activation energy. B_1 was predicted to have the form aq/kT where a is an activation distance for the field-assisted displacement of an ion of charge q.

For the thicker anodic films, Setoh and Miyata[44] presented the model of a duplex structure-active (barrier) and porous layer. This idea was accepted later by Rummel,[45] Baumann,[46] and Cuthbertson[47] and dramatically confirmed by the well known cylindrical pore model of the ALCOA group, Keller, Hunter, and Robinson.[48] Recently, Murphy and Michelson,[49-51] object, from the chemical point of view, to the cylindrical pore model as being "too physical."

From a practical point of view, the invention of electro- and chemical polishing in the 1930's and 1940's promoted the development of bright anodizing for use for reflectors[52-55] and decorative parts. The development, greatly due to the Swiss group, of coloring techniques by dyestuffs and inorganic pigments and of printing techniques of labels and nameplates, widely expanded the use of aluminum in decorative products and in the printing industry. The hard coating proposed by Tomashov[56,57] in Moscow in 1946 opened new applications in engineering, although such a process, it seems, had been proposed in Europe some years before.

The architectural use of aluminum was greatly expanded by the improvement of normal anodizing techniques, and therefore of extrudable alloys and dyestuffs, and recently by the introduction of integral color (self color) processes which give excellent resistance to weathering deterioration. The pioneering contribution of the Kalcolor process by the group of Kaiser Aluminum and Chemical Corp.[58] has been extremely important to the self coloring of films, although the self-color effect has been known for a number of years for films produced in oxalic acid. Continuous anodizing of wire and sheets is of recent origin, and here the rate of anodizing plays an important role.

Anodizing in nonaqueous environments provides an important process for producing thick films. The author has obtained extremely hard corundum (α alumina) films from bisulfate melts[59] and hard, smooth, and integrally colored films from a boric acid–formamide system which requires no sealing treatment.[60] These films are finding new applications in precision industries.

ANODIZING PROCESSES OF CURRENT IMPORTANCE AND INTEREST

Outline of Anodic Oxidation of Aluminum

When aluminum is the anode in a suitable electrolyte, it reacts with the nascent oxygen ($H_2O \rightarrow H_2 + O$) supplied from the electrolyte to form Al_2O_3, and a compact and adherent film is produced. The oxidation proceeds strictly according to Faraday's law, and a definite amount of oxide is produced by a given quantity of electricity. In addition the reaction is substantially exothermic.

$$\tfrac{1}{3} Al^{3+} + \tfrac{1}{2} O^{2-} \rightarrow \tfrac{1}{6} Al_2O_3 \text{ (1 Faraday)}$$

$$2Al^{3+} + 3O^{2-} \rightarrow Al_2O_3 + 400 \text{ kcal}$$

With the passage of 1 Faraday, $\tfrac{1}{3}$ mole of Al is oxidized to produce 17 g of Al_2O_3. However, alumina is an electric insulator without free electrons in the conduction band. Thus, there is a possibility of the reaction terminating after only a monomolecular film is formed. In order for the film to grow, the ions will move under the high electric field across anodic films ("high-field" ionic conduction). While quantum mechanical tunneling of electrons is probable when the anodic film is very thin (within 20 Å), it is not a significant consideration. Under the high-field ionic conduction mechanism, the growth rate was determined to be 13.7 Å/V by Walkenhorst;[61] this corresponds to a high electric field in the range of about 10^7 V/cm. When the film becomes too resistive (thick) to permit anodic current to flow, its growth terminates. Therefore, it is necessary to raise the anodic potential (and, accordingly, the bath voltage wherein the cathodic polarization is negligibly small) in order to maintain film growth. If there is no effect of the solvent upon the film, viz., in the case of ammonium borate solution, the voltage can be raised to about 700 V and the resulting film thickness will be 14 Å $\times 700 \approx 1\ \mu$, i.e., an anodic film can be formed which can sustain about 700 V. When the voltage exceeds a certain limit, which depends upon the nature, concentration, and temperature of the electrolyte, then random arcing occurs on the film which moves from one point to another, that is, the film breaks down. The electronic current now predominates but it makes no contribution to the film growth (but promotes the crystallization of amorphous alumina, as described later, by its local heating action).

The anodic film which thickens linearly with the applied voltage is called the "barrier layer." This layer is electrically insulating, its dielectric constant being 8–10. The value is not high compared with other dielectric materials, but the film thickness can be controlled within narrow limits and a substantial electrostatic capacity can be obtained. This is the basis for the electrolytic capacitor.[4]

On the other hand, when the film tends to dissolve in the solvent, the cell attempts to restore the barrier layer to the limiting thickness corresponding to the applied voltage. This results in simultaneous formation and solution of oxide and the eventual development of a compact array of oxide cells, each containing a cylindrical pore. During this process, a balance is established such that the barrier layer attains a constant thickness, and continued film formation only produces an increase in the thickness of the porous layer. With this type of coating (duplex film) the outer region is truly porous since it contains a multitude of minute pores which can be observed with the electron microscope.

In this type of anodic oxide film, the porous layer grows at a constant voltage, determined by the steady-state relationship between formation and dissolution at the pore base area. Here, the potential drop is that across the duplex structure of the barrier layer and its associated pores which contain the electrolyte. Porous layers with thicknesses up to 100–200 μ

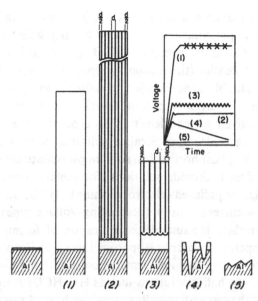

Fig. 1. Voltage–time characteristics at constant current density and corresponding surface structures. 1) Barrier-layer-forming type. x indicates sparking by electric breakdown of the anode film. 2) Electropolishing type (fluctuation type). 3) Duplex-film-forming type (stationary type). 4) Anodic pitting type. 5) Anodic etching (general corrosion) type (low-voltage stationary type). Left: original aluminum surface with about a 20-Å barrier layer naturally formed in atmosphere. (After Tajima, Itoh, and Fukushima.[62,63])

can be obtained. For Al, this characteristic occurs only among the valve metals and widens the scope of its use.

The corrosion resistance of these films containing cylindrical pores is often unsatisfactory because they permit the passage of electric currents. Fortunately, a sealing treatment has been invented; this involves steaming the filmed articles in autoclaves and subsequently in boiling water or salt solutions, which acts to plug the pores with hydrated alumina or with metal hydroxide or salts.

A schematic classification of the voltage–time curves of Al at a constant current density and the corresponding surface structures are shown in Fig. 1.[62,63] There are five types of curves: 1) For a given formation voltage the current rapidly approaches to nearly zero and only a small current, the so-called "leakage current" (electronic current through impurities, etc.), flows. In order to maintain a constant current (constant rate of film formation), the anodic potential must be raised linearly with time (ascending-voltage type—barrier layer formation). 2) The voltage becomes constant for a given current density (the stationary type). This type of curve is seen when there is a suitable concentration of sulfuric and oxalic acid at the proper temperature (barrier + porous layer). 3) The voltage fluctuates periodically for a given current density. This type of curve is to be seen in concentrated strong acids or in alkaline solutions with a strong solvent action—sulfuric and phosphoric acids, sodium phosphate solutions,[52,53] and sodium or potassium hydroxide solutions (for metal). The anodic surface in this case is brightly polished (electropolishing). 4) The voltage gradually drops for a given current density (descending-voltage type). This type of curve is seen when there is a suitable concentration of formic or acetic acid at the proper temperature (barrier + porous layer). 5) For a given current, the initial voltage is very low and remains constant; the anode surface is roughly etched (acidic halide solutions such as HF, HCl, strong alkalis, etc.).

In the case of the descending-voltage type the heat of reaction and Joule heating accumulate as the film becomes thicker, thereby increasing the dissolution. The film cannot be expected to grow infinitely; there is a maximum thickness which can be obtained. According to the statistical calculations of Tajima and Shimura,[66] the growth rate of sulfuric acid films (15 vol. % H_2SO_4, 1 A/dm^2, on rocking anode of 99.99% Al) is

$$\dot{d} = 0.347 - 10^{0.025T-1.943}t^{0.34} \qquad (1)$$

where $\dot{d}$ is the thickening rate in μ/min, T is the temperature in °C, and t is the anodizing time in min.

In words, Eq. (1) states that the film grows by 0.347 μ/min (20 μ/h) but the rate decreases when the temperature is higher and the time of electrolysis is longer.

If we integrate Eq. (1), we have

$$d = 0.347t - 10^{0.025T-2.0708}t^{1.34} \qquad (2)$$

When $T = 48$°C, the time needed for $d = 0$ is 16.5 min, and at $t = 10$ min, $d = 0.52 \mu$. That is, under these conditions ($T = 48$°C), the anodic oxide film first grows, reaching the maximum thickness, and in 16.5 min, the thickness becomes zero, the barrier layer remaining.

The growth rate for oxalic acid (0.7 wt. %) is

$$\dot{d} = 0.347t - 10^{0.008T-1.170} \qquad (3)$$

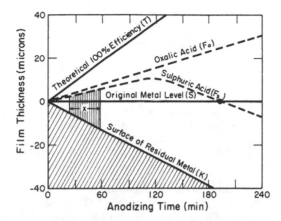

Fig. 2. The change in level of base metal and outer surface of film during growth of anodic oxide films in oxalic acid and sulfuric acid solutions (current density approx. $1.6\,A/dm^2$; x is the commercial range for sulfuric acid). Line through origin is the original level of metal. Line K is the boundary surface between oxide film and the residual metal core. Line T is the boundary surface between oxide film and electrolyte with theoretical 100% efficiency. Curve F_0 is the boundary surface between oxide film and oxalic acid electrolyte. Curve F_s is the boundary surface between oxide film and sulfuric acid electrolyte. The difference in ordinates between the curves F_0 and F_s and the line K gives the effective thickness for the two processes. The thickness of the film formed in oxalic acid increases proportionally with current density. (After Hübner and Schiltknecht.[6])

Commercial sulfuric acid baths use a 10–15% concentration, and unless the temperature is kept below 25°C, sufficiently dense films with corrosion resistance and dye affinity will not be obtained. Thus, the thickness can vary within a wide range,[67] depending upon the process used, the electrolyte, the anodizing time, and the temperature (Fig. 2).

It is possible, most readily by the oxalic acid process, that anodization will completely consume the aluminum. The metal completely disappears and a sheet may become transparent (Fig. 3). For normal applications, the thicknesses of both films range from 5 to 25 μ.

Figure 4 shows three schematic diagrams of corrosion and oxidation domains in various electrolytes, where voltage and electrolyte concentrations are plotted at a constant current density of 1 A/dm² at normal tem-

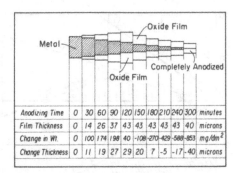

Anodizing Time	0	30	60	90	120	150	180	210	240	300	minutes
Film Thickness	0	14	26	37	43	43	43	43	43	40	microns
Change in Wt.	0	100	174	198	40	-108	-270	-429	-588	-853	mg/dm²
Change Thickness	0	11	19	27	29	20	7	-5	-17	-40	microns

Fig. 3. Influence of anodizing time on film structure. Pure aluminum foil, 0.12 mm thick, was anodized in sulfuric acid for varying periods of time by the dc (20°C, 14 A/ft²) process and the cross sections arranged in a row. The oxidation can be continued until all the metal has been converted into oxide. Such so-called "through anodized" test pieces are transparent, since the metal core has disappeared. The trend of the curve F_s in Fig. 2 may be seen from this figure. (After Hübner and Schiltknecht.[6])

perature.[62,63] It is interesting to note that monobasic acids always cause pitting corrosion without any film formation. This was noted early by Jenny[2] and by the author[62,63] and recently stressed by Kissin.[16] The causes of this behavior will be discussed later.

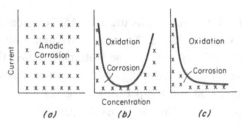

Fig. 4. Schematic representation of the domains in which anodic oxidation or corrosion occurs. a) Formic and acetic acids. b) Propionic acid (sulfuric acid, sodium bisulfate). c) Oxalic, malonic, succinic, adipic, lactic, citric, tartaric, and maleic acids (sulfamic acid). (After Tajima, Itoh, and Fukushima.[62,63])

Scope of Anodizing Electrolytes and Their Characteristics

Sulfuric Acid

Sulfuric acid [64-74,401] and 10–15 vol. % sulfuric acid are the most widely used electrolytes for anodizing. Sulfuric acid produces a transparent and thick film which is excellent in dye adsorption. This sulfuric acid is the basic electrolyte for special anodizing processes such as bright anodizing, hard coating, and integral color anodizing (which will be dealt with later on). Historically, the bath was well investigated and commercially applied in the USA as the Alumilite process and in Germany as Eloxal (Elektrolytisch oxidiertes Aluminum) where the process was termed as GS (dc sulfuric acid anodizing). The German operating standards for sulfuric acid and oxalic acid processes are summarized in Table 1.[3,72]

Kissin, Deal, and Paulson[16,73] detail the anodizing characteristics of commercial alloys in 15% sulfuric acid at 12 A/ft² and 20–25°C (Fig. 5). Tyukina[74] discusses specifically anodizing Al–Cu alloys in sulfuric acid. Bogozavlevskij[75] studied the effects of ultrasonic agitation on films in sulfuric acid baths and found that they become harder, less porous, and more corrosion resistant. Unpublished work in the author's laboratory gave negative effects due to local heating of the surface. Moreover, when sulfuric acid is added to electrolytes which form barrier layers it is possible to form a duplex film which has the property of the barrier layer. The film so formed has wide uses in architectural anodizing processes such as the integral color (self color) process.

Oxalic Acid

The oxalic acid process was originally developed in Japan in 1923[29-33] and, with the invention of the steam-sealing process in 1928,[36] was used extensively before World War II for kitchen utensils and electric insulation. Before and during World War II, the oxalic acid process was almost the sole process adopted in Japan. Although in Germany the GX, GXH, WX, and WGX processes with dc, ac, and dc + ac using oxalic acid had been known (Table 1), it seems this acid was not used as much in the past as the GS process. Therefore, the author believes that the brief description of the Japanese oxalic acid process (Alumite process) will be of interest to the reader.[77] (A brief introduction to the history of the process has already been given.)

In Japan, the oxalic acid anodizing process is carried out mainly by superimposing ac on dc, and the resulting anodized color is yellowish gold.

Table 1. German Processes (ELOXAL) for Normal Commercial Anodizing

Name	Electrolyte	Current form	Bath voltage, V	Current density, A/dm²	Temperature, °C	Time, min	Energy consumption, kwh/m²
GS	sulfuric acid (150–275 g/liter)	dc	ca. 12–20	ca. 1–2	20–22	30–40	ca. 0.5–2.5
GX	oxalic acid (5 g/liter)	dc	ca. 20–60	ca. 1–2	18–20	40–60	ca. 3–12
GXH	oxalic acid (5 g/liter)	dc	ca. 30–35	ca. 1–2	35	20–30	ca. 1–3.5
WX	oxalic acid (5–10 g/liter)	ac	ca. 20–60	ca. 2–3	25–35	40–60	ca. 3–9
WGX	oxalic acid	1. dc 2. ac	ca. 30–60 ca. 40–60	ca. 2–3 ca. 1–2	20–30	15–30	total 2–10
GSX	{ sulfuric acid oxalic acid }	dc	ca. 20–25	ca. 1–2	20–27	—	—

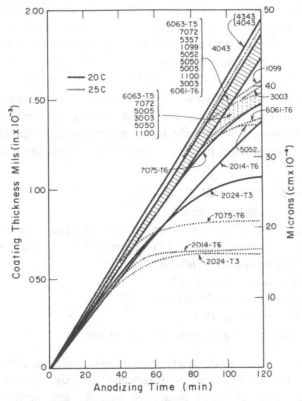

Fig. 5. Anodic coating thickness of aluminum alloys (15% H_2SO_4, 12 A/ft², 0–120 min). (After Kissin, Deal, and Paulson.[16,73])

The reasons for the popularity of anodizing for kitchen utensils in Japan were, in the author's opinion, the following:

1. The development of anodizing process kept pace with the production and popularization of aluminum articles.

2. The anodized color produced on aluminum by the oxalic acid process was largely to the liking of the Japanese people (aided by timely propaganda of the manufacturer) in addition to its improved properties of hardness and corrosion and abrasion resistance.

3. Anodizing processes were monopolized by the Institute of Physical and Chemical Research, Tokyo, and the oxalic acid process was used exclusively.

4. People had become accustomed to its color and tended to think of aluminum articles with the "Alumite" color.

The patents expired in 1943, and many of the anodizing shops which started after the war adopted the sulfuric acid anodizing process on account of the high cost of oxalic acid and higher electric power consumption of the process. However, to comply with the desires of the people, a similar color was obtained by dipping anodized articles in a hot saturated solution of Japanese tea. But the tea dyeing, it was said, degraded the quality of the film, and oxalic acid anodizing and steam sealing recovered its former popularity. The process is especially indispensable for electrical insulations.

The Japanese standard process for oxalic acid anodizing has been and is as follows:

Electrolyte: oxalic acid 3% (1–5%).

Cathode: graphite or lead.

Voltage: 60–100 V. (ac is often applied, but the superimposing of ac on dc is more commonly employed—the dc voltage is 15–30 V and the ac voltage is 60–100 V.)

Current density: 10–20 A/ft^2.

Time of electrolysis: 30–40 min.

Temperature: 75 to 95°F.

Steam sealing: 70 psi steam for 30–60 min.

The articles are rinsed in water, dyed in some cases, and finally sealed in high-pressure steam. The steam-sealing process improves the transparency, corrosion resistance, electrical insulating quality, and hardness of the surface film.

The changes in bath voltage with time, when the current density is kept constant (5, 10, 20, 50 mA/cm^2), are shown for 2% oxalic acid at 7°C (44.6°F) in Fig. 6. The voltage rises steeply at the start, reaches a maximum value, and then decreases slightly, maintaining nearly a constant value during 2–3 h of electrolysis. It is to be noticed that the products of current density and time required to reach the maximum voltage (total electric quantity, A/cm^2·sec) are nearly constant (about 0.5 C in this case) over a fairly wide range of current density, as shown in Table 2.

The critical voltage varies with the concentration and current density of the solution. Figure 7 shows the change of critical voltage with concentration and current density in 2% oxalic acid at 18–19°C.

When the anode is taken from the bath at the maximum voltage, the surface is covered with a very thin film with interference color. It is concluded that the maximum voltage and the total electric quantity were required to form this thin film and, after its formation, only a relatively lower voltage is necessary to build up a dense film.

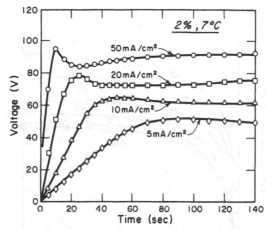

Fig. 6. Voltage–time relationships and the critical
voltage in 2% oxalic acid at 7°C. (After Setoh
and Miyata.[400,77])

Oxalic acid anodizing by straight dc sometimes causes a local attack
on aluminum due to the accumulation of oxalate ion, and/or of chloride
ion. The superimposing of alternating current on direct current improves
this condition.

In recent years, hard anodizing for engineering uses and integral color
anodizing for architectural uses have been developed and oxalic acid plays
an important role in these processes. Thus, the current trend of the world
anodizing industry is toward the "revival" of oxalic acid and other related
organic acid electrolytes such as malonic and maleic acids. As seen in

Table 2. Total Electric Quantity Required to Reach Maximum Voltage for 2%
Oxalic Acid at 7°C (After Setoh and Miyata[77,400])

Current density, mA/cm^2	Time, sec	Total electric quantity, C/cm^2
1.0	600	0.60
2.0	300	0.60
5.0	100	0.50
10.0	50	0.50
20.0	25	0.50
50.0	10	0.50
100.0	7	0.70
200.0	4	0.80

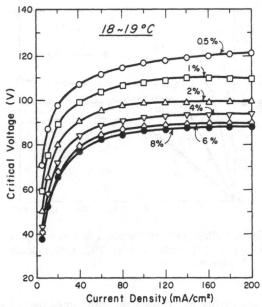

Fig. 7. Critical voltage plotted against current density for various concentrations of oxalic acid. (After Setoh and Miyata.[400,77])

Table 1, oxalic acid requires about a 3 to 4 times higher voltage than sulfuric acid, which is due to the more barrier-forming nature of oxalic acid.

Oxalic acid with or without additions is very extensively reviewed and studied by Setoh and Miyata,[76] and Mason and Slunder.[67]

A sulfuric–oxalic acid mixture (for example, 180 g/liter H_2SO_4 and 4 g/liter oxalic acid) is currently popular in Europe. The merits and the control of solutions are mentioned by Zurbrügg,[80] Neunzig and Röhrig,[81] Sautter,[82] and Kape.[137]

The Ematal process developed by Schenk[3,79] in Switzerland and the related process studied in detail by Hübner[78] is based on oxalate. The Ematal electrolyte consists of 40 g potassium titanium oxalate, 8 g boric acid, 1.2 g citric acid, 1.2 g oxalic acid, and 1 liter water. Anodizing is carried out at 55–60°C with an initial voltage of 80–90 V raised finally to 120 V in order to maintain the current density of 30 A/ft². Titanium can be replaced by thallium or zirconium. The film is opaque and can be dyed to give pastel shades. The process is described in detail by Schenk[3] and Hübner and Schiltknecht.[6]

Malonic acid gives a similar film to oxalic acid and the process was early patented by Kujirai and Ueki.[32] One example is 1.0–3.0% malonic acid with a temperature under 86°F using a current density of 0.45–0.27 A/dm² by applying 60–100 V ac or dc.

Anodic behaviors of aluminum in mono, di, and oxycarbonic acids, including malonic acid, were studied by Tajima, Itoh, and Fukushima.[62] Kape[137,208] describes the merits of malonic acid anodizing as giving a deep yellow or sepia color of extreme hardness and abrasion resistance.

Chromic Acid

The chromic acid process, originally developed by Bengough and Stuart[34,35] in 1923, found applications in the UK and USA for the improvement of corrosion resistance of Al and its alloys, particularly of Duralumine for aircraft and marine vehicle parts.[83] The process has the advantage of inducing no corrosion even when the electrolyte remains in recesses. The film is grey and, when dyed, gives nonmetallic, plasticlike shades which find some uses in decorative ware. Another outstanding property is its ductility, which is such that articles can be formed from anodized and dyed sheets with negligible signs of crazing of the film. The original chromic acid process is 2.5–3.5% CrO_3 at a temperature of $40 \pm 2°C$ and a voltage of 0–50 V. The voltage is kept at 0–40 V for 10 min raised to 40 V for 40 min, and then to 40–50 V for 10 min (total time, 1 h). The resulting current density is 1–3 A/dm².

A modified high-temperature process was first proposed by Lewsey[84] in 1952 and later by Brace and Peek[85] with particular attention to opaque coatings. This process and the mechanical and corrosion properties of the resulting film were recently described in detail by Modic.[86] According to Brace, the opaque color comes not from the film, but from the etching of the aluminum surface while anodizing.

Modic[86] states that the acid concentration in the Bengough–Stuart process is unreasonably low and he ascribes it to the high cost of the acid

Table 3. Film Thickness (μ) and Chromic Acid Concentration at 50°C after 35 min (After Modic[86])

Voltage, V	CrO₃%					
	2.5	3.8	6.4	10.0	15.0	28.0
15	1.5	2.1	4.0	5.5	6.5	6.5
40	7.0	7.5	8.0	8.0	7.5	7.0

Table 4. Film Thickness and Temperature at 40 V after 40 min (After Modic[86])

Temperature, °C	30	35	40	45	50	55	60
Thickness, μ	1.2	2.0	3.5	5.0	8.5	8.0	7.0

at that time. He explained the correlation of acid concentration, film thickness, voltage, and temperature which is given in Tables 3 and 4.

From these tables, the optimum condition is determined to be 10–15 wt. % CrO_3 at 50–55°C and 40 V. Below 6% acid, only dark film is obtained and above this concentration, clear, opaque, enamel-like films are obtained. Up to 35°C, quite clear and transparent films or slightly turbid films (like sulfuric acid films) are obtained; above 40°C fully opaque films are produced. This process is described in Fig. 8.

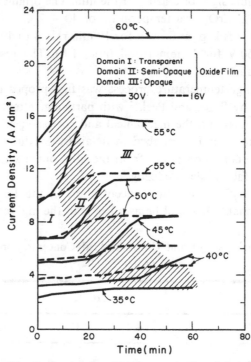

Fig. 8. Current density–time curves in relation to temperature and voltage in chromic acid anodizing. (After Modic.[86])

Sulfamic Acid

The sulfamic acid anodizing process was first suggested by Gordon and Cupery[87] and Piontelli[88,89]; a detailed study was published by Tajima, Fukushima, and Kimura.[90]

The solubility of sulfamic acid is 21.32 g in 100 g water at 20°C. For anodizing, a 5–10% concentration at 1 A/dm² is recommended. It seems that sulfamic acid gives intermediate properties between barrier and duplex films and the resulting voltage is higher than that of sulfuric acid anodizing (Fig. 9). The film is not so easily dyed as sulfuric acid film, but the abrasion resistance is superior (Fig. 10). The mixture of the two acids, or the consecutive two-step anodizing with sulfamic and sulfuric acids, improves the anodizing conditions and dye affinity.[90] However, sulfamic acid has no remarkable merit as compared with conventional sulfuric and oxalic acids except that the process is particularly suited to use without refrigeration in the tropical zone where the room temperature is 30°C or higher. Sulfamic acid found some uses as a component of mixed electrolytes for integral color anodizing. Schweikher[91] patented a process involving the use of sulfamic acid at 170°F in continuously anodizing foil and sheet to give corrosion and abrasion resistance and good dye absorption. Fukushima[92] studied the cause of local corrosion frequently experienced by sulfamic acid anodizing;

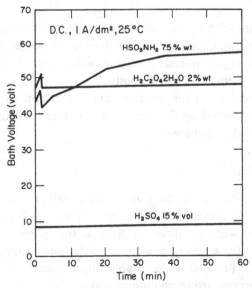

Fig. 9. Comparison of bath voltages of sulfuric, oxalic, and sulfamic acid anodizing; dc, 1 A/dm², 25°C. (After Tajima, Kimura, and Fukushima.[90])

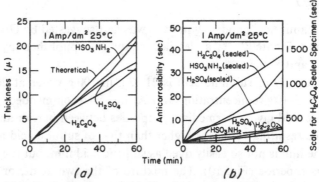

Fig. 10. Comparison of (a) film thickness and (b) corrosion resistance (JIS-8601 alkaline-corrosion test) of sulfuric (15 vol. %),
oxalic (2 wt. %), and sulfamic (7.5 wt. %) acid films; 25°C, dc,
1 A/dm², 60 min, unsealed and steam-sealed at 70 psi. (After
Tajima, Kimura, and Fukushima.[90,77])

he proposed to first anodize at low current density (e.g., 0.2 A/dm² for
6 min) followed by a higher current. The proceedings of the recent symposium concerning applications of sulfamic acid supplies considerable
relevant data.[423,424]

Phosphoric Acid

Oxalic and phosphoric acid solutions are most frequently employed in
commercial preplating anodic oxidation since these two acids, particularly
the latter, give a highly porous structure which permits the mechanical
bonding of the subsequent electrodeposits.

Phosphoric acid anodizing of aluminum and its application to electroplating was first patented by Fischer[94] and described in detail by Spooner
and Seraphim[95] and elsewhere.[96-99]

Bright Anodizing

Bright anodizing originated during the development of an electrochemical brightening process for aluminum reflectors. After the brightening
operation, objects were anodized to form a transparent and protective
film, and although these processes were not called such, they were definitely
"electropolishing." (The term electropolishing appeared some years later,
in relation to a process using perchloric acid.)

Pullen[52,53] was the first to produce a bright surface on a high-purity
aluminum by using electrochemical means. He used an $Na_2CO_3 + Na_3PO_4$

mixture for electrobrightening, followed by anodizing in an $NaHSO_4$ solution which gives a transparent and protective thin coating. This process was called the Brytal process.

The Alzak process, developed by Mason and Tosterud[54] of ALCOA, used HBF_4, $H_2SO_4 + HF$, or $CrO_3 + HF$ solutions to produce anodically bright surfaces, followed by anodizing in H_2SO_4. The process was introduced by Dickinson[100] and Edwards.[101]

The Illuminite process developed by Nakayama[55] in Japan used phosphoric acid to brighten aluminum for reflectors, followed by anodizing in H_2SO_4. In the author's opinion, he was the first to use phosphoric acid to electropolish aluminum.

Bright anodizing is now widely used not only for reflectors but also for decorative parts, particularly for automobile parts, such as trims, grilles, hub caps, bumpers, etc.,[102,103,115,121] and parts for space vehicles.[357-359,422] The process of bright anodizing is the combination of electropolishing in phosphoric or phosphoric–sulfuric acids (referred to later in this paper) with or without additions, or chemical polishing in phosphoric–nitric acid[105-107] or hydrofluoric acid (ammonium bifluoride)–nitric acid (Erftwerk process of VAW),[108-114] and anodizing in sulfuric acid or in barrier-forming electrolytes.[422]

The technical advance of bright anodizing has been made by the improvement of chemical or electropolishing processes[114-120] and by the improvement in alloying.[110] High-purity Al (99.85%), Al–Mg alloys, Al–Mg–Si extrudable alloys, and Al–Mg–Zn alloys of higher strength are most fitted for chemical and electropolishing followed by protective and transparent anodizing. To secure uniform and transparent films, alloys must be mechanically worked and heat treated very carefully. Iron contamination must be less than 0.005% for nitric–bifluoride chemical polishing solutions, and below 0.06% for phosphoric–nitric acid solutions. An increase in iron content affects negatively the polishing and the transparency of anodic oxide films. The behavior of copper ($\leq 0.14\%$) in the alloy and in the polishing solution is interesting; it promotes dissolution of Al by galvanic action to improve brightness.

Recently, Morris,[406] Terai,[121] Sundberg and Samuelson,[122] and Teubler[123] discussed in detail the metallurgical aspects of bright anodizing aluminum and its heat treatment. The addition of 0.1% Cu to Al–Zn–Mg alloys favors the chemical and electropolishing effect, just as the addition of copper salt to chemical polishing baths improves the polish. The alloy developed by Sundberg and Samuelson of the Svenska Metallverken contains 4.4% Zn, 1.1% Mg, 0.03% Cr and Cu (Fe + Si + Mn $\leq$ 0.20%).

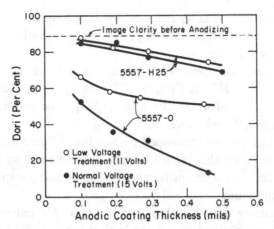

Fig. 11. Effect of anodic coating thickness on image
clarity (DORI, distinctness of reflected image) of
5557-H25 and 5557-O specimens. (After Cochran
and Keller.[17])

Cochran and Keller[17] studied in detail the effect of voltage during anodizing
of aluminum. The voltage employed in sulfuric acid solutions has an important influence on the structure, properties, and appearance of the oxide
coating. Higher voltage favors high anodizing efficiency, higher density,
and higher abrasion resistance but produces lower image clarity. The better
anodic finish clarity obtained with lower voltage treatments is attributed
largely to a decrease in the surface roughness developed at the metal–oxide
interface. This is associated with a reduced attack on microconstituents and
less development of grain relief, during anodizing, of the metal surface. A
low-voltage anodizing procedure is suggested in order to obtain finishes
having the highest image clarity and least amount of reflection haze. This
is an advantage in applications for bright decorative trim or specular re-
flections. This procedure is most effective on annealed metal but also yields
improvements in anodic finish clarity for some lots of strain-hardened alloys.
It has also been found effective in producing improved image clarity of
anodized, heat-treatable Mg–Si alloys as, for example, 6463 and 6563 extrusions (Fig. 11).

Hard Anodizing

Particularly thick and hard anodic films produced by oxalic acid were
reported early by Miyata[124] in Japan (65 μ), Rummel[125] in Germany (600 μ),
Schenk[3] in Switzerland (100 μ), and Elssner[126] in Germany. However, To-

mashov[128],[129] in the USSR was the first to actually propose hard coating by sulfuric acid for the engineering purposes of increasing hardness and wear and corrosion resistance. Research in this field has been very active since then, and various processes of applying dc, ac, or dc + ac to sulfuric, oxalic, and other organic acids and their mixtures[143] have been proposed and practiced.

As described before, in normal anodizing a maximum film thickness is reached. When the film approaches this maximum ($>30\ \mu$) and achieves a sufficient compactness, the maximum hardness is obtained (VHN 350–1200). Thus, hard anodizing may be said to be a process of thickening the anodic film. The so-called barrier layer, of course, is compact, but it can be made only as thick as 1 μ. In order to establish a hard-anodizing process, it is necessary to find intermediate electrolytic conditions between the barrier- and duplex- (barrier and porous) film-forming conditions. Such conditions are favored by keeping the temperature as low as possible, selecting an electrolyte low in solvent action, and providing strong agitation to eliminate the heat evolved in the film due to reaction and Joule heating. Hard anodizing usually requires high formation voltages (up to 100–150 V), depending on the thickness required and the nature of the electrolyte. The film growth rate and hardness of alloy species differ remarkably, even under the same electrolytic conditions.

In Table 5 various proposals for thick and hard anodic coatings are listed. (The details of these processes are produced in the original literature referenced in the table.) Most of these processes use constant current. However, Kape[137] recommended a constant-wattage process since the current reduces as the film thickens; thus, the local heating at the anode can be moderated. Csokán[138] proposed a constant-voltage process.

Csokán's work[138–147] at the Research Institute for Nonferrous Metals (Fémipari Kutató Intezet) in Budapest is interesting and worth mentioning here. He used a very dilute sulfuric acid (1%) at about 0°C, applying constant voltage to obtain very hard coatings, contrary to the popularly used concentration of about 10% for hard coating. Csokán chose 5, 2.5, and 1% sulfuric acid to anodize, for example, pure Al and to obtain a lower specific conductivity. In these baths anodizing was carried out at initial voltages of 10, 20, 30, 40, 50, 60, 70, 80, and 90 V. Figure 12 shows that lowering the concentration causes a decrease in current density. The effect is more pronounced than the effect of cooling, with the result that anode overheating is slight, even at high voltages. The current density in a 5% electrolyte at room temperature and 50 V is 112 A/dm², and even at −10°C it is about 90 A/dm². In a 2.5% electrolyte, the initial current density is 40 A/dm²;

Table 5. Hard-Anodizing Processes

Source (process)	Electrolyte	Temperature, °C	Current density	Voltage, V	Time, min	Thickness, μ	Remarks
Tomashov[46,47]	H_2SO_4, 200 g/liter	1–3	2.8 A/dm²	23–100, finally 150	240	130–175	20% porosity for 2S Al 35% porosity for AK4 (Y alloy)
G. L. Martin[130] (M.H.C.)	H_2SO_4, 15% (saturated with CO_2)	0	20–25 A/ft²	20–60			constant current Alumilite 225,226 for wrought alloys Alumilite 725,726 for cast alloys (8–10°C)
Campbell[131–135] (Hardas)	H_2SO_4, 10 vol. %	0–5	10–35 A/dm²	ac 10–70 dc 20–140	15	50–75	ac + dc, the proportions varying with the alloy. ac makes easy the high-current use and favors alloy anodizing
	oxalic acid	0–4					
Sanford[136]	H_2SO_4, 7 vol. %; peat extract, 3%; nonyl alcohol, 0.02%; polyethylene glycol, 0.02%; CH_3OH, 7 vol. %;	−10	1.2–2.2 A/dm²	15–60, finally 100		1.2–2.5 per min	
Kape[137] (Hiduran)	H_2SO_4, 10 vol. %	10					constant wattage
	malonic acid, 12 wt. %	55					
Csokán[138–147]	H_2SO_4, 1%	0		40–60		maximum 200–220 120–150	constant voltage high-purity Al 2S

Reference	Electrolyte	Temperature	Current density	Voltage		Time	Remarks
Stalzer[160]	dil. H_2SO_4	20	50 A/dm² (200 A/dm³)			50/4 min	5 cm/sec, anode rocking 20°C (boiling under reduced pressure)
Wiesner and Meers[161]	H_2SO_4, 12 wt. % oxalic acid, 1 wt. %	48–52°F	36 A/ft²	10–75	100	25/20 min	
	H_2SO_4, 385 g/liter oxalic acid, 11–15 g/liter	50°F					for high Cu–Al alloys
(GL Process—Coloral, S.A.)[162]	oxalic acid	normal	3–6 A/dm²	45–90		5–6	smooth and wear resistant, suited for watch and precision machine parts, made of Anticorrodal (Al–Mg–Si)
Lelong, Segonds, and Herenguel[163]	oxalic acid, 80 g/liter formic acid, 55 g/liter	15	60 A/ft²	25–60	100	100–250	even high current densities are possible
	$NaH_2SO_4 \cdot H_2O$, 240 g/liter citric acid, 100 g/liter	15	60 A/ft²	50–100	100		applicable to various alloys
Jogarao et al.[164,165]	oxalic acid, 5 wt. % CaF_2, 0.1 g/liter H_2SO_4, 0.5 g/liter $MnSO_4$ or $Cr_2(SO_4)_3$, 1 g/liter	5–10	4.4 A/dm²	55–100	45	25–50	VPN 1000–1200. Addition of CaF_2 gives VPN max. 1400
Shenoi[166]	oxalic acid, 5% malonic acid, 4%	20–24	4 A/dm²	71–150	63	58	VPN 470. Addition of 0.3% $MnSO_4$ and 0.15% $NiSO_4$ gives VPN 560

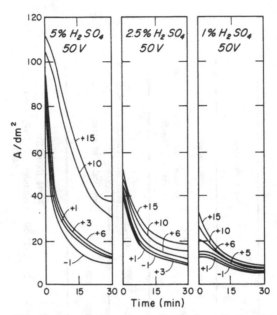

Fig. 12. Change in current density in sulfuric acid
at different concentrations and temperatures and at
50 V. (After Csokán.[141])

in a 1% electrolyte it is only 12–13 A/dm², and this is decreased to 5–7 A/dm²
in 30 min.

In a low-concentration bath the dissolution of the oxide layer is negligible, even for longer anodizing times. This favors oxide growth. A change
in oxide formation is noticed here. In a 5% electrolyte, with a 30-V initial
voltage, a normal oxide layer is formed at all temperatures. In colder baths,
a thicker and harder film may be obtained as the dissolution of the oxide
is retarded by the decrease in temperature. At initial voltages of 30–50 V
Csokán observed a particular phenomenon which is different from those of
normal anodizing. In normal anodizing scattered, small, dark gray or black
spots appear and where they are more numerous, such as at edges and
corners, and where the current density is higher, they form a very hard
and slightly uneven surface. At 50–60 V or higher, Al is dissolved in the
electrolyte. A decrease in temperature reduces the dissolution of the oxide
so that an oxide coating formed at −1°C and 50 V has an even yellowish
gray color over the whole surface and is very hard. At voltages exceeding
60 V the color becomes pale, and the structure becomes more porous; at
the edges of the sample, signs of metal dissolution can be detected.

In 2.5% and 1% sulfuric acid up to 35 V, the formation of the oxide layer is similar to that in more concentrated solutions. At higher voltages, however, striking phenomena are observed. It is characteristic that in the very thin but extremely close-packed layer formed in the first seconds of anodizing at voltages exceeding 35 V, breakdown can only be observed at a small number of spots. Oxidation is not uniform over the whole surface but spreads readily from distinct active spots. Finally the oxidation areas are merged.

At 45–50 V, these active centers become so numerous at the beginning of anodizing, that the aluminum surface is completely covered and the formation of the close-packed oxide layer proceeds evenly over the whole surface. The current density remains low even at a 50–60-V initial voltage, local overheating of the electrodes is less likely, and dissolution of the oxide is nil. At 70 V and 10–15°C, a porous layer is formed as a result of the dissolution of the oxide, but in colder baths, +1 to −1°C, a more compact layer is formed. The coating produced by anodizing with voltages above 80 V is spongy and powdery, and scales off easily from the surface. From these experimental results, Csokán states that the rate of formation of the oxide layer is at a maximum and that of dissolution at minimum in 1% sulfuric acid at temperatures of about 0°C and at cell voltages of 40–60 V. In these circumstances, the formation of a hard coating starts at scattered centers, which are then surrounded by gradually spreading secondary oxidation zones; in between these zones new centers are formed continually. The oxide nuclei do not all grow at the same rate nor are they initiated at the same time. As a result of this, Csokán states that the compact oxide coating will display a peculiar prismlike structure as shown in Fig. 13.

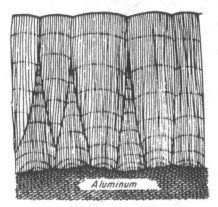

Fig. 13. Irregular prismatic macrostructure due to uneven growth of the hard oxide coating. (After Csokán.[141])

Doyle, Harris, and Kape[137] recommend the constant-wattage procedure, since the current reduces as the film thickens and the local heating of the anode can be moderated.

Kape,[137] Isawa,[148] and others have also observed the local islands of oxide pointed out by Csokán. The author[149] believes that this usually happens by anodizing in low-concentration and low-conductivity electrolytes such as in very dilute sulfuric acid or in higher carbonic acids. Preventing the occurrence of these local islands is an important factor in securing high-quality uniform hard coating. Furthermore, on thick films crazing occurs; this is slight during electrolysis, but increases when the anodized parts are withdrawn from the bath due to the difference in the expansion coefficients of metal and oxide. The crazed film has an oil retention property which finds uses in frictional parts requiring lubrication.

Lakshminarashimhan, Narasimhan, and Shenoi[156] recently reported hard anodizing in an oxalic–malonic acid mixture at normal or higher temperatures. The results are shown in Fig. 14 and the optimum condition is given in Table 6.

The thick and hard anodic films are usually colored brown to black. This property was used at one time in the construction of curtain walls and exterior panels. However, due to the high cost of using low-temperature

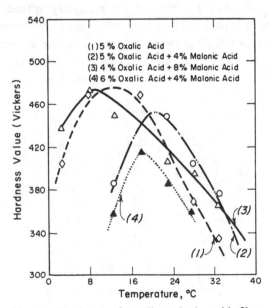

Fig. 14. Hardness of oxalic–malonic acid films (4 A/dm², 1 h). (After Lakshminarashimhan, Narasimhan, and Shenoi.[156])

Table 6. Effects of Addition Agents on Hardness (After Lakshminarashimhan, Narasimhan, and Shenoi[156])

No.	Addition agent added, %	Hardness	Surface appearance
1	0.5% tartaric acid	415	green
	1.5% tartaric acid	425	dark olive
2	1.5% titanium potassium oxalate	385	light olive
	3.0% titanium potassium oxalate	415	green
3	0.2% manganese hydrogen phosphate	405	olive
	0.5% manganese hydrogen phosphate	460	dark olive
	1.0% manganese hydrogen phosphate	475	dark gray
	1.5% manganese hydrogen phosphate	500	dark gray
4	0.7% manganese sulfate	426	dark olive
	0.14% manganese sulfate	450	dark gray
	0.21% manganese sulfate	500	black
	0.28% manganese sulfate	500	black
	1.00% manganese sulfate	500	black
5	0.07% nickel sulfate	425	dark olive
	0.21% nickel sulfate	450	dark gray
	0.28% nickel sulfate	460	dark gray
	0.71% nickel sulfate	505	black
	0.95% nickel sulfate	525	black
	1.60% nickel sulfate	505	black

baths and the monotonous shade it affords, this use gave way to the newly developed integral color processes to be described in the following pages. Hard anodizing, its properties and uses are reviewed in numerous references.[157-168]

Integral Color Anodizing

After World War II, a large quantity of aluminum sheets and extrusions were being used in the architectural field, which owes much to the technical progress in anodizing. The transparent film produced by anodizing in sulfuric acid is used directly after being colored by dyes or inorganic pigments.

Organic dyes have been used for many years to produce virtually any hue or shade and the contributions of the Swiss chemists in this field should be noted. However, organic dyes tend to fade at varying speeds when exposed to sun and weather (although there are a few exceptions).

Another coloring technique is the ferric ammonium oxalte hydrolysis process (called "Elangold" in Europe) which imparts various shades of gold by depositing ferric oxide or hydrate in the pores of the anodic coating before sealing. The color produced has good light fastness, but a separate dyeing operation is required.

A third coloring technique is the double decomposition process designed to precipitate colored inorganic compounds in the pores before sealing. The so-called "Elanbronze" process, still widely used in Europe, consists of dipping alternately in $KMnO_4$ and in cobalt acetate solutions to produce MnO_2 or hydrate precipitates dispersed over the anodic oxide film. The color of the resulting film is yellow to brown and has good weather resistance. However, it also requires additional processing steps.

Electrolytic color-anodizing processes have been developed which require no separate dyeing and the films are extremely weather resistant, hard, and light fast. As the various shades of color, associated with thicker films, became popular, the "integral color" and "self-color" products were widely used.[168-176]

The yellow or golden colored film produced in oxalic acid by applying ac + dc or ac is very stable and has been used extensively in Japan, mainly for kitchen utensils, for 40 years. The process has also been used for architectural purposes but only to a limited extent. This process is definitely the "integral color" process in the present-day term.

The "Kalcolor" process developed by the Kaiser Aluminum and Chemical Corporation toward the end of the 1950's is an epoch making one in architectural anodizing. This process makes use of sulfosalicylic acid with a small addition of sulfuric acid and gives yellow through brown-to-black colored anodic films which are very stable and hard, and are suited for curtain walls, spandrels, solar screens, windows, etc.

The old patents of Schering-Kahlbaum AG[177] loosely covered[178] the organic aromatic sulfonic acid and, further, thymol sulfonic, phenol sulfonic, and cresol sulfonic acids with additions of small amounts of sulfuric acid or sulfate. However, the Kalcolor process is nonetheless the more popular process.

The color is also influenced by the alloy. Some examples from the Kalcolor process[176] are shown in Table 7.

In the Kalcolor process the pieces are anodized in an electrolyte consisting of 10 wt. % of sulfosalicylic acid, 0.5 wt. % sulfuric acid, and the balance water. The temperature is 77°F. Anodizing starts at 24 A/ft² dc and continues at this current density until the voltage reaches the value indicated in Table 1, under "Maximum volts"; it is held at this voltage for

Table 7. Kalcolor Process Variables Used to Produce Colors on Aluminum Alloys (After Diel and Swanson[58], and Church[179])

Alloy	Color range	Initial current density, A/ft²	Time to maximum V, min	Maximum V	Total anodizing time, min
5005 clad with 5005 (sheet)	amber gray	24	20	50	30
	charcoal brown	24	35	60	45
1100 clad with 1100 (sheet)	tan	24	20	50	30
	olive	24	30	60	45
2024-T3 (sheet and extrusion)	light blue	12	15	65	45
3003 clad with 3003 (sheet)	dove gray	24	10	50	20
	charcoal gray	24	20	65	40
5086 (sheet)	black	24	25	65	40
	gray	24	15	50	20
5052 (sheet)	light bronze	24	15	40	30
	Golden brown	24	35	60	45
5357 (sheet)	light brown	24	20	50	30
	brown	24	30	60	45
6061-T6 (sheet and extrusion)	antique bronze	24	10	50	20
	jet black	24	30	65	40
6063-T5 (extrusion)	amber	24	20	50	30
	light brown	24	35	60	45

the remainder of the specified anodizing time. The samples are then rinsed and sealed in a 195°F solution consisting of 1 g/liter cobalt acetate, 5 g/liter boric acid, 5 g/liter desugared calcium lignosulfonate, and the balance water, at pH = 5.7 for 15 min for coatings produced in 15–30 min, and 20 min for those produced in 30–45 min, and finally rinsed in water (120°F) and dried. Other similar integral color processes are listed in Table 8.

The confused state of Japanese integral color processes is due to the extensive use of oxalic acid coloring processes over the past 40 years. These processes are known by several proprietary names (all produce nearly the same colors), viz., Colorfine (alloys, electrolytes; Kobe Steel Co.), Nikcolor (alloys; Nikkei Aluminium), Sumitone (alloys, electrolytes; Sumitomo

Table 8. Integral-Color-Anodizing Processes (Except Japan)

Process	Electrolyte type	Usual operating conditions			
		Temperature, °C	Voltage range, V	Current density, A/ft^2	Film thickness range, μ
Kalcolor (Kaiser)	5-sulfosalicylic acid + sulfuric acid	25	35–70	24–30	20–40
Duranodic (ALCOA)	4-sulfophthalic + sulfuric acid	25	35–100	15–40	20–40
	5-sulfoisophthalic + sulfuric acid	25	35–100	15–40	20–40
Permalux (ALUSUISSE)	5-sulfosalicylic acid, sulfuric acid, and maleic acid	25	35–70	24–30	20–40
ISML	Oxalic acid, tartaric acid, and sulfuric acid	20–30	40–80	15–24	20–40
Veroxal (VAW)	Oxalic acid, maleic acid, and sulfuric acid	20–30	40–80	15–24	20–40
Alcandox (ALCAN, Banbury)	Oxalic acid (saturated)	18–22	40–80	14–18	

Lightmetals), Hishicolor (electrolytes; Mitsubishi–Reynolds), Enalmite (electrolytes; Nippon Aluminium), Colormite (electrolytes; Osaka Aluminium), Rikcolor (electrolytes; Riken Alumite), etc.

The most important step in the Kalcolor process is the addition of a small amount of sulfuric acid. Sulfosalicylic acid alone gives only a light yellowish barrier film, the voltage rising steeply, but the addition of sulfuric acid converts the barrier film into a growing duplex film. This effect is very distinct, and the integral color processes proposed later mostly include the addition of a small amount of sulfuric acid or oxalic acid. Even boric acid, a typical electrolyte giving a barrier film, gives a duplex film with a small addition of sulfuric acid.

Another type of integral color is produced by alloying (Table 9). Sulfuric acid gives a transparent film on high-purity Al, which becomes

Table 9. Alloys and Colors Anodized in H_2SO_4 (After Nakayama[88])

Alloy	Element %	Hot-working temperature	Color
AlAg	10–30	$\leq$ 400	brown
AlAu	0.3–5	$\leq$ 400	rose
AlB	0.8–5	$\leq$ 400	white
AlBe	0.5–2.5	$\leq$ 400	white
AlBi	0.5–4	$\leq$ 270	purple gray
AlCd	1–6	$\leq$ 320	gray
AlCo	0.5–2	$\leq$ 400	blue gray
AlCu	0.5–10	$\leq$ 400	yellowish brown
AlFe	0.5–3	$\leq$ 400	yellowish gray
AlMg	4–8	$\leq$ 400	white
AlMg$_2$Si	0.5–8	$\leq$ 400	white
AlMgZn$_2$	8–20	$\leq$ 400	gray
AlMn	0.5–3	$\leq$ 400	rose
AlMo	0.3–1	$\leq$ 400	yellow
AlNi	0.5–3	$\leq$ 400	blue gray
AlZn	10–25	$\leq$ 25	gray

turbid or opaque as a result of alloying impurities, such as Fe and Cu. Initially, the phenomenon was looked upon as a disadvantage[2] as evidenced by Brenner and Vogel[180] and Keller and coworkers,[181,182] who studied in detail the effects of constituents on anodized color, and by Morris[406] who recently studied the effects on haziness and defects. However, by intentional addition of appropriate elements, light-fast colors can be obtained by anodic oxidation. The most typical alloy is 3–5% Al–Si,[183–187] which gives a gray-to-black color by anodizing in sulfuric acid. The ALCOA headquarters building, constructed in 1952 in Pittsburgh, may be said to be an historical monument to this alloy. Its color is derived from dispersed Si particles which remain unoxidized during anodic film formation.

Two alloys give a gold color by anodizing: the Al–Cr alloy (Cr, 0.2–0.5%) of Kaiser Aluminum and the Al–Cr–Mg–Si alloy (Mg$_2$Si, 0.75–2.0%; Cr, 0.2–0.4%) of ALCOA.

According to Zurbrügg,[186] an effect similar to Cr is obtained with the addition of about 1% Cu. The color intensity of the anodic film does not increase with the increase of Cu above 2%, and becomes zero at 4%. The combination of Cu and Cr gives a more intense color than single alloying. The alloy AlMgSi 0.5 developed at the Research Institute, ALUSUISSE, Neuhausen, contains 1% Cu, 0.4% Cr, and 0.1% Fe (0.04% for bright anodizing).

Table 10. Effects of Various Alloying Elements on the Color of Anodic Films
(After Kape[171])

Primary alloying elements, %	Color produced in 25-μ film	
	Oxalic acid; 9.0%, 20°C	Sulfuric acid; 7.5%, 20°C
Super purity aluminum Trace of iron	pale straw	clear transparent
Approx. 0.4 iron 0.4 silicon	pure gray–gold	silver gray
1.25 manganese	brown	pale brown
2.2 magnesium 0.2 manganese	golden bronze	silver, opaque
0.7 magnesium 1.0 silicon 0.7 manganese	bronze brown	pale brown–gray
0.7 magnesium 0.5 silicon	pale gold	silver, opaque
1.0 magnesium 0.6 silicon 0.25 chromium 0.25 copper	dark brown	yellow
4.0 copper 0.6 magnesium 0.5 manganese	bluish white	yellow
5.0 magnesium 0.5 manganese	brownish yellow	silver gray
5.5 silicon 0.5 magnesium	gray	gray
3.0 copper 5.0 silicon 0.5 manganese	bluish gray	yellow brown

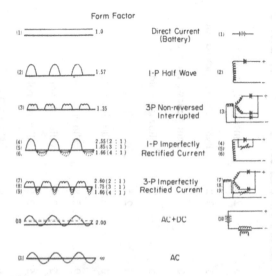

Fig. 15. Various wave forms applied and the circuit diagram. (After Tajima *et al.*[189,63])

The alloying approach is influenced not only by the component elements and their structures, but also by mechanical working, heat treatment, and particularly by homogenization and precipitation treatments. The results given in Table 10[171] are, therefore, only representative of the general shades which can be obtained from different alloys on anodizing in two standard electrolytes.

Integral color is also obtainable by the use of modulated currents, the role played by the current being, mainly, to increase the color shades. Tajima and coworkers[189,63] applied various forms of current (Fig. 15) to sulfuric and oxalic acid anodizing. Some results for oxalic acid are shown in Table 11.

Various opinions exist as to the origin of the self color (the phenomena are not yet completely clarified, Fig. 16):

1. Color by excess aluminum dispersed in the film.
2. Color by alloying constituents.
3. Color by the polymerization of anions embedded in the film by a reaction similar to the Kolbe reaction.
4. Color by the reduction products of sulfur (included in the film in the case of ac anodizing).
5. Color by the dispersion of carbon (in the case of organic acid, by reduction or decomposition).

Table 11. Properties of Oxalic Acid Films [$(COOH)_2 \cdot 2H_2O$; 3 wt. %, 30°C, 1 A/dm² (Mean Positive Current), 30 min]

Current form	Voltage, V		Coating ratio	Thickness, μ	Density, g/cm³	Appearance
	dc	ac				
Direct current (battery)	42.0	44.0	1.27	8.5	2.6	noncolored, transparent
1-P halfwave	42.0	45.0	1.35	7.5	3.2	light yellow, transparent
3-P Nonreversed, interrupted	42.0	45.0	1.29	7.0	2.7	light yellow, transparent
1-P imperfectly rectified (2:1)*	12.5	34.0	1.22	6.5	2.3	yellowish gold
(3:1)	14.0	34.0	1.20	6.5	2.5	yellowish gold
(4:1)	14.5	36.0	1.23	7.0	2.5	yellowish gold
3-P imperfectly rectified (2:1)	18.0	33.0	1.20	5.0	2.8	yellowish gold
(3:1)	23.0	35.0	1.21	5.5	2.8	yellowish gold
(4:1)	24.0	36.0	1.23	6.5	2.6	yellowish gold
Superimposed (ac + dc)	24.0	25.0	1.25	7.0	3.0	light yellow
Alternating current	4.0	40.0	1.14	4.5	2.9	yellowish gold

* Ratio of positive (1 A/dm²) and negative currents.

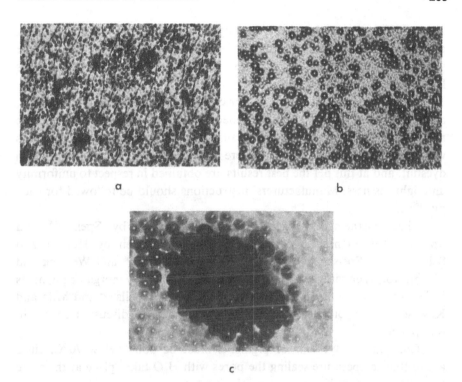

Fig. 16. Electron micrograph of integral color films formed in (a) malonic acid, (b) sulfosalicylic acid (plus sulfuric acid), and (c) boric acid–formamide nonaqueous electrolyte. Circular cells are partly colored with transparent cell centers. As the films grew thicker, the cells became larger and the color gradually predominated over the whole surface. (After Tajima et al.[255,256])

Coloring by Dyestuffs and Pigments

The duplex film formed on Al is very reactive and adsorptive for dyes and pigments. Bengough and Stuart[35] were the first to dye anodic film. They dyed chromic acid film with Anthracene Blue, Anthraquinone Blue, Alizarin Red S, etc. Flick[190] anodized Al in an ammonium sulfide solution containing Benzopruprine to obtain colored film. He also tried to dye after anodizing. The dyes employed were said to "lake," that is, to form an Al salt by reaction with the oxide or hydroxide. Gower[39] tried to anodize in sulfuric acid containing potassium bichromate, lead acetate, and barium sulfide to get colored film. Bengston[191] combined sulfuric acid anodizing and dyeing with organic dyes and inorganic pigments. Shirotsuka,[192] in Japan, applied for a patent to a process involving anodizing Al (or its alloys), dyeing them, and keeping them in boiling water or superheated steam.

Tosterud[193] proposed the double decomposition process of dipping anodized aluminum in lead acetate and then in potassium bichromate solutions. The successful development of dyeing, particularly in the use of dyestuffs, is largely due to Swiss chemists.

The dyestuffs most often used for coloring are known as the acid type, and their pH is usually 4.5–6.5. Generally, a dye with a pH outside the range of 4–8 is not recommended because strongly acid or alkaline solutions will attack the anodic coatings. There is an optimum pH range for each dyestuff, and at this pH the best results are obtained in respect to uniformity and light fastness. Manufacturers' instructions should be followed for each dye.[194]

Light fastness of organic dyes is described by Speiser[195] and Spooner.[196,197] General processes for dyeing are given by Hübner and Schiltknecht,[6] Speiser,[198–201] Mita,[203,204] Vanden Berg,[205] and Wernick and Pinner,[8] together with theory and references. Coloring by inorganic pigments is described by Hermann and Hübner,[207] Kape and Mills,[208] and Mita and Kawase.[209] Elangold and Elanbronze processes were discussed earlier in this paper.

Coloring by dyestuffs or pigments must be done below 70°C, since above that temperature sealing the pores with H_2O takes place at the same time. Sealing must be done after coloring.

Asada[210] patented a process of precipitating metal oxides, hydroxides, and salts on the conventional anodic coatings by successive treatment with ac in a bath containing suitable metal salts. For example, Al–Mg alloy is anodized in sulfuric acid for 40 min, electrolyzed in a mixture of 2% sulfuric acid and 0.2% nickel sulfate (pH 1.0–1.1), with 40 V (ac), and finally sealed. A blue color is obtained. In a similar manner, light-fast blue, green, brown, reddish violet, and yellow can be obtained. The process is said to be more economical than the integral color processes now in use in the architectural field in Japan. ALCAN has taken a world licence under the name of the Anolok process.

Special Anodizing Processes

Anodizing in Molten Salts

According to the recent publication by Ginsberg, Neunzig, and Sautter,[42] a German patent was issued to VAW (H. Ginsberg[211]) in 1938, for anodizing in molten salt; the process was named Pyroxal but found no practical use.

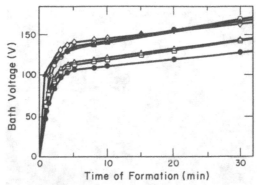

Fig. 17. Voltage–time curves at a constant current of 1 A/dm² in bisulfate melts. ×) NH₄HSO₄; □) KHSO₄; ○) NaHSO₄–NH₄HSO₄ 1:1 mole; △) KHSO₄–NaHSO₄ 2:1 mole; ●) NH₄HSO₄–KHSO₄ 1:1 mole; ▲) NaHSO₄. (After Tajima et al.[59])

Schaaber[212] in Germany in 1949 patented a process of anodizing Al (and its alloys) in a nitrate–nitrite mixture, producing a transparent, glass-clear coating by applying dc, interrupted current, or reverse current. One percent to 50% alkali metal chromate can be added to produce a light-yellow-to-deep-brown coating, and a 0.1% to 2.5% citric acid addition will produce a black coating.

Tajima, Soda, and Mori[59] anodized Al in the low-temperature melts of bisulfates, such as NaHSO₄, KHSO₄, and NH₄HSO₄. The single salt and the mixtures melt at between 100–160°C, and anodizing is done, in these melts, at 1 A/dm² to form a porcelainlike corundum (α alumina) film. The film can be made as thick as 100 μ or more. Voltage–time curves are shown in Fig. 17. Sparking and luminescence is evidenced during anodizing. The oxide films consist mainly of α alumina, which has been converted from the initial amorphous alumina, through γ, to α alumina. This is a typical example of electrochemical oxidation and electrothermic transition at the localized area of the aluminum anode surface. Although the film is porous due to sparking, it is very hard and dramatically resistant to tangential abrasion; it is not as resistant to normal shock (for example, to sand blasting). The film is very stable and attacked by neither alkali nor acid, nor even by fluoride ion. Hydrofluoric acid penetrates the film through the pores to attack the substrate aluminum, and the film is separated.

The film can retain oils and be impregnated with lacquers, etc., to improve corrosion resistance and gloss. The corundum coating is finding

Fig. 18. Heat-resistant, ceramic insulated CI coil made of α-alumina film anodically formed in bisulfate melt. (After Tatsuta Electric Wire and Cable Co., Japan.[213])

applications in Japan in heat-resistant wire for magnetic coils (500°C) and CI coils (ceramic insulated coils),[213] and business machine parts where extreme wear resistance is required, such as drying rolls and winding rolls (Fig. 18). An electrolytic capacitor was made with this corundum coating as dielectric.[214]

Anodizing in a Nonaqueous Solvent System

Tajima and Baba obtained a thick, colored film on Al by anodic oxidation in boric acid–formamide solution (270 g boric acid in 1 liter formamide at 2–4 A/dm², room temperature to 60°C). The film is pore free but can be dyed. No sealing treatment is required (dipping in boiling water does not increase the weight) and the resulting film is extremely corrosion resistant and hard. The film formed at 0.5 A/dm² is soft and coarse and that formed at 2–4 A/dm² is very hard, compact, and smooth and consists of Al_2O_3 and B_2O_3. The film gives integral color, and the color is related to thickness as follows:

$$0–10\ \mu \quad \text{light gold}$$
$$10–20\ \mu \quad \text{gold}$$
$$20–40\ \mu \quad \text{light brown}$$
$$40–60\ \mu \quad \text{brown}$$
$$>60\ \mu \quad \text{dark brown}$$

It is interesting that aqueous borate solutions or nonaqueous borate-polyalcohol solutions give only a thin barrier layer, while nonaqueous solutions give a thick, colored, noncrazing film. (The mechanism of formation is discussed later in the paper.) The film is particularly smooth and lustrous, and it is finding applications in precision machine parts.

Anodizing in molten salt systems and in nonaqueous systems is a new and promising field both from scientific and practical points of view. In the author's opinion, molten salt systems will usually give a corundum coating in any of the oxidizing baths and nonaqueous systems have the possibility of giving oxide films of various properties, depending on the combination of electrolyte and solvent.

Kape[215] recently published a paper on anodizing in an oxalic acid-formamide–water system; some of the films obtainable in this system are claimed to be promising as an integral color process for architectural use.

Continuous and High-Current Anodizing

Continuous anodizing of wire and strip is of rather recent origin. This process permits the use of a higher current density (20–30 A/dm²) due to the continuous movement and cooling of the anode wire or strip. The coating must be sufficiently flexible to be capable of further mechanical working. The electrolyte commonly used is warm sulfuric acid containing magnesium sulfate or chloride, with dc or ac. Oxalic acid may also be used with dc or ac; a better electric insulation is secured by oxalic acid coating. After continuous anodizing, the wire or strip is sealed by hot water or lacquer.

One-side anodizing of strip is done while an electrical contact is made on the other side of the strip, or by using the bipolar electrode principle.

Continuous anodizing is described by Nickelson,[216] Kape,[217] Csokán (with a patent and literature list prior to 1962),[218] Johnston,[219] Church,[220] Hermann,[221] Richaud,[222] and Barkman,[223] who used the multicathode system to distribute the current evenly.

Church[220] and Sautter[62] illustrate the Lloyd process of producing 4 μ thick films in warm sulfuric acid at 11 A/dm².

Litchtenberger-Bajza and Jagandha-Raju[225] reported rapid anodizing of aluminum in oxalic–formic acid at constant voltage. This was effective for Al and its alloys, except for those containing Cu.

Sealing

The sealing process is the final chemical operation in anodizing. Sealing plugs the reactive and adsorptive cylindrical pores formed in the oxide layer and increases the resistance to corrosion, staining, and to damage in handling. During sealing, the oxide film is converted to AlOOH or Al_2O_3 $\cdot H_2O$ (boehmite). This reaction occurs above 80°C; below this temperature $Al_2O_3 \cdot 3H_2O$ (bayerite) is formed, degrading the quality of the oxide film. The reactions are as follows:

$$Al_2O_3 + H_2O \rightarrow 2AlOOH \rightarrow Al_2O_3 \cdot H_2O \quad \text{(boehmite above 80°C)}$$

$$2AlOOH + 2H_2O \rightarrow Al_2O_3 \cdot 3H_2O \quad \text{(bayerite below 80°C)}$$

Sealing is often accompanied by adsorption of inhibitors, such as acetate, bichromate, or molybdate, which may be added in small amounts to the sealing solution. These inhibitors improve the corrosion resistance of the film and the fastness of the dye adsorbed by reacting with the $Al-Al_2O_3$ system or with the dye. However, in the presence of phosphate or silicate in the sealing water, the uptake of H_2O by the oxide film is inhibited and the weight gain of the film is smaller. Phosphate is said to disturb the hydration reaction and to be harmful to sealing, but this explanation is subject to some dispute.[344]

Typical sealing operations are described by Spooner,[226] Sacchi and Paolini,[227,232] Richaud,[228] Neunzig and Röhrig,[229] Cooke and Spooner,[230] Elze,[231] Darnault,[233] Birtel,[234] Pake,[403] and Barkman[235] and listed in Table 12. The mechanism of sealing will be discussed later.

MECHANISM OF ANODIC OXIDATION

Stability and Corrosion of Aluminum (Pourbaix Diagram)

Before entering into the mechanism of anodic oxidation, the fundamental properties of aluminum oxide will be related to the Pourbaix diagram.[236,237]

Figure 19a and 19b shows the theoretical domains of corrosion, immunity, and passivation of aluminum at 25°C, in the absence of substances with which aluminum forms soluble complexes or insoluble salts. Figure 19a refers to passivation by the formation of a layer of hydrargilite, $Al_2O_3 \cdot 3H_2O$. Figure 19b refers to passivation by the formation

Table 12. Typical Sealing Processes

Composition	pH	Temperature	Time	Notes
Water, preferably distilled or demineralized	5.5 (adjust with acetic acid or caustic soda)	not less than 98°C	as anodizing cycle	very popular; some dyestuffs may bleed
Saturated steam	—	100°C	15–45 min	very efficient and clean, avoids bleeding of dyestuffs, but relatively expensive
70-psi steam (in autoclave)	—	100–160°C	30–60 min	popular in Japan; particularly effective for oxalic acid film
5.5 g Ni(CH₃COO)₂ 1.0 g Co(CH₃COO)₂·4H₂O 8.5 g H₃BO₃ 1 liter distilled or demineralized water	5.6–5.8	98°C	3–30 min.	satisfactory for dyed films
50 g Na₂Cr₂O₇ 1 liter water	6–8	90–100°C	10–30 min	produces yellow discoloration; widely employed for service and aircraft specifications
Lacquers, various types (directly or by electrophoretic deposition)	—	air dried or stoved	—	useful for temporary protection

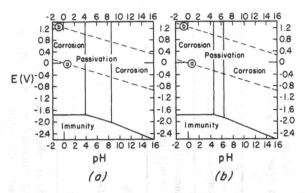

Fig. 19. Theoretical domains of corrosion, immunity, and passivation of aluminum at 25°C. a) Passivation by a film of hydrargillite, $Al_2O_3 \cdot 3H_2O$. b) Passivation by a film of boehmite, $Al_2O_3 \cdot H_2O$ (or AlOOH). (After Deltombe and Pourbaix.[236,237])

of a layer of boehmite, $Al_2O_3 \cdot H_2O$. In noncomplexing solutions of pH roughly between 4 and 9, aluminum tends to become covered with a film of oxide.

Aluminum oxide (or alumina, Al_2O_3) occurs in various forms, viz. corundum or α alumina (rhombohedral), β alumina (hexagonal), γ alumina (cubic), δ alumina (rhombohedral), and amorphous alumina. The physical and chemical properties of alumina depend to a large extent on the temperature reached during its formation; thus, when heated to a high temperature, alumina loses its hygroscopic property and at the same time becomes practically insoluble in acids and bases (α Al_2O_3).

Aluminum hydroxide gels, produced by adding acid or alkali to solutions containing Al, are not stable. They crystallize in the course of time to give first the monohydrate γ $Al_2O_3 \cdot H_2O$ or boehmite (rhombohedral), then the trihydrate $Al_2O_3 \cdot 3H_2O$ or bayertie (monoclinic), and, finally, another trihydrate, hydrargillite, crystallizing in the same system. This development of aluminum hydroxide is known as "ageing."

The corrosion resistance of aluminum is determined, essentially, by the behavior of its surface oxide film. Practically neutral solutions are, in general, without effect, except in the particular cases when there is danger of pitting due to the presence of halide ions. In acid or alkaline solutions, aluminum is attacked as soon as the oxide film is removed. The reaction is slower in acid solutions than in alkaline solutions. For a detailed description of the modifications of alumina and the various hydroxides the Symposium Proceedings edited by Karsulin[404] is very useful.

Classification of Anodic Films on Metal

The anodic films which are produced on metals can be classified, simply, as follows:

Noncontinuous Films. These films have various pores or defects and have low electrical resistance during film formation. They grow under an anodic potential near the equilibrium potential of the metal. Usually, they are crystalline ($AgCl$, $PbCl_2$), precipitated from the reaction of ionized metal and electrolyte. Sometimes the films are produced independent of the anode metal. MnO_2 deposition on a stainless-steel anode from manganese sulfate electrolyte is a typical example. Very often these films are semiconducting.

Continuous Film. These films may be either electronic conductors or insulators: (A) Passive films on Ni, Cr, stainless steel, etc.; these films are usually semiconducting and cannot be made thick. (B) Anodic films on valve metals (Al, Ta, Nb, etc.); these films are continuous. In these metals very thin films of 50 Å or so cover the metal surface completely and are usually amorphous oxide. The ionic conduction under high field is the sole mechanism of growth. These films can be made as thick as 1 μ, which requires a very high anode overpotential of several hundred volts. Sometimes they show rectification (resistant to anodic current but not to cathodic current)[407,417] and electroluminescence.[417,418]

Duplex Film (Continuous and Porous Film). When a continuous film is gradually transformed into a noncontinuous film, it takes a duplex structure. A typical example is the thick anodic coating on aluminum which is the main subject of this review. Also, electropolished surfaces of aluminum belong to this type.

Chemical Composition of Anodic Oxide Films

According to Setoh and Miyata,[44] Plumb,[239] Tajima and Shimura,[66] and others, the loss of aluminum from the metal substrate during anodic oxidation is strictly faradaic and the anodic reaction proceeds in the following way:

$$Al \rightarrow Al^{3+}, \quad 2Al^{3+} + 3O^{2-} \rightarrow Al_2O_3$$

Thus, it is reasonable to assume that Al_2O_3 is produced first and further dissolves if there is any solvent action of the electrolyte. Therefore, the role of the anion is important in the dissolution process of Al_2O_3. At high overpotentials in chloride-containing solutions and where the Cl^-/Cl_2 equilibrium potential is exceeded, the discharge of Cl^- predominates over that of oxygen, and aluminum never forms Al_2O_3 but is dissolved into the solution. In general, it is the usual case that in the anodic oxide film the anions, their polymers, or other derivatives are usually incorporated.

The term "coating ratio" (C.R.), that is, $Al_2O_3/2Al$, was introduced by Mason and Slunder[67] as a measure of the overall anode efficiency with respect to coating efficiency. If all the aluminum ionized was converted to Al_2O_3 without loss into the electrolyte, the C.R. value will be 1.889. They reported that for many of the duplex film-forming electrolytes, such as oxalic, sulfuric, and oxalic + sulfuric acids, the C.R. ranges from 1.35 to 1.46. It is known that lower temperature, lower concentration of the electrolyte, and higher current density favor the C.R. value.

The coating ratio can be determined by weighing the original aluminum, the film produced, and the aluminum weight loss. All these measurements can be made by using the "stripping solution," invented by Mason, of 35 ml of 85% phosphoric acid and 20 g CrO_3 per liter at about 95°C, which dissolves the anodic coating only; the substrate is, practically, not attacked.

The coating ratio obtainable with oxalic acid is, according to Mason and Slunder,[67] only a little lower than that obtainable with sulfuric acid, but the anode current efficiency based on aluminum removed is only 92.5%. This indicates that 7.5% of the current must have been consumed in some electrochemical reaction, possibly oxidation of oxalic acid.

Mason and Fowle[405] state that with a sulfuric acid coating containing 14% SO_3, the postulated theoretical coating would be 2.20 and not 1.89 (Al_2O_3 only) when the efficiency is 100%. In practice, this is not reached because the volume of the coating represented by the pores will dissolve, as well as some of the outer surfaces.

The coating ratio of the film formed in bisulfate melts is nearly 1.76–1.81,[59] indicating their barrier nature, and that of the film formed in nonaqueous systems of boric acid–formamide is unusually high, 2.3 to 4.0, due to the presence of B_2O_3 or organic compounds according to the electrolytic conditions.[60]

The main constituent, Al_2O_3, in the anodic oxide film, exists mostly in the amorphous form, and in the case of barrier films, for example, the borate film, some amount of γ Al_2O_3 and γ' Al_2O_3 (γ Al_2O_3 of low crystallinity) also exists.

According to Verwey,[240,241] O^{2-} ions in γ' Al_2O_3 are arranged in a face-centered cubic lattice, and the smaller Al^{3+} ions are distributed statistically in the interstices between the O^{2-} ions, with about 70% of the Al^{3+} ions having a coordination number of six, and 30% having a coordination number of four. There are some unoccupied lattice positions.

Taylor, Tucker, and Edwards[242] reported that Al_2O_3 crystallization is promoted if formed in a dilute solution at higher temperature and voltage (200 V). Tajima and Soda[59] found that the anodic films formed in bisulfate melts (120–180°C) consist mainly of extremely hard and adherent α Al_2O_3, which can be made as thick as 100 μ or more.

It is known that barrier films, as well as duplex films, contain water when formed from aqueous solutions. Norden,[27] as mentioned previously, early in 1899, found the water content to be 14.8% in sulfuric acid films, which is equal to the values obtained later by Jenny[2] and other researchers. Güntherschulze[1] found 2% H_2O in boric acid films formed at 750 V. Pullen[243] reported that in chromic acid films, no water is included, but in sulfuric acid and oxalic acid films, the composition is equivalent to monohydrate. Baumann[46] noted 5–7% water in sulfuric acid and oxalic acid films formed at 2 A/dm² and 15–20°C. Edwards and Keller[244] found 1–6% H_2O in sulfuric acid films.

The structure and composition of various films formed on Al are summarized in Table 13 according to Domony and Lichtenberger-Bajza[245]; the barrier film thickness was measured by the Hunter–Fowle method[246] of determining the voltage at which the current suddenly increases and calculating the thickness based on 13.7 Å/V.

Ginsberg and Wefers[247] noted that in anodic oxide films water exists in a molecular state, being adsorbed on the active alumina as in the case of a zeolyte. By differential thermal analysis, the bond was found to be fairly strong, but in the case of a sealed film, water is fixed as OH radicals. Karsulin and Lahodny[404] have studied the dehydration of alumina hydrate.

It is well known that the newly formed anodic film is very reactive, adsorptive, and hygroscopic. Thus, sealing with steam or boiling water is a widely practiced process in commercial anodizing. However, once sealed, dried, or left in normal atmosphere for a long time, the film loses the reactivity.

Dorsey[248] showed that the infrared absorption of the barrier layer distinguishes it from the porous layer even when the two layers are present together within the same film (duplex film). Experiments were performed for boric, chromic, sulfuric, oxalic, malonic, succinic, tartaric, fumaric, glutaric, and adipic acids and the correlation chart given in Table 14 was obtained. Table 15 shows the values found by Dorsey for the highest absorption frequency ranges for each type of bond.

Table 13. Thickness of Barrier Films and of Total Films Produced under Various Conditions. (After Domony and Lichtenberger-Bajza[245])

Medium	Temperature, °C	Barrier thickness, μ	Total thickness, μ	Structure and composition of protective films
In dry air atmosphere	20	0.0010–0.0020	0.0010–0.0020	amorphous Al_2O_3
In dry oxygen atmosphere	20	0.0010–0.0020	0.0010–0.0020	amorphous Al_2O_3
In dry air atmosphere	500	0.0020–0.0040	0.04–0.06	amorphous $Al_2O_3 + \eta\ Al_2O_3$
In dry oxygen atmosphere	500	0.0100–0.0160	0.03–0.05	amorphous $Al_2O_3 + \eta\ Al_2O_3$
In humid atmosphere	20	0.0004–0.0010	0.05–0.10	boehmite and hydrargillite
In humid atmosphere	300	0.0008–0.0010	0.10–0.20	????
Boiling water	100	0.0002–0.0015	0.50–2.00	boehmite
High-temperature and -pressure water	150	ca. 0.0010	1.00–5.0 according to purity of Al	boehmite
Chemical oxidation (conversion)	70–100	0.0002–0.0008	1.0–5.0	boehmite + anions of the solution (viz., CrO_3, PO_4)
Chemical polishing	50–100	ca. 0.0005	0.01–0.1	boehmite + anions of the solution
Anodic oxidation	18–25	0.0100–0.0150	5.0–30.0	amorphous Al_2O_3 + anions of the solution
Hard anodizing	+6 to −3	0.0150–0.0300	150–200	amorphous Al_2O_3 + anions of the solution
Electropolishing	50–60	0.0050–0.0100	0.1–0.2	unknown Al_2O_3 + anions of the solution
Anodic barrier formation	50–100	0.0300–0.0400	1.0–3.0	crystalline and amorphous Al_2O_3 + anions of the solution

Table 14. Infrared Data for Anodic Aluminas. (After Dorsey[248])

Electrolytes and anodizing parameters	Infrared absorptions (wavenumbers)			
	AlO↔H stretch	Al⇔O stretch	Al↔OH bend	Al↔OAl stretch
A. 2.2M orthoboric acid				
100°C, 250 V, unsealed	none	none	955 (M,s)	none
100°C, 450 V, unsealed	none	none	955 (M,s)	none
(Orthoboric acid films were examined over the 4000–33 cm^{-1} region, using IR-7 and IR-11 instruments)				
B. 0.3M chromium (VI) oxide				
41°C, 0.216 A/dm^3, 1.7 μ, unsealed	3360 (S,b)	none	1100 (W,b) 1060 (S,s) 925 (S,b)	670 (S,b)
41°C, 0.216 A/dm^3, 1.7 μ, sealed	3360 (S,b)	none	1100 (W,b) 1060 (S,s) 925 (S,b)	670 (S,b)
C. 3.1M othophosphoric acid				
25°C, 0.756 A/dm^2, 2.0 μ, unsealed	3000 (W,b)	none	1150 (S,s)* 1140 (S,sh,s)* 908 (M,b)	700 (S,b)
25°C, 0.756 A/dm^2, 2.0 μ, sealed	3000 (M,b)	none	1150 (S,b)* 908 (M,b)	700 (S,b)

Table 14. (*Continued*)

Electrolytes and anodizing parameters	Infrared absorptions (wavenumbers)			
	AlO↔H stretch	Al⇔O stretch	Al↔OH bend	Al↔OAl stretch
D. 1.5M sulfuric acid				
25°C, 1.30 A/dm², 1.8 μ, unsealed	3400 (M,b)	1325 (M,b)	1150 (S,b)[†] 1090 (S,sh,s) 900 (M,b)	650 (M,b) 610 (M,s)
25°C, 1.30 A/dm², 1.8 μ, sealed	3450 (S,s) 3420 (S,s) 3390 (M,sh,s)	1325 (M,b)	1125 (S,b)[†] 1090 (S,sh,s) 900 (M,b)	650 (M,b)
E. 0.6M oxalic acid				
3°C, 1.08 A/dm², 1.3 μ, unsealed	none	1480 (S,s)	940 (S,s)	610 (S,b)
10°C, 1.08 A/dm², 2.8 μ, unsealed	none	1470 (M,s)	935 (S,s)	740 (W,b) 650 (W,b)
F. 0.6M malonic acid				
68°C, 1.08 A/dm², 2.5 μ, unsealed	none	1470 (M,s)	1130 (M,s) 960 (S,s)	610 (S,b)
90°C, 1.08 A/dm², 1.7 μ, unsealed	3400 (W,b)	none	933 (S,s)	675 (W,b)
G. 0.6M succinic acid				
50°C, 1.08 A/dm², thickness not measurable, unsealed	none	1470 (W,b)	960 (S,s)	755 (W,sh,b)
85°C, 1.08 A/dm², thickness not measurable, unsealed	none	none	965 (S,s)	755 (W,sh,b)

H. 0.6M tartaric acid				
25°C, 1.08 A/dm², thickness not measurable, unsealed	none	1450 (W,b)	955 (M,s)	610 (S,b)
75°C, 1.08 A/dm², 2.8 μ, unsealed	none	1475 (M,s)	970 (S,s)	610 (S,b)
I. 0.6M fumaric acid				
10°C, 1.08 A/dm², thickness not measurable, unsealed	none	1468 (M,s)	965 (S,s)	610 (S,b)
100°C, 1.08 A/dm², thickness not measurable, unsealed	none	none	960 (S,s)	725 (W,b)
J. 0.6M glutaric acid				
98°C, 1.08 A/dm², thickness not measurable, unsealed	none	none	957 (M,s)	none
K. 0.6M adipic acid				
95°C, 400 V, thickness not measurable and C.D. not constant, unsealed	none	none	970 (S,s)	none

The following notations are used to describe the intensity of the infrared bands: S, strong; M, moderate; W, weak; b, broad; s, sharp; sh, shoulder on another band.

* This band may be due to aluminum phosphate. The 2.0 μ film thickness contained 6%, by weight, PO_4^{3-}.

† This band may be due to aluminum sulfate. The 1.8 μ film thickness contained 13%, by weight, SO_4^{2-}.

Table 15

Alumina system, chemical bond	IR absorption, wave number range	Interpretation
AlO ↔ H (stretch)	3660–2940 cm⁻¹	presence of water either free or adsorbed, or combined as hydroxide
Al ⇔ O (stretch)	1696–1345 cm⁻¹	presence of double bond, indicating an adsorptive or reactive material that probably does not have much crosslinking
Al ↔ OH (bend)	1162–900 cm⁻¹	presence of aluminum hydroxides; monohydrates are characterized by bands near 1070 cm⁻¹, while normal trihydrates show bands below 1025 cm⁻¹
Al ↔ OAl (stretch)	below 900 cm⁻¹	high degree of crosslinking in the material; strong bands in this region are often associated with an absence of Al=O bonds; as the degree of crosslinking, i.e., molecular weight, increases, the bond absorption frequency will shift to lower values

Using these interpretations Dorsey analyzed his data. The most common absorption observed is in 1000–900 cm⁻¹ range indicating the existence of a barrier layer. In sulfuric, oxalic, malonic, succinic, tartaric, and fumaric acids, the adsorptive and reactive Al ⇔ O bond was noted, which disappeared by sealing treatment. The barrier-layer-type oxide was produced in boric acid.

Dorsey compared the anodic films formed in H_2SO_4 and D_2SO_4 solutions by infrared absorption analysis. A comparison of the infrared spectrum of the protonated material with that of the deuterated compounds reveals those bands that involve structural hydrogen; these will have shifted to lower frequencies as a result of the isotopic mass effect. The deuteration of the alumina lowered the absorption frequency of the 900–1000 wavenumber barrier-layer band and also reduced the amount of both the porous layer Al ⇔ O and barrier layer. He concluded that the barrier layer is a trihydrate. The primary phase of the barrier layer, as obtained initially or by boric acid anodizing, may be a cyclic aluminic acid trihydrate. A sec-

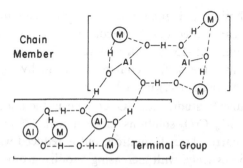

Fig. 20. Primary barrier consisting of cyclic aluminic trihydrate with divalent aluminum. (After Dorsey.[248])

ondary barrier phase may be a decyclized form of the primary, having more terminal groupings and a lower polymer weight. A proposed structure for the primary anodic barrier layer is shown in Fig. 20, where Al is present both in divalent and trivalent states, which is in conformity with the established fact of oxygen deficiency of the anodic barrier layer. A second type of bonding Al ↔ O forms as anodizing progresses further to produce the porous layer. It was noted that the IR absorption frequency of the barrier-layer bond increased as this occurred. This means that the barrier layer underwent a change from high to low weights as anodizing proceeded to the porous-layer stage.

The other constituents come from the electrolyte. Norden[27] published a paper only a few years after electrolytic aluminum became available in commercial quantities, stating that dilute sulfuric acid ($d = 1.050$, 20–30°C) anodic coating contained about 13% SO_3 (Al_2O_3, 69.8%; SO_3, 13.2%; H_2O, 14.8%; SiO_2, 2%). The film was separated by current reversal. Liechti and Treadwell[249] separated the film from the Al substrate by Werner's bromine–ethanol solution, and by chemical analysis and ignition loss, and noted that in sulfuric acid film there exists 1–2% tightly bound water, 3–5% loosely bound water, and 8–13.5% SO_3. In oxalic acid film there was 5–6% water and 3–3.3% oxalate ion. The latter coincides with the results of Pullen.[243] Generally, the SO_3 content of sulfuric acid film increases with current density and decreases with increasing temperature of the solution. According to Mason,[250] a general formula is $Al_2O_3 \cdot x H_2O$ ($x = 0.18–0.26$). The SO_3 content is reduced to 2% by immersion in boiling water, by electrodialysis, or by precipitation as $BaSO_4$.

Lichtenberger-Bajza, Domony, and Csokán[251] concluded from x-ray and DTA analysis that the hard coating from dilute sulfuric acid consisted of bayerite, boehmite, and anhydrous aluminum sulfate. Domony and Lichtenberger-Bajza[245] noted that hard coatings and normal coatings from sulfuric acid contain 13.5–15.2% SO_3 and 12.8–13.7% SO_3, respectively. It should be noted that the sulfate content is unusually higher than other anions (oxalite, chromate, etc.) in each case.

Brace and Baker[252] studied the SO_3 content in the anodic films formed in H_2SO_4 and H_2SO_4–CrO_3 solutions labelled with S^{35} and found that it fluctuates during barrier layer formation, but is reduced with the initiation of pores and, progressively, with increasing anodizing time. Provided there are two adjacent vacancies in the lattice, a sulfate ion will fit readily into the oxide lattice, since the distance between the O atoms in the Al_2O_3 is similar to that between the O atoms in the SO_4^{2-} ion. In addition, SO_4^{2-} is symmetrical and can fit with ease whichever pair of O atoms reaches the surface first. Since an oxide film at the oxide–electrolyte interface can be most easily considered as a two-dimensional lattice, a reduction by one half of the number of available ions for incorporation into it, makes it probable that the incorporation of oxalate will be $(\frac{1}{2})^2 = \frac{1}{4}$ of that of the sulfate ion. The fact that oxalate contents of anodic coatings are about 3% compared with about 15% for SO_4, supports this reasoning.

$$
\begin{array}{ccc}
O^- & & O \\
\diagup & & \diagdown\!\!\!\!\diagdown \\
C\!\!=\!\!\!\!\diagdown & \!\!\!\!-C & \\
& & \diagdown \\
O & & O^-
\end{array}
$$

Raub, Kawase, and von Krusenstjern[253] measured, by S^{35}, the SO_3 content in sulfuric acid films; it ranged from 10–17%. Higher current combined with higher voltage and lower temperature increases the SO_3 content.

Lewis and Plumb[238] indicated that in the barrier film produced in dilute sulfuric acid solution marked with S^{35}, Al_2O_3 is not in a stoichiometric proportion to Al^{3+}, which is in excess (oxygen deficiency). McNeil and Guss[254] identified 46% phosphate in the barrier layer formed in $1N$ NaH_2PO_4 marked with P^{32} and confirmed that the anion is concentrated at the anode, where dehydration or crystallization occurs, leading to oxide formation. In the case of an insulating barrier layer, like that of Al, deposition of anions occurs. On the other hand, when the anode dissolves, the deposition of anions is completely supressed. In addition, in electrolytes which tend to produce anodic films polymerization of anions occurs rather easily at the anode.

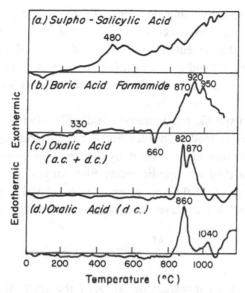

Fig. 21. DTA curves of various oxide films.
The reaction is exothermic at points above the
x axis in each case, and endothermic below it.
(After Tajima et al.[255,256])

Tajima and coworkers[255,256] investigated the structure of the typical
integral color films. Integral color films formed in sulfosalicylic acid
($+H_2SO_4$), in the nonaqueous system of boric acid–formamide and in
oxalic acid, were subjected to differential thermal analysis. The curves
obtained are shown in Fig. 21.

The exothermic maxima existing between 800 and 900°C are due to the
transition of alumina from γ Al_2O_3 to η Al_2O_3 as shown by Ginsberg and
Wefers.[247] The curve (a) for the anodic film formed in sulfosalicyclic acid
($+H_2SO_4$) shows, additionally, an exothermic reaction at about 480°C due
to the decomposition of organic components derived from sulfosalicylic
acid. The heat of combustion of salicylic acid is greater than that of oxalic
acid or formamide. Curve (b) was obtained from an anodic film heavily
formed in the nonaqueous system of boric acid–formamide. The composition of the film was 80 wt. % Al_2O_3 + 20 wt. % B_2O_3. It is strongly colored
(yellow through brown to black, depending upon film thickness), was
crackfree and required no sealing treatment for practical use. It was thought
that the small endothermic maximum at 660°C on the curve was possibly
due to the melting of excess Al or boron oxide.

Anodic films have been formed in oxalic acid using dc with superimposed ac, and also by using dc alone. There were no appreciable differences in the other curves and curve (d), obtained from differential thermal analysis, although the current form used (ac, dc, ac + dc, interrupted dc, or other current forms) had a remarkable influence on the color of the film as reported by Tajima and coworker[63,189] and referred to earlier in this paper.

The change in specific resistance of anodic films (with barrier layers of a nonideal nature) formed in duplex-film-forming electrolytes was measured at the initial stage of formation by Tajima and coworkers.[255,256]

The thickness d of the anodic oxide film formed by anodizing in a duplex-film-forming electrolyte is given, approximately, by the current through the "pure" barrier layer and is an exponential function of field strength:

$$d = \frac{M}{6F\varrho} (i - i_e - i_R)t \qquad (4)$$

where M is the molecular weight of Al_2O_3, i the total current density in A/cm^2, i_e the electronic (leakage) current density, i_R the current density equivalent to the rate of dissolution of the oxide film, t the time in sec, F the Faraday constant, and ϱ the density of the film. The ohmic drop across the film is given by

$$V = rdi \qquad (5)$$

where r is the specific resistance of the oxide film.

By combining Eqs. (4) and (5), and differentiating with respect to time, the following equation is obtained:

$$\left(\frac{\partial V}{\partial t}\right)_i = \frac{M_i}{6F\varrho} r(i - i_e - i_R) \qquad (6)$$

Neglecting both i_e and i_R, which are small compared with i, the specific resistance is given by

$$r = \left(\frac{\partial V}{\partial t}\right)_i \cdot \frac{6F\varrho}{M} \cdot \frac{1}{i^2} = 1.93 \times 10^4 \left(\frac{\partial V}{\partial t}\right)_i \cdot \frac{1}{i^2} \qquad (7)$$

where the density of the film is assumed to be independent of the nature of the film and time of anodizing, and of value $3.4 \ g/cm^3$. The specific resistance thus can be evaluated using Eq. (7) by measuring the voltage–time transients at constant current by a transistorized galvanostat and using an electromagnetic oscillograph with an automatic photographic recorder.

Figure 22 shows the relationship between barrier film thickness and

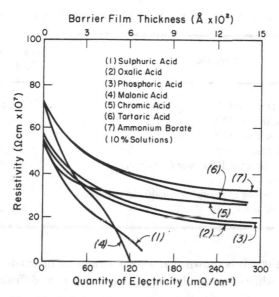

Fig. 22. Relationship between differential specific resistance and barrier film thickness at the initial stage of anodizing in various solutions at 25°C. (After Tajima et al.[255,256])

differential specific resistance in various solutions at 25°C. The current density used was 0.015 A/cm² and the barrier layer thickness was calculated from the total current passed through the electrolyte. In each case, the specific resistance of the film was initially high at about 6×10^8 $\Omega \cdot$cm (specific resistance of pure Al_2O_3 is reported to be 10^9–10^{11} $\Omega \cdot$cm). The value of the film decreased with increasing thickness and this decrease was greatest in films formed in sulfuric acid and malonic acid. However, if organic compounds, such as acridine or thiourea (an effective inhibitor for aluminum corrosion with a positively charged group), were added to the anodizing solution, then within experimental error the specific resistance was unaffected by thickness. The initial value of the specific resistance depended on the current density: the higher the current density, the lower the value. At the higher current densities, there was no change in specific resistance with increasing film thickness.

Values of i_R, the current density equivalent to the dissolution rate of oxide formation, were determined from the current–time transients obtained by anodizing at a constant voltage, where $(\partial V/\partial t) = 0$ and $i = i_R$. The temperature dependence of i_R values led to the calculation of the activation energy of the dissolution of the film. The results are shown in Table 16.

Table 16. Activation Energy of the Film Dissolution. (After Tajima et al.[255,256])

Electrolyte	Activation energy, kcal/mole
Sulfuric acid 10%	14.6
Oxalic acid 10%	14.3
Phosphoric acid 10%	19.3
Chromic acid 10%	15.4
Sulfosalicylic acid 120 g/liter ($+H_2SO_4$)	9.47
Malonic acid	8.21
Electropolishing in phosphoric–sulfuric–chromic acid solution	4.87

From these values, it is shown that anodic oxidation proceeds under activation control, while electropolishing occurs under a diffusion-controlled process.

Infrared spectroscopy was used to analyze anodic films and also the electrolytes in which the films were formed. The results are shown in Fig. 23. Curve A is the spectrogram obtained for sulfosalicylic acid anodically oxidized with smooth platinum and curve B that for the oxide film formed in sulfosalicylic acid ($+H_2SO_4$). Both curves show the presence of phenol and the $-SO_2$ and $-SO_3$ radicals. Curves C and D are, respectively, the spectrograms for the boric acid–formamide electrolyte anodically oxidized with platinum and the anodic film formed in it. Incorporation of the anions $-CONH_2$ and $-CN$ is clearly demonstrated. Similarly, curves E, F, and G, for the films formed in conventional surfuric acid and in oxalic acid using dc and ac, show that anions are incorporated into the anodic film. For the films formed in oxalic acid, there is not so much difference in the spectrogram for the film formed with dc (F) and for the film formed with ac (G), although there was a marked difference in color shades and mechanical properties.

Tajima and Mori[257,258] measured by the Becke-line method the refractive indices of various anodic oxide films and of chemical conversion films (chromate–phosphate and chromate–carbonate) on 99.99% Al, separated from Al by iodine–methanol or mercurous chloride solutions. The results are shown in Fig. 24. Alpha-alumina films can be formed on aluminum by anodizing for short times in bisulfate melts; it is even possible to form them in aqueous ammonium borate solutions by anodizing for rather longer times (15 h) at sparking voltages. These extreme conditions cause

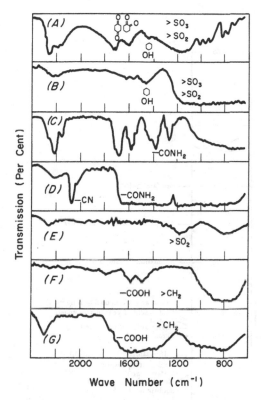

Fig. 23. Infrared spectroscopic analysis of various oxide films and their electrolytes. (After Tajima *et al.*[255,256])

the formation of α alumina by combination of electrochemical oxidation and electrothermic transition reactions at localized areas on the electrode surface. The results show the variation in the value of the refractive index of anodic films formed in different electrolytes, and the marked influence of the anion upon the refractive index.

Normally, the anodic oxide films formed in aqueous solutions show lower refractive indices than those of pure amorphous Al_2O_3 or of α Al_2O_3. In addition, hard coatings formed at low temperatures (0 to 10°C) and those films formed by sealing treatments generally give lower refractive index values than normal or as-anodized films. Sulfuric acid hard coatings show local double refraction which is believed to be due to γ $Al(OH)_3$, converted from the amorphous state by the high stress normally existing in thick hard coating.[258] Barrett[261] made a precise determination of the refractive index of thin barrier film (2000 Å) from ammonium borate solution,

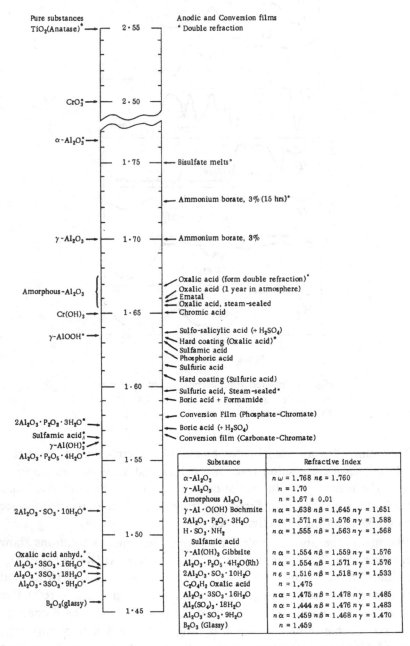

Fig. 24. Refractive indices of pure substances compared with those of various anodic oxide films and chemical conversion films. (After Tajima *et al.*[256-258,63])

by ellipsometry and by the Becke-line method. The values coincide with 1.665 ± 0.05. Ginsberg and Wefers[247] made similar measurements on selected films.

Barrier Film

Barrier oxide films are produced from electrolytes without solvent action, such as boric acid, ammonium borate, ammonium tartarate, ammonium citrate, very dilute sulfuric acid, and other very dilute oxyacids.

The bottom layer of duplex films on Al is also a barrier layer with a thickness of about 100–150 Å or more, which is linearly dependent on the anodizing voltage and is responsible for the anodic potential drop of the film. The barrier layer is compact and insulating, and little electronic conduction is expected except for a very thin layer of less than 20 Å through which quantum-mechanical tunneling is possible. Therefore, ionic conduction is the predominant means of electric conduction across the barrier layer, in spite of the high dimensional resistance which requires a high electric field across the layer. The electric conduction across the barrier layer is called the high-field ionic conduction, the phenomenon peculiar to the so-called "valve metals."

Ionic conduction in barrier layer films is due to metal cations, oxygen ions, other ions (impurities, alloying components), protons, or their combination. There is a charge-transfer and at the same time a mass-transfer process at the metal–oxide and oxide–electrolyte interface leading to chemical reaction. The substrate is consumed and the anodic film grows.

Whether or not aluminum, during film formation, moves outward through the film, or oxygen moves inward, or both, is and has been[262] a problem of active discussion. The reaction site depends on which of the three possibilities takes place; the oxide film could be growing at the oxide–solution interface, at the metal–oxide interface, or somewhere between the two. In the past, it was commonly believed that oxygen moved across the film and the oxidation reaction took place at the metal–oxide interface. (Rummel,[45] Schenk,[3] Anderson,[263] and Hass and Bradford.[264])

To assess these possibilities Schenk considered the atomic and ionic volumes of Al and O. The volume of Al^{3+} is about 1/23 of Al, while O^{2-} will be greater by nearly the same ratio than atomic O. Atomic diameters for these ions are: O^{2-}, 1.4 Å; Al^{3+}, 0.5 Å; Ta^{5+}, 0.7 Å. The discharge potential of O^{2-} is only about 1.23 V. Therefore, oxygen ions first discharge at the solution–oxide interface and the neutral oxygen atoms diffuse into the already formed Al_2O_3 layer, and near the oxide–metal interface are again ionized to react with Al^{3+}. The volume of the ionized Al^{3+} is reduced to accept two O^{2-} ions in the space next to the Al^{3+}, but there is no room for the third oxygen ion. The original face-centered cubic lattice of Al will be deformed, thus leading to the formation of a loose γ Al_2O_3 (γ' Al_2O_3):

$$2Al_{13}^{2+8+3} + 3O_8^{2+8} \rightarrow 2Al_{13}^{2+8} + 3O_8^{2+8} + 6e$$

Hass and Bradford[264] have attempted to determine the site at which oxide growth takes place. They first preevaporated TiO_2 on the surface and then anodized it in a tartaric acid solution. New alumina formed at the Al–TiO_2 interface can only be explained by oxygen migration.

Davis and coworkers[265,266] recently found, by radioactive tracer methods, that, in the current density range 0.1–10 mA/cm^2, the cation transport fraction was 0.6 and independent of current density for a non-aqueous glycol electrolyte and increased with current density from 0.4 to 0.7 for an aqueous electrolyte. Transmission electron microscopy has been employed by Doherty and Davis[267] to determine the kinetic mechanisms of the thermal oxidation of Al. These studies showed that above $500°C$, "crystalline" oxide forms at the oxide–metal interface by anion transport through the existing amorphous oxide film. Evidence was presented indicating that the thermal "amorphous" film grows largely by cation transport controlled at the oxide–metal interface.

Whitton,[268] using a radiotracer technique, concluded that oxygen migration along lattice defects is predominant in anodically formed tantalum and zirconium oxides (n^+, 0.28 ± 0.03 in tantala). Francis[269] used transmission and replica electron microscopy of the oxide in the vicinity of vacancy pits on Al to investigate the growth mechanisms of anodized amorphous oxide films on (100) and (110) planes of 99.992% single crystals.

The basic experimental technique used by Francis is illustrated schematically in Fig. 25. The initial oxide film (a) was either the as-electropolished oxide or a thicker film grown either thermally in air or anodically. When the crystal was heated to an elevated temperature (ranging from 400 to $575°C$) and cooled (b), micron-sized pits bounded by low-index crystallographic planes nucleated and grew under the oxide in the metal by the condensation of lattice vacancies. Oxide was then grown anodically (c) under various controlled voltage conditions in 1% ammonium citrate and 1% citric acid in distilled water. At the edge of a vacancy pit, a step is produced at the oxide–metal interface whose height is equal to the thickness of oxide grown by anion transport, and a step of height equal to the thickness of oxide grown by cation transport is produced at the oxide–electrolyte

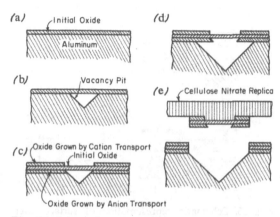

Fig. 25. Schematic representation (not to scale) of experimental technique. a) Initial surface. b) After first heating. c) After anodizing. d) After second heating and cooling. e) Removal of oxide fragment. (After Francis.[269])

interface. Then the crystal was heated a second time, and, on cooling, a fraction of the vacancy pits grew to a larger size (d). An adherent cellulose nitrate replica was used then (e) to remove the oxide over the pits, and these oxide fragments were viewed in normal transmission in the electron microscope. The topography of the oxide–metal interface was observed by shadowing the fragments. The topography of the oxide–electrolyte interface was observed either by shadowing the surface of the crystal prior to application of the replica or by using conventional techniques to replicate the surface before the second heating without removing any oxide.

Another experimental technique used by Francis for studying films is shown in Fig. 26. Both an initial high-temperature oxidation and a subsequent anodic treatment are used. On the oxide produced by high-temperature oxidation crystallites are nucleated at the oxide–metal interface and these then grow into Al as shown in Fig. 26a. At room temperature the anodic oxide will grow only if the field strength exceeds 10^7 V/cm. Thus, if a crystal with this initial oxide–metal interface topography is subsequently anodized by slowly increasing the voltage, growth will not occur at the crystallite sites until the oxide thickness becomes uniform (Fig. 26b). Growth by cation transport will produce depressions over the crystallite sites at the interface between the electrolyte and the final surface. Growth by anion transport will tend to "swallow up" the crystallites.

By combining these two methods, Francis concluded that both cation and anion transport takes place simultaneously. There is no influence by

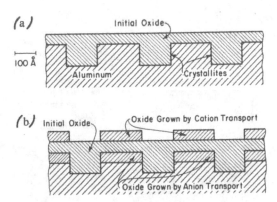

Fig. 26. Schematic representation of surface after nucleation of crystalline oxide. a) Before anodization. b) After anodization. (After Francis.[269])

the surface orientation on the transport, but it is possible that the nature of the electrolyte could influence the structure of the oxide–electrolyte interface or the structure of the cationically grown oxide since its growth occurs at the oxide–electrolyte interface. If a porous layer exists, anion transport might be limited by oxygen diffusion through the pores to the surface of the barrier layer.

Despite the above observations most of the recent research indicates that cations, usually smaller than oxygen ion, are the mobile species. Vermilyea (on tantalum),[270] Dekker and van Geel,[271] Lewis and Plumb,[238] Amsel and Samuel,[272] Bernard,[273] Berry, Kennedy, and Waggener,[274] and Tajima and coworkers (in molten salts)[283] support the oxide–electrolyte-interface reaction.

When we consider the more detailed mechanism of the formation of an oxide film, there are two basic probable transport mechanisms shown schematically by Amsel and Samuel in Fig. 27.[272]

Interstitial Transport. Oxygen or metallic atoms diffuse through the lattice in interstitial positions, the atoms in the formed layer remaining at rest.

Substitutional Transport. Either oxygen or metal jumps from one substitutional position to another, which includes "vacancy diffusion" and "exchange interstitial diffusion" in which a moving atom hits an atom at rest and takes its place in a substitutional position, the process being repeated by the atom initially at rest. Thus, at any moment, there is an atom in excess in an interstitial position, but it is a different atom at each jump.

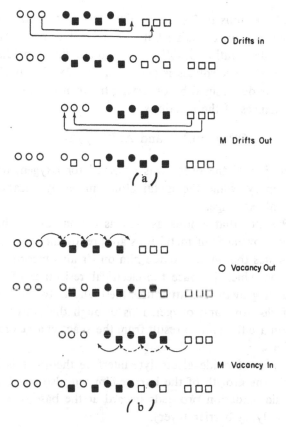

Fig. 27. Symbolic representation of the basic ion-transport models. a) Direct movement. b) Propagation. ○) Oxygen; ●) labeled oxygen; □) metal; ■) labeled metal. (After Amsel and Samuel.[272])

McMullen and Pryor[275] showed the presence of 8×10^{14} vacancies/cm^2 having an activation energy of 0.17 ± 0.04 eV in γ-Al$_2$O$_3$–boric acid film.

Since the mechanism of ion transport is quite controversial, a variety of novel methods have been employed to discern the detailed process. One method involves using two kinds of anodic layers produced by successively anodizing in two different electrolytes or conditions. The one electrolyte is, for example, labeled by S^{35} or O^{18}. Which of the films is upper or lower in the barrier-layer system is then determined by checking the intensity or concentration of the labeled atom or by comparing certain chemical properties of the two layers.

Amsel and Samuel[272] used the tracer (O^{18}) method to determine the order of atoms in Al$_2$O$_3$ and Ta$_2$O$_5$. Films were obtained by anodic oxidation at 40 V of the foils in 3% ammonium tartarate containing 90 at. % labeled O^{18}. The foils were oxidized in normal solution (0.204 at. % O^{18}) at 240 V dc. The distribution of O^{18} in the film anodically oxidized in both enriched O^{18} and O^{16} electrolytes is shown in Fig. 28. The position of the labelled atoms was determined by observing the forms of excitation curves near sharp resonances of the reactions

$$O^{18}\,(p,\,\gamma)\,N^{15} \text{ and } Al^{27}(p,\,\gamma)\,Si^{28}$$

Perfect conservation of the order was observed for oxygen, its sublattice remaining stationary, while the metal atoms move by vacancy diffusion and by interstitial exchange.

Bernard[273] states that anions as well as cations are mobile through the film but the movement of metal ions are predominant.

Banter[276] states that anodic oxide film on Zr also incorporates anions from the solution which increase the electrical resistance of the films as well as limit their growth. He attributes these effects to the lowered rates of diffusion of electrons and oxygen ions through the films. The lowered rates of diffusion are thought to result from the space charge created by the incorporated ions.

With the help of oxide–electrolyte interface theory it is possible to explain not only the growth of the barrier film but also the duplex film in which the anodic oxidation proceeds inward at the base of pores (at the top of the underlying barrier layer).

The foregoing studies tell nothing about the possible role played by protons or hydroxyl ions. Hoar and Mott[277] concluded that the movement of the oxygen ion is difficult because of its larger volume. However, smaller OH^- ions would migrate more readily through the oxide to the metal–oxide interface where they decompose into O^{2-} and H^+ (bare proton), and mobile protons come back, toward the cathode, to the oxide–electrolyte

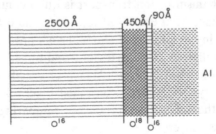

Fig. 28. Final structure of the oxygen distribution in the film. (After Amsel and Samuel.[272])

interface to combine with O^{2-} ion in the electrolyte to form again OH^- ion as shown below (after Hoar and Mott[277]):

$$O^{2-} \rightarrow OH^- \rightarrow H^+ + O^{2-}$$

(electrolyte)

(oxide) $Al_2O_3 \leftarrow Al^{3+}$ (aluminum)

This mechanism is very interesting since the well known valve action of the aluminum oxide barrier layer may also be explained very simply by the proton migration through the (cathodized) anodic barrier oxide film. However, the Hoar–Mott theory is still controversial in view of subsequent studies which have demonstrated that both cationic and anionic ions are mobile and in view of the evidence that barrier films have been found to grow to significant thicknesses in nonaqueous systems where no OH^- ions exist.[59,60]

Vermilyea,[270] by measuring the stresses produced in anodic oxide films of various metals, postulated that a new layer formed during anodizing is hydrated, and is subsequently dehydrated by outward migration of protons.

Heine and Pryor[278] measured the resistivity of anodic film on aluminum by an ac bridge during film thinning in a chromate solution. They observed that the resistance of the oxide adjacent to the electrolyte was somewhat less at 1 kc/sec than that of the bulk of the oxide film, which they ascribed to the hydration of the former due to the inward migration of OH^- ions with the electric field. Substitution of OH^- ions in the oxide lattice produces a more n-type structure of lower resistance.

Brock and Wood[279] anodized aluminum in both a substantially nonaqueous 20% ammonium pentaborate–ethylene glycol electrolyte and a 3% ammonium tartarate aqueous electrolyte. The former film was essentially free from hydration and the ac resistance at 1 kc/sec was independent of the forming current density; the latter film was progressively more hydrated as the forming current density was lowered. No hydration in the nonaqueous electrolyte film indicates that the film growth was occurring mainly by aluminum and oxygen ion migration. In the film formed in aqueous solution hydration is evident and is most extensive in the outer-film regions as indicated by inward movement of OH^- ions. The duplex film formed partly in borate and partly in tartarate indicates that O^{2-} ions move inwards during borate anodizing, and both O^{2-} and OH^- ions during tartarate anodizing. Nothing definite could be said about cationic mobility.

The growth of barrier films proceeds linearly with the applied anode potential, and the growth rate is, according to Walkenhorst,[41] 13.7 Å/V.

This value is independent of the anodizing time and, to a limited extent, of the electrolyte concentration and temperature. It is generally valid for the barrier layer of the barrier-plus-porous-duplex films, although in this case the value is somewhat erratic according to the nature of the electrolytes. Values of 10 Å/V for 2% oxalic acid and 12.5 Å/V for 3% chromic acid are reported.

Until now we have considered the phenomenology of film growth processes. It is now appropriate to consider specific mechanistic aspects of atomic movements through the films. The ionic conduction model proposed early by Frenkel assumes that the mobile species are defects which may be either interstitials or vacancies. Our attention will be directed primarily to mobility of vacancies. A mobile ion in the interstitial position will move when it acquires enough energy by thermal fluctuation to pass over a potential energy hump to a neighboring site where it is immediatedly deactivated by the loss of energy due to collisions. The ion is still oscillating in this new site with an average energy of the order kT and at a frequency of about 10^{13} cps, until it acquires the activation energy much greater than kT needed to again exceed the potential energy barrier. The electric field interacts with the ion to change its potential energy, with the result that the ion moves in the direction of the field. In the case where the work done by the field F is much greater than kT, the movement against the field is negligible and the mean velocity will be $2a \exp[-W(F)/kT]$, where a is the distance from the potential energy minimum to the peak. The exponential term is the probability that an ideal linear harmonic oscillator has an energy $W(F)$; the dependence of $W(F)$ on F is usually written in the form $W_0 - qaF$, where W_0 is the activation energy for ion migration at zero field and q is the charge on the ion. Using the assumptions outlined above the ionic current under high electric field F is expressed as

$$I_i^+ = q^+ 2a^+ n^+ \nu + \exp\left[\frac{-(W_0^+ - qaF)}{kT}\right] \qquad (8)$$

$$I_i^- = q^- 2a^- n^- \nu - \exp\left[\frac{-(W_0^- - qaF)}{kT}\right] \qquad (9)$$

where $n^+ q^+ = n^- q^-$ (electroneutrality condition), $2a = 2.95$ Å, $W_0 = 1.3$ eV and $n^+ q^+ = 3e$.

The $\log I_i - F$ plot gives a linear, Tafel relationship. That is, when the field (applied voltage) is increased at a steady rate greater than that needed to maintain constant current density, a Tafel-type relationship exists between current density and voltage. This is one of the characteristics of

Table 17. Values for the Constants A_i and B_i

A_i, A/cm²	B_i, cm/V	Thickness measurements	Authority
3.6×10^{-23}	4.2×10^{-6}	capacity	Güntherschulze[1] and Betz
10^{-18}	3.0×10^{-6}	above	Charlesby[282]
	4.8×10^{-6}	weight	
10^{-23}–10^{-24}	5.4×10^{-6}	capacity	Bernard and Cook[298]
	4.4×10^{-6}	$(1/i)(dV/dt)$	
10^{-20}	4.3×10^{-6}	capacity	Nagase[295,296]
10^{-19}–10^{-20}	4.7×10^{-6}	capacity calibrated by ellipsometry	Videm[300]
4.5×10^{-20}	4.7×10^{-6}	ellipsometry	Barrett[261]

barrier films. These relationships are simply expressed as

$$I_i = A_i \exp(B_i F) \tag{10}$$

$$F = \frac{V}{d} \tag{11}$$

where I_i is the ionic current density in A/cm², d is the thickness of the barrier layer, V is the applied voltage across the layer, F is the field strength, A_i is the parameter 10^{-22} A/cm², and B_i is the parameter 4×10^{-6} cm/V.

A number of authors have evaluated Eq. (10) and their results are summarized in Table 17. Nagase[295,296] and Bernard and Cook[298] prepared barrier film of Al_2O_3 free of other anions and impurities by anodizing a 99.99% Al foil in a carefully selected electrolyte, i.e., 30 wt. % of ammonium pentaborate in ethylene glycol, up to 150 V, at 100% efficiency, and measured the values of parameters. Charlesby[282] and Van Rysselberghe et al.[299] obtained somewhat different values. Barrett[261] and Videm[300] of the Trondheim school gave the newest data by combining capacity measurements and ellipsometry (Table 17). From Eq. (1) the thickness of the oxide film is dependent upon the ionic current density I_i and applied voltage V:

$$d = \frac{B_i V}{\log I_i - \log A_i} \tag{12}$$

If $\log I_i$ is negligibly small as compared with $-\log A_i$ (this is actually the case) and the current I_i after film formation is not so much dependent upon V, then

$$d = KV \quad (K \text{ is a constant}) \tag{13}$$

$$\frac{d}{V} = K \tag{14}$$

that is, the film thickness is the function of applied voltage only. According to Walkenhorst,[61] $d/V = K = 14$ Å/V. Thus, the field strength is $F = V/d = 1/K = 7 \times 10^6$ V/cm. This value gives the maximum field strength for the barrier film on Al. If the field strength for the barrier layer exceeds this value, the film growth will proceed until the equation is satisfied.

Charlesby[282] showed that the electronic current (or leakage current), which makes no contribution to the growth of oxide film, is expressed by the equation

$$I_e = A_e \sinh B_e F \tag{15}$$

where A_e and B_e depend on temperature. He attributes this current to the impurities within the film which cause semiconduction.

At a constant voltage, the anode current decreases rapidly to a small, steady value, which is known as the electronic current (or leakage current). Usually high purity Al gives a typical barrier layer with the least electronic current. In the case of Al alloys, the current leakage through heterogeneous phases is not negligible and this limits the voltage rise.

The barrier film consists of amorphous $\gamma' \, Al_2O_3$ or $\gamma \, Al_2O_3$. The former differs from the latter by the order of cations in the spinel systems, $\gamma' \, Al_2O_3$ being in disorder and $\gamma \, Al_2O_3$ in order. The thinner film is mostly amorphous and the thicker one is in a crystalline state. Young[10] has summarized the literature in his excellent monograph.

According to Verwey[260] O^{2-} in $\gamma \, Al_2O_3$ are arranged in a cubic face-centered lattice and the smaller Al^{3+} are distributed statistically in the interstices between O^{2-} ions, with about 70% of Al^{3+} ions having a coordination number of 6, and 30% with a number of 4. There are some unoccupied lattice positions.

Dignam[284] showed that the barrier film thickness (observed as voltage) is different between as-anodized oxide film and annealed anodic film, the latter being thicker as shown in Fig. 29. When the anodic layer is annealed, its ionic conduction decreases. Dignam interprets the phenomenon, based on the theory of Bean et al.[285] and Dewald,[286] by the decrease of Al^{3+}

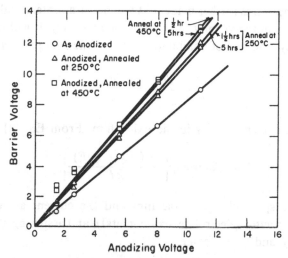

Fig. 29. Influence of annealing on the "barrier voltage"
values for anodic films. (After Dignam.[284])

ions in the amorphous film, which can become current carriers, rather than
by a change in activation energy or activation distance. Annealed anodic
films and amorphous oxide films formed in dry oxygen could not be distinguished. For amorphous films formed or annealed at high temperatures, the
concentration of interstitial vacancies will be small due to the high recombination rate caused by thermal diffusion of the ions within the film.
For anodic barrier films formed in aqueous ammonium carbonate electrolyte, the structure of which is amorphous, Dignam calculated the effective
activation distance and compared the values.

It is assumed that during anodic oxidation, metal ions (cations) are
mobile species and the rate-determining step is the passage of ions through
the films. In the bulk of the film, every lattice ion is a potential source of
an ion–vacancy pair, the rate of production of ion–vacancy pairs being
given by

$$\frac{dn^+}{dt} = \frac{dm^+}{dt} = (N_0 - m)\nu \, \exp\left[\frac{-(W - q\lambda F)}{kT}\right] \tag{16}$$

where n is the carrier concentration, m is the vacancy concentration, N_0 is
the concentration of lattice sites, W is the activation energy for the creation
of an ion–vacancy pair, λ is the equilibrium distance to the position of
the maximum in the potential energy diagram of an ion around a lattice
site, q is the charge of the interstitial ion, and F is the electric field in
the film.

The assumption is further made that whenever an interstitial ion passes within an area σ in the vicinity of an immobile vacancy, it will be captured by that vacancy; the rate of annihilation of ion–vacancy pairs is, therefore, given by

$$\frac{dn^-}{dt} = \frac{dm^-}{dt} = \frac{i'\sigma m}{q} \tag{17}$$

where i' is the current density in the oxide film. From Eq. (8) we have

$$i' = q2an\nu \, \exp\left[\frac{-(W_0 - qaF)}{kT}\right] \tag{18}$$

Neglecting space charge in the film and the current associated with charging the capacitor (electrolyte–oxide–metal), and assuming $n = m < N_0$, Eqs. (16), (17), and (18) become

$$\frac{dn}{dt} = N_0\nu \, \exp\left[\frac{-(W_0 - q\lambda F)}{kT}\right] - \frac{i\sigma n}{q} \tag{19}$$

and

$$i = q2an\nu \, \exp\left[\frac{-(W_0 - qaF)}{kT}\right] \tag{20}$$

where i is the external, observed current density.

The solution of Eqs. (4) and (5) for steady-state conditions (i.e., $dn/dt = 0$), can be written as follows:

$$q(\lambda + a) = \frac{kT}{2F} \ln\left[\frac{i^2(\sigma/a)}{q^2\nu^2 2N_0}\right] + \frac{W + W_0}{2F} \tag{21}$$

During the formation of the anodic films under discussion, steady-state conditions prevailed, permitting the use of Eq. (21). The samples were anodized at a constant current, about $i = 7 \times 10^{-5}$ A/cm². By comparing the voltage to which samples are anodized with the corresponding barrier voltage (Fig. 29) one obtains $1/e = d/V = 0.83 \times 12.7 = 10.5$ Å/V.

Since the activation energy for anodic oxidation of aluminum determined by Charlesby,[232] 1.6–1.8 eV, is the same for high-temperature oxidation, the value $(W + W_0)/2 = 1.76$ eV may be used here. Setting $N_0 = 4.27 \times 10^{22}$, the concentration of aluminum ions in Al_2O_3 of a density of 3.6 g/cm², $\sigma/a = 10^{-7}$ cm (an estimate), $\nu = 10^{12}$, and $kT = 0.02$ eV, one obtains

$$\frac{q(\lambda + a)}{2} = 10.5 \text{ e}Å \tag{22}$$

While the solution of Eqs. (19) and (20) for the case of a linearly increasing applied voltage (transient state during an anodic polarization measurement) is very complicated, a useful solution can be obtained for certain limiting conditions. For amorphous films either formed or annealed at high temperatures, the concentration of interstitial ions and vacancies will be small due to the high recombination rate caused by thermal diffusion of the ions within the film. During the early stages of a polarization measurement, the rate of annihiliation of ion–vacancy pairs may therefore be neglected by comparison with their rate of formation, giving

$$\frac{dn}{dt} = N_0 \exp\left[\frac{-(W - q\lambda F)}{kT}\right] \tag{23}$$

Finally Dignam obtained

$$\frac{q(\lambda + a)}{2} = 11.3 \text{ eÅ} \tag{24}$$

by substituting the available data including those of Dewald and Vermilyea. What he has attempted is to determine which of the parameters is primarily responsible for the decrease in conduction properties of anodic oxide film on annealing. The values of $q(\lambda + a)/2$ are nearly the same in both cases, which shows that the change in conduction properties by annealing can be ascribed to the preexponential factor, rather than to either the effective activation energy $(W + W_0)/2$ or the effective activation distance $(\lambda + a)/2$.

For films formed by anodizing at 7×10^{-5} A/cm², N_0 appears to be nearly equal to the total number of aluminum ions in the film, that is, about 4×10^{22} ions/cm². On the other hand, for annealed films, $N_0 = 10^{11}$ ions/cm³ [from $vv\,(2N_0a/\sigma)^{1/2} = 7.3 \times 10^{-2}$ cm·sec; v is the volume per metal in the oxide]. Annealing, therefore, appears to reduce greatly the number of lattice sites which are potential sources of ion–vacancy pairs. Thus, the number of current carriers is reduced.

Wilsdorf's model for the amorphous oxide shows that in a randomly oriented molecular array certain cations are expected to be in relatively high potential energy sites which should be excited into interstitial positions, thus, leading to cationic conductivity. On annealing, some ordering of the molecular group could take place, causing a reduction in the number of cations in high potential energy sites. It is probable that annealing the barrier films expels water incorporated in the film. However, further anodizing an annealed film recovers the conduction properties of a freshly anodized sample, implying a rehydration of the dehydrated film. Dignam considers that such a rapid recovery in a few minutes is difficult. The transport numbers for hydroxyl ion would be insignificant compared with that for aluminum ion, and protons would hardly diffuse against a potential gradient of about 10^7 V/cm. He assumes that the anodic barrier layers are essentially amorphous, differing only slightly from films formed in dry oxygen. Dignam has further refined the theory of conduction mechanisms and kinetic processes in high fields for anodic oxidation and dry oxidation of aluminum in a series of papers.[287–294]

For the electrolytic conduction and electrolytic rectification mechanisms in anodic oxide films, the recent papers by Schmidt,[407] Vermilyea,[408] Doremus,[409] Smith,[410] Lehovec,[411] Boddy,[412] Huber,[413] Hale,[414] Michelson,[415] and Dunn[416] at the Symposium in Dallas, 1967, are of considerable value.

The oxide film anodically formed from bisulfate melts consists mainly of $\alpha\ Al_2O_3$. Aluminum melts at 660°C, and transition from η to $\alpha\ Al_2O_3$ occurs above 1100°C. This is a typical example of the combination of electrochemical oxidation and electrothermic transition at a localized area of anode surface

$$2HSO_4^- \rightarrow H_2SO_4 + SO_3 + O + 2e$$
$$2Al^{3+} + 3O + 6e \rightarrow Al_2O_3$$

Thermodynamic calculations indicate the validity of this reaction.[283,63] Electron micrographs of films formed in the bisulfate melt are compared with the typical porous structure of the film formed in sulfuric acid in Figs. 30 and 31. The former shows the film from sulfuric acid and the latter shows that formed in bisulfate.

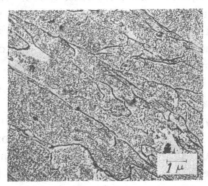

Fig. 30. Surface structure of as-anodized film; 10% H_2SO_4, 23–24°C, 1 A/dm², 18–20 V, 30 min. (After Tajima et al.[283,63])

Fig. 31. Surface structure of films formed in
sodium bisulfate melt; 190–198°C, 1 A/dm²,
4 min. (After Tajima *et al.*[283,63])

Figure 31 shows the topology of α-alumina film formed in bisulfate
melts. The craters formed are produced by sparking. Numerous small
craters are healed in the course of electrolysis while a smaller number of
new larger craters are formed. This continuous healing may be explained
by assuming that the oxidation reaction occurs at the oxide–electrolyte
interface. Figure 32 shows the Al–oxide interface without any intermediate
layer. The hexagonal crystals of α-alumina grow directly on the cubic lattice
of Al. Therefore, the film can be said to be the barrier type despite the fact
that the film can be made thicker than 100 μ. This film includes many macro-
pores produced by sparking which results from the electric breakdown
during anodizing.

Huber's well known work[303–306] and that of Lacombe *et al.*[302] show
that, though the anodic oxide films formed in aqueous electrolyte with

Fig. 32. Cross section (cutting angle, 45°) of the oxide film and substrate aluminum.
Sodium bisulfate melt; 186–188°C, 1 A/dm², 160–198 V, 30 min. (After Tajima *et al.*[283,63])

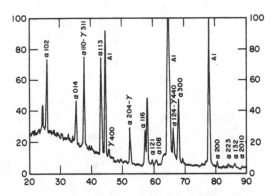

Fig. 33. X-ray diffraction chart of anodic oxide film formed in $NaHSO_4$–NH_4HSO_4 melt (1:1 mole); 120°C, 1 A/dm², 60 min. (After Tajima *et al.*[59,63])

solvent action are optically isotropic, amorphous, or of fine-grain γ alumina, they show double refraction; the films are microporous and can be regarded as a mixed body consisting of parallel cylinders of an isotropic material penetrating through the whole thickness of another isotropic material. The phenomenon is called "form double refraction." The bisulfate-melt film shows, in spite of the porous nature, no form double refraction, as the pores produced by sparking are larger than the wavelength of the light applied and this result indicates that the film is of the barrier type (Fig. 33).[59]

The anodic oxide film formed in boric acid–formamide nonaqueous system can be made as thick as 100 μ, and the resulting film is generally of the barrier type.[59] At lower current density, 0.5 A/dm², the film is of organometallic nature, soft and coarse, while at higher current, 2–4 A/dm², it is compact, very hard, and integrally colored, as mentioned earlier. From thermodynamic calculation of the decomposition potential of form-amide from available data, and from other physical, optical, and chemical experiments, the following reaction mechanism was proposed:

$$HCONH_2 \rightarrow HCONH^- + H^+$$

$$Al^{3+} + 3HCONH^- \rightarrow Al(HCONH)_3$$

$$2Al(HCONH)_3 \rightarrow Al_2O_3 + 3HCONH_2 + HCN$$

as the film contains about 20% B_2O_3, the direct reaction

$$Al^{3+} + BO_3^{3-} \rightarrow AlBO_3$$

is also probable.

Duplex Film

Anodic oxidation of Al at suitable conditions under a rather low and nearly constant anode potential in sulfuric, oxalic, or phosphoric acids usually produces the films with double structure which grow to $100\,\mu$ or more. Sulfuric acid requires 15–20 V (dc) while oxalic acid requires 70–80 V indicating more of a barrier (compact) nature to the film than with the sulfuric acid film. The main part of the duplex film is the porous region which is the remainder of the initial oxide film subjected to the solvent action of the electrolyte. Distinct duplex-type films are only to be seen in aluminum among the so-called "valve metals."

A paper by Setoh and Miyata[44,76] proposing the duplex structure first appeared in 1931. The summary of this historically important paper is reproduced here:

1. The active layer, composed mainly of Al_2O_3 and covering the metal anode as a very thin layer, is more or less porous. These pores come from the uniform destruction of the active layer all over the surface by the attack of the electrolyte anions, and are of different character than those such as caused by local electric breakdown of dielectric at the weakest points; oxygen gas in the nascent state, produced by the electric discharge in these pores, violently combines with aluminium and seems to be absorbed into the electrode, thus forming a new active layer which is supplied to the old active layer continuously. The porosity of the visible film, which is supposed to be derived from the active layer, is considered to be essentially inherent from the first instant of formation of the active layer itself.

2. During electrolysis at constant current density, there is a rapid rise of terminal voltage and thickening of the active layer. This initially rapid process does not depend significantly on the kind of electrolyte used. This is clearly shown by an elementary mathematical analysis of the voltage–time characteristics. The tendency for the film to thicken is opposed by its dissolution in the electrolyte; and the rate and mode of dissolution depends specifically on the electrolyte. Thus, it may be concluded that the field-controlled growth kinetics are independent of the electrolyte unless some impurities are introduced which change the electrical properties of the film.

3. Oxygen gas is occluded in the pores of the active layer and its presence may be confirmed by rapid expansion when the terminal voltage is suddenly varied. The presence of this occluded oxygen prevents direct contact between the base metal and electrolyte, and prevents the usual anodic dissolution of aluminum. This isolation of the anode from the direct wetting of the electrolyte is of great importance for the anodic behavior, especially in oxalic acid.

4. The gas evolved from the anode consists mainly of oxygen which contains small amounts of CO_2 and H_2. CO_2 is produced by the oxidation of oxalic acid at the anode and H_2 by the dissolution of aluminum. The hydrogen cannot be oxidized because it remains isolated in the structure of the visible film already oxidized. These results give evidence of pure oxidation which takes place almost exclusively at the anode.

Thus, for the formation of an anodic film on aluminum, we ought to reject the theories based upon the assumption that bulky colloidal aluminum hydroxide precipitates at the anode and hardens there by aging, or burns off into a hard and compact compound by electric sparks. It is thought most plausible that electrolytically produced oxygen evolves because of the preferable protection of the anode metal with the active layer against the attack of anions, and that it combines directly with aluminum through the pores, which is the result of the destruction of the active layer by anions uniformly all over the surface and permits easy passage of current. The active layer, although it has many pores filled with oxygen gas, plays an important role in isolating the anode from the electrolyte, and it not only guards the base metal from anodic corrosion but also facilitates the decomposition of water in solution into oxygen gas which combines with the anode aluminium and forms a new active layer in turn. The previous discussion indicates that 100% of the anodic current is used to form the surface film; however, the film dissolves without evolution of hydrogen so that the effective efficiency with respect to film growth is 75–80% at most. Our experiment in this paper shows that the anodic behavior of aluminum in various electrolytes depends wholly on the delivery of electrons reacting in a specific way with the anion in solution. The same proposition was suggested some time ago by Güntherschulze, whom we support here by our experiments. If the anion contains an element which is an electric conductor in itself, such as oxalic acid, chromic acid, etc., the active layer is liable to be destroyed with comparative ease and, consequently, the critical voltages become low and indefinite. The current-choking properties in these electrolytes are less. In short, it might be thought that an aqueous solution of oxalic acid produces the anodic behavior of aluminum, which is quite similar to that of other electrolytes having far higher and more definite critical voltages, and forms an anodic film having the composition Al_2O_3.

This active and porous-layer theory was later accepted by Rummel (Fig. 34)[45] who thought that the oxide film grows from the boundary into the metal, Baumann (Fig. 35),[46] Fischer and Kurz (first by electron micrograph),[301] by Lacombe and Beaujard[302] and Huber and coworkers[304–306] (measurement of form double refraction), and finished up with the well known cylindrical pore model of the Alcoa group—Keller, Hunter, and Robinson.[48]

Duplex films are partly hydrated and include anions of the electrolyte. The pore number calculated from Fig. 30 is about $174 \times 10^9/cm^2$. This

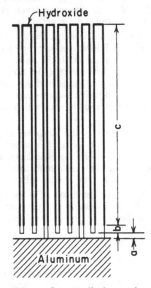

Fig. 34. Scheme for anodic formation of oxide films. a) Initial oxide film, partly broken. b) Oxide film produced from the initial oxide, which is not yet attacked by the electrolyte. c) Oxide–hydroxide film which has been penetrated by the electrolyte. (After Rummel.[45])

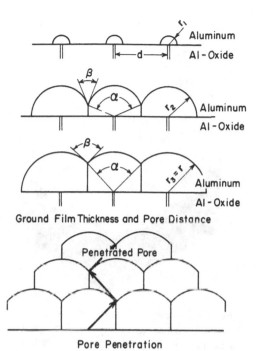

Fig. 35. Growth process of Al_2O_3 film. Relation between the initial film thickness r (radius of attack point) and pore distance d is given by $d = r_2$. r_1, r_2, r_3 are the radii of attack points at various stages of growth. (After Baumann.[46])

value is nearly identical with those calculated by Rummel (by optical microscopy),[44] Keller and coworkers (by electron micrograph),[48] and Burwell, Smudsky, and May (ethylene and nitrogen gas adsorption).[307]

According to the Alcoa model, the oxide film is composed of hexagonal columnar cells with a cylindrical pore of star-shaped section in each center. Beneath and in contact with the porous layer, is a barrier layer. Figure 36 shows a unit cell of a 120-V phosphoric acid coating visualized by the Alcoa group. They proposed the following general formulas:

$$R = \frac{H}{2} + \frac{W^2}{8H} \qquad (25)$$

$$C = 2WE + P \qquad (26)$$

$$V = \frac{78.5P^2}{C^2} \qquad (27)$$

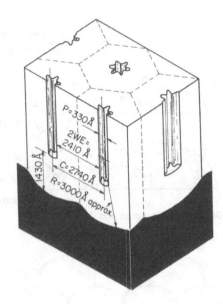

Fig. 36. Structure of 120-V phosphoric acid coating constructed on cross section of cell base pattern. Dimensions of pore, cell, cell wall, barrier, and radius of curvature are shown at 65,000×. (After Keller, Hunter, and Robinson.[48])

where R is the radius of curvature of the cell bottom (3000 Å for 120-V phosphoric acid coating), H is the height of the scallop, C is the cell size, W is the wall thickness (Å/V), E is the forming voltage, P is the pore diameter, and V is the pore volume. Some data derived from Eqs. (25), (26), and (27) are listed in Table 18.

Table 18. Pore Diameter and Wall Thickness of Typical Duplex Films. (After Keller, Hunter, and Robinson[48])

Electrolyte	Temperature, °F (°C)	Pore diameter P, Å	Wall thickness W, Å/V
15% sulfuric acid	50 (10)	120	8.0
2% oxalic acid	75 (24)	170	9.7
3% chromic acid	100 (38)	240	10.9
4% phosphoric acid	75 (24)	330	10.0

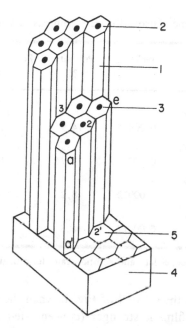

Fig. 37. Cross section of porous anodic oxide
film formed in $4N$ H_2SO_4. 1) Oxide cell. 2)
Pore. 3) Cell wall. 4) Aluminum. 5) Ground
cell pattern. (After Thomaschov, Tyukina, and
Zalivalov.[308])

Since the ALCOA model has been frequently described in the literature,
no further explanation will be given here.

Figure 37 shows the model of Zalivalov, Tyukina, and Tomashov.[308]

Booker and Wood[309],[310] were the first to take an electron micrograph
of the cross section of the duplex-type sulfuric acid film which showed clearly
the aggregate of cylindrical and vertically stretched cells.

Cosgrove[312] measured, by n-butane (at $0°C$) and krypton (at $-195.8°C$)
adsorption and by applying the classical BET equation, the surface area,
pore volume, and diameter of porous films formed from sulfuric, oxalic,
and chromic acid solutions. The data for n-butane are shown in Table 19.

Paolini and coworkers[313] performed comparative calculations of the
parameters of sulfuric acid films formed by dc and ac, using BET adsorption,
gravimetry, and the electron micrograph. They pointed out that a roughness
factor should be introduced in order to obtain an agreement between data
from different sources.

Grubitsch, Geymer, and Burik[311],[404] proposed the use of pore structure
for the filter (Eloxalfilter) for virus filtrations, isotope separations, etc.

Table 19. n-Butane Adsorption Experiments. (After Cosgrove[312])

Electrolysis		Thickness, in.	Porosity, %	Pore diameter, Å	Estimated,* Å
15% H_2SO_4;	70°F, 15 V, 30 min	0.00035	14.9	166	141
15% H_2SO_4;	70°F, 15 V, 60 min	0.00060	17.9	161	162
2% oxalic;	80°F, 30 V, 30 min	0.00012	8.2	203	200
3% chromic;	100°F, 40 V, 60 min	0.00017	16.3	290	270

* Based on electron-microscope data and solution rate determination.

As pointed out in the curve 4 of Fig. 1, when the solvent action of the electrolyte on anodic films is stronger (concentrated acids or alkali) than that of the normal electrolytes which produce duplex film, the aluminum anode is electropolished. In this case too, the aluminum surface takes a very thin duplex structure, the barrier layer being so thin as to permit the transmission electron micrograph. Lichtenberger and Hollo[167,419] showed clearly the hexagonal pore structure of a surface electropolished in a phosphoric acid–butyl alcohol electrolyte and in a perchloric acid electrolyte (pore size, 400–800 Å; cell size, 550–1300 Å). A typical structure is shown in Fig. 38. Raub and Baba[420] confirmed the same structure and Akahori[320] has taken a similar picture for a film anodized briefly in sulfuric acid.

As Kissin[17] pointed out, according to the ALCOA model, a key factor is the balance between the barrier layer formation and its dissolution at the base of the pores. At the steady-state condition, the anodizing voltage (and thus the potential drop across the barrier layer) must be constant. The ALCOA model assumes that the nature of the barrier layer and the pore wall is the same. Hunter and Fowle[314] deduced the following equation expressing the dissolution rate of barrier film by simple immersion in 15% sulfuric acid used as anodizing solution:

$$\log R = 0.0196T - 1.45 \qquad (27)$$

where R is the dissolution velocity in Å/min and T is the temperature in °F. R is 0.84 Å/min at 21°C. The value of R is also valid for the dissolution rate at the top surface of the porous oxide during anodizing since the

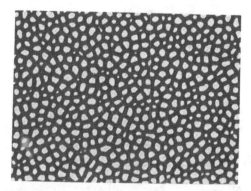

Fig. 38. Oxide film detached from a piece electropolished at 35 V in a phosphoric acid–butyl alcohol electrolyte. Direct transmission electron micrograph. (After Lichtenberger and Hollo.[167,419])

anodizing voltage is predominantly impressed upon only the barrier layer between the pore base and the aluminum substrate. On the other hand, the forming rate of the barrier layer is calculated to be 3725 Å/min at 12 A/ft² (1.29 A/dm²). In equilibrium, the dissolution rate v_s must be equal to 3725 Å/min. Thus, v_s: $R = 1:4000$. They concluded that in order for $v_s/R = 1$, in a 15% sulfuric acid at or near room temperature, the concentration of the acid at the pore base must be about 50% and the temperature about 125°C (nearly in a boiling state). However, it is well known that slight changes in temperature and concentration of the bulk electrolyte greatly affect parameters of film formation. It seems difficult to reconcile the results of Hunter and Fowle[314] with the purely physical cylindrical pore model.

Skulikidis and his school[315–319] studied in detail the formation mechanism of anodic oxide films from 15% sulfuric acid at various current densities and temperatures. According to them, a dense and compact γ Al_2O_3 film of up to 500 Å thickness, which has no affinity for dyes, is produced first. The mechanism of formation is of the Wagner type (Al^{3+} ion diffusion). Above this film is an oxygen gas layer which is a precursor to oxygen ions and which causes the "luminescence" phenomenon peculiar to Al. As a result of breakdown of the γ Al_2O_3 layer, oxidation proceeds at the metal-oxide interface. From the value of the activation energy (6.46 kcal/mole), the reaction may be controlled by the diffusion of sulfate ion through the pore. The film growth is linear with time [$Y(\mu) = kt$ (min)] and current density. This type of film (γ_1 Al_2O_3) grows up to 36 μ, where the formation of γ_2 Al_2O_3 begins. The formation mechanism of γ_2 Al_2O_3 is proportional

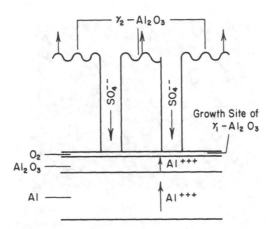

Fig. 39. Schematic model for Al_2O_3 growth. (After
Skulikidis, Paraskevopoulis, and Argyriou.[318])

to time but independent of the current density. The activation energy
(20.84 kcal/mole) relates to the diffusion of Al^{3+} through γ_1 Al_2O_3 lattices
outwards. Important is the fact that with the initiation of the formation
of γ_2 Al_2O_3 (total current 8530 C/dm²), regardless of the current density
and time applied, the luminescence phenomenon disappears. γ_1 Al_2O_3 and
γ_2 Al_2O_3 differ from each other in their capacity for uptake of dyestuff.
In the presence of γ_2 Al_2O_3, γ_1 Al_2O_3 cannot be dyed. From these experimental facts, Skulikidis proposed the model shown in Fig. 39.

Akahori[320] has taken various excellent electron micrographs of sulfuric
and oxalic acid films. According to him, the size of a unit cell is uniform
from the outside to the bottom; the funnel-like entrance of the pore with
a wide opening, as proposed by Setoh and Miyata[44] and Baumann,[46] could
not be observed. Figure 40 shows that linear unit cells and aluminum
matrix are separated by a spherical boundary which forms the barrier layer.

The insert in Fig. 40 shows clearly the light and thin arclike structure
which is the barrier layer. The thickness of the barrier layer is only 30–50 Å,
being much less than the calculated value of 800 Å based on 14 Å/V. With
the other observations, Akahori concluded that the zigzag pore model of
Baumann may explain the formation of the discontinuous growth of the
barrier layer, while the layer actually observed grows continuously over
the substrate, probably by the reaction of molten Al and oxygen. The thin
molten layer under the barrier layer leads to the uniform growth of the
layer irrespective of the crystal orientations of the substrate aluminum. The
barrier growth rate is proportional to the current density. The value 14 Å/V
is erroneous and the real thickness of the barrier layer is less than 100 Å,

Fig. 40. Cross section of oxalic acid film near the aluminum matrix. Between the parallel patterns are micropores. The dark layer consisting of circular arcs is the barrier layer. Left corner: barrier layer was sandwiched by carbon film. (After Akahori.[320])

with no relation to the forming voltage. The value 14 Å/V may be the one including the thickness of the gas layer.

Kaden[321,322] proposed the normal temperature theory based on the thermal balance calculations. He assumed that the total heat evolved (Q) is caused by electric energy and by the heat of formation of Al_2O_3 at the pore base, which amounts to a maximum of 0.15 cal/cm^2·sec when anodized at 4 A/dm^2 and 12.5 V. Q is balanced by the heat conduction from one side of the barrier to the other, where the specific heat conductance of the barrier (as corundum) $\eta = 5.6 \times 10^{-3}$ cal/cm·sec·°C, and the thickness $d = 140$ Å.

$$Q = \eta \frac{T}{d}, \qquad T = 4 \times 10^{-5} \,°C \qquad (27)$$

He concludes that the temperature at the film-growth site is nearly equal to that of the bulk of the electrolyte.

Barkman[224,324] presented a schematic concept of the preferential epitaxial growth, which is shown in Fig. 41. The orientation of the pores indicated by the slanted position has been verified by careful cross-sectional studies. This indicates that the epitaxial growth of the oxide depends on the grain orientation of the substrate aluminum. One of the pores is discon-

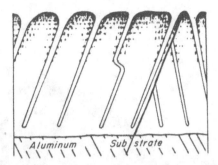

Fig. 41. A schematic concept of the structure
of the growth of anodic oxides on aluminum
showing epitaxial pore growth and dissolution
of the outermost portions of the film. (After
Barkman.[224,324])

tinuous or is changing its path or growth due to some obstacle, such as an insoluble impurities or an inoxidizable element like Si. The outermost part of the oxide layer loses its structural characteristics, being softer and subjected to chemical attack. This property favors dye adsorption.

A dynamically growing model, as pictured by Gorbev,[14] is shown in Fig. 42.

Ginsberg and Wefers[247] state that they agree in principle with the classical pore model, but that it must be modified. They refer not to the "pores through the oxide films," but to "the fiber structure of the oxide." They have taken many excellent electron micrographs of the oxide films by mechanically deforming them and have confirmed the existence of the fiber structure growing vertically on the cubic crystal of Al. Each fiber is tubular, the wall being x-ray amorphous and in disorder, and includes ion groups from the electrolyte. The wall becomes chemically active toward the inner surface. By water sealing, the film is converted to AlOOH which crystallizes in needles parallel to the fiber orientation. This process begins at the entrance of the tuber (top layer) and develops inward in the course of sealing.

Murphy and Michelson[49] have disputed the "too physical" cylindrical pore model and have proposed a new and highly controversial model based on a colloid chemical viewpoint. They pointed out that the evidence for the existence of "pores" has been based entirely upon experiments with dried oxide. These researches involve working with the oxide film under low pressures. They argue that the thin spots observed in transmission electron micrographs of separated films are the results of a less dense structure of the original oxide base which does not have in itself a hole or pore in the wet, as-anodized state. They state that this is reasonably

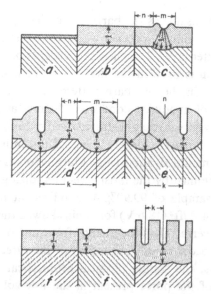

Fig. 42. Growth model of anodic oxide film.
a) Initiation of oxide formation. b) Barrier
layer thickness *l*. c) Initiation of a pore at *m*.
d) Growth model when the pore distance is 3*l*.
e) Growth model when $k = 2l$. f) Continuous
growth model of duplex film. (After Gorbev.[14])

well substantiated by the results of Franklin[325] and Stirland and Bicknell[326]
on the dissolution rate of oxide films. Their work indicated that even the
anodic film formed in boric acid (typical "uniform" barrier-type film) has
shown a cell structure and was not uniform in its solubility in acid solution.
They point out that the film subjected to the transmission electron micrograph are thin because of brief anodizing times or because the thicker films
have been thinned by dissolution. Thus, pores possibly do not exit originally
but may be artificially produced by the dissolution or shrinkage processes.

Diffusion studies by Burwell and May[327] have demonstrated that anodic
films provide a substantial barrier to the diffusion of ions, and this result
is consistent with the barrier layer proposed by Keller and coworkers,[48]
who state that conduction through this oxide occurs by ionic diffusion
through the electrolyte in the pores. The mechanism also requires continuous
dissolution of oxide at the pore base in order to propagate the pores. As
already described, Hunter and Fowle[314] concluded that the temperature of
the acid at the pore base must be about 125°C and the acid concentration
about 50% to maintain the "hole-drilling" activity. On the basis of the
straight cylindrical pore model, the dielectric material between the pores

must essentially be the same as the barrier layer. The porosity is apparently produced by a "hole-drilling" process which converts the dielectric barrier layer to a porous outer layer. The classical concepts of straight-sided pores, the location of which is fixed in the first few moments of anodizing, does not permit a change in the cell base pattern or the development of new pores by variations of anodizing conditions. Hunter and Fowle referred to the work of Renshaw,[323] who demonstrated the development of new colonies of pores and presumably the development of an entirely new pore base pattern by further modified anodizing. Murphy observed that the oxide base pattern is determined by the most recent anodizing conditions. He demonstrated that a sample of 99.99% Al anodized at 6 A/ft² (11.8 V) for 20 min and then at 24 A/ft² (18.6 V) for 2 min showed the oxide base pattern of 18.6 V, and vice versa. Thus the change from one oxide base pattern to another appears to occur readily regardless of the direction of change. To explain these results, the simple physical geometrical pore model would require that some of the pores present in the low-voltage coating would cease to function when the voltage is raised and, conversely, that new pores would have to be produced when the voltage was decreased from 18.6 V to 11.8 V. It is difficult to visualize the mechanism by which such events occur on the basis of the physical geometrical pore model.

The existence of boiling concentrated acid at the pore base as required to obtain the necessary dissolution rate should not lead to the development and propagation of a straight pore, but rather should cause enlargement of the pore base area. This, in turn, would lead, in contradiction with general observation, to an oxide of lower density and lower hardness near the metal than at the outside surface.

On the basis of the existing model, there is no mechanism to provide for the growth of oxide in the regions between the pores. The resistance of the pore walls would seem to be sufficiently great to prevent the migration of as much aluminum into them as migrates through the barrier layer at the pore base. Similarly, if the pore wall is of a quality similar to that in the barrier layer, as assumed, it is difficult to account for the required migration of the oxygen species into this region to form the oxide.

The Murphy–Michelson model (Fig. 43) is outlined as follows:

1. The anodic film is considered an agglomeration of submicrocrystalline particles of (probably) anhydrous Al_2O_3 surrounded by, and held one to another by, submicroscopic regions containing electrolyte anions, H_2O, OH^-, and H^+, in a relatively complex hydrogen-bonded system. The "internal surfaces" surrounding the submicrocrystallites are considered to be

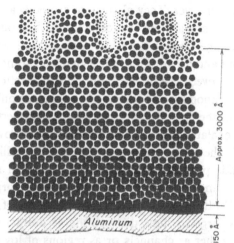

Fig. 43. Schematic representation of possible structure of "porous" anodic coating within about 3000 Å of aluminum surface. White areas between particles represent hydrous and hydrated oxide which provide for electrolytic conduction between electrolyte and barrier layer. Bottom of pore provides singularity in conduction required for radius of curvature of scallops in metal surface. (After Murphy and Michelson.[49])

the essential structural feature of the outer portion of anodic films formed in highly acid electrolytes. In this model the regions of lesser density, which show in electron microscopy as pores, may be explained as resulting from dissolution effects promoted by the convergence of current paths through the internal surfaces in the anodic film with resultant increased heating of the electrolyte along these paths. This heating increases the rate of conversion of anhydrous to hydrous oxide, and ultimately to solution of the oxide, forming a pore.

2. Barrier layer is formed at the oxygen-rich outer zone of the barrier layer by Al^{3+}-ion migration from the metal to react with O^{2-} or OH^- to form, essentially, anhydrous Al_2O_3.

3. Interpore material has been formed from the barrier layer by a process in which the barrier layer is selectively dissolved to produce particles of anhydrous materials separated by a hydrous, anion-containing region of high electrolytic conductivity.

4. Oxygen carriers (OH^-, H_2O) are brought to the barrier layer by an electrolytic conduction, through or along the internal surfaces of the porous

layer above the barrier layer. They also cause a modification of previously formed barrier layer into a more protonated and hydrous oxide, thereby tending to decrease the electric thickness of the layer, while protons are transported outward and dehydration of the oxide is promoted. The balance is thus established between the forward and reverse reactions.

The reason that anodizing of porous films proceeds only in certain acid solutions may lie in the special hydrogen bonding properties of such acid anions as sulfate, oxalate, and phosphate, which are known to prevent or decrease drastically the rate of rearrangement and crystallization of gels of aluminum hydroxide precipitates formed from chemical reagents. This is the reason why the anodic films in such acids are "amorphous" by x-ray or DTA analysis. On this basis, the submicrocrystallites of crystalline oxide are not sufficiently large to show distinguishable diffraction patterns.

5. The pores, either as channels or as regions of lower oxide density, are the result of dissolution effects from the convergence of current paths through the hydrous "mud" which fills the interstices of the anhydrous alumina "rock pile."

Criticisms raised against the Murphy–Michelson model[20] are summarized in the following:

1. Cross-sectional electron microscopic pictures and replicas of the top surface (not in vacuum) clearly confirm the existance of the cylindrical pores (R. C. Spooner).

2. Murphy attributes lack of crystallinity to the presence of sulfate ions. However, oxalic, phosphoric, and chromic acid films are amorphous, too, while their amounts of inclusion are known to be far smaller than 15% of sulfate (R. C. Spooner).

3. The hardness change in the film from inside toward outside, as measured on the cross section, can be explained well by the Murphy model but not by the Alcoa model (J. M. Kape).

4. Is it really true that the internal surface region must be dehydrated to become a part of the barrier layer by evacuation (Y. Yahalom)?

5. Can the valve action of the anodic film be explained by this model? (Murphy: this is a problem in the barrier oxide). Why does the chromic acid film reach a very high sealing weight gain as compared with the sulfuric acid film (A. W. Brace)?

6. From the measurement of sealing weight gain, there must be a considerable traffic of ions within the walls of the pores. The fact is better understandable by the new model than by the classical model with pore walls similar in nature to the barrier layer. The existence of pores can be

easily confirmed by dyeing, without evacuating and drying the film (F. Sacchi).

7. Densely dyed (more than grey) surface and a much more evenly distribution of coloring matter can be explained easily by the new model (G. H. Naylor).

8. Is the phenomenon of luminescence often seen in oxalic, sulfuric, and tartaric acid the problem of the barrier layer beyond the "mud"—internal surface (author)?

Murphy, in 1967,[50,51] further advanced his theory to construct the two possible ionic arrangements on the following basis:

Proton Space Charge. Under the driving force of high anode potential, anions and water penetrate into the coating, resulting in the accumulation of proton space charge, suggested by the work of Onada and De Bruyn on ion oxide

$$2Al^{3+} + 3OH^- \rightarrow Al_2O_3 + 3H^+$$

$$2Al^{3+} + 3H_2O \rightarrow Al_2O_3 + 6H^+$$

$$2Al^{3+} + 3HSO_4^- \rightarrow Al_2(SO_4)_3 + 3H^+$$

These protons are neutralized by field-assisted migration of protons out of the film and by field-assisted migration of anions (OH^-, HSO_4^-).

Colloid Chemical and Electrokinetic Considerations. The isoelectric point (zero point of charge, ZPC) of various aluminas is on the alkaline side (pH = 9) of neutrality. This means that a high concentration of OH^- is necessary to neutralize the inherent capacity of adsorbed H^+. Certain specifically adsorbed anions, such as sulfate and phosphate, cause a shift in the ZPC in the direction of more acid solutions. Since the two models are quite similar, only one is shown here (Fig. 44).

Murphy referred to Kissin's observation[12] that duplex films are produced only by dibasic or tribasic acids (sulfamic acid?). A knowledge of colloid chemistry tells us of the high degree of adsorption of divalent ions relative to monovalent ions. For example, the concentration of electrolyte necessary to cause flocculation of a positive colloid is ten times as high for a monovalent anion as for a divalent anion. In addition, for the anodic coating, there is evidence of strong specific adsorption of phosphate, oxalate, chromate, and sulfate. It is likely that the adsorption of the divalent anions by the anodic coating will influence the coating properties. This influence may be discussed by visualizing what would happen in the absence of strong electrolyte–anion adsorption. In this case, the hydroxyl ion would

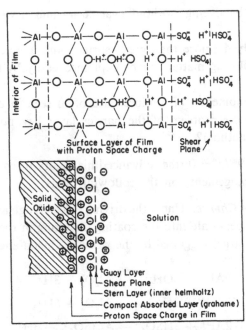

Fig. 44. One of the two possible ionic arrangements in the anodic film–electrolyte interface showing proton space charge in the film, balanced by anions in the compact adsorbed layer and in the solution. Top, chemical; bottom, schematic. (After Murphy[50,51])

adsorb to neutralize the contained and adsorbed protons, with the result that something very much like hydrated alumina with low foreign anion content would be produced. This is probably approximately what happens in the slightly basic or neutral solutions which produce barrier films of low anion content and low conductivity.

The high hydrogen concentration of porous-coating electrolytes materially changes the reactions of anodizing from those which occur in barrier-forming electrolytes. First, as previously noted, the hydrogen ion probably enters the oxide and may also form a compact adsorbed layer on the oxide surface. The high hydrogen-ion concentration in the solution also decreases the concentration of hydroxyl ion in solution and, to a lesser extent, the concentration of the free anion of the acid used (sulfate, phosphate, or oxalate).

At the pH of anodizing electrolytes, the hydroxyl ion concentration is decreased to such an extent that adsorption of the electrolyte–anion can compete favorably with the hydroxyl ion, which is normally much more strongly adsorbed. The presence of the anion in and on the oxide makes possible a higher concentration of adsorbed and included hydrogen ion, which in turn modifies the properties of the barrier film and porous coating material. The high "foreign" ion content of the barrier film formed in porous-coating electrolytes can account for its higher and nondecreasing conductance. The specific adsorption power of anions of alumina probably decreases in the order: phosphate, oxalate, sulfate. However, because of the differing dissociation constants of the acids, for the same hydrogen-ion concentration, the anion concentration increases approximately in the same order. Thus, qualitatively the anion adsorption can be in the same range for the three acids, which may account for the similarities. The observable difference in their anodizing behaviors can be attributed to differences in the degree and strength of adsorption, the properties of the adsorbed anions, and the concentrations and temperatures of the acids.

Murphy's novel approach, introduced here, is still in part speculative, but it is of great significance in that he has first applied the already established colloid chemical and electrokinetic double-layer structure information to the anodized boundary.

The specific resistance calculation of the initial anodic films (referred to earlier) using Eq. (7) has been further extended by Baba, Muramaki, and Tajima[421] for various organic acids (concentration, 1 mole/liter) and the results are summarized in Fig. 45, where the four curves, from upper to lower, indicate current densities of 0.5, 1.5, 3.0, and 6.0 A/dm^2, Δ indicates solubility of the acid below 1 mole/liter (saturated solution), and X indicates that no anodic films were formed (pitting type, e.g., pelargonic acid). It was demonstrated that the organic acids in which thick films were formed are polybasic acids, lactic, glycolic, salicylic, O-hydroxybenzoic, and acrylic acids and the Na salts of aromatic acids. The following conditions may be deduced, with some exceptions, to grow duplex films:

1. High specific conductance of the electrolyte.

2. At least two hydrogen atoms are indispensable which are easily ionized. Their dissociation constants, indicated in the figures, suggest that the primary dissociation of the acid is responsible for porous film formation without pitting.

3. Chlorine-containing acids give pitting due to the adsorption and replacement of oxygen by chloride, leading to dissolution of aluminum.

Hoar and Yahalom[328] studied the initial film formation in H_2SO_4, $H_2SO_4 + NaHSO_4$, $NaHSO_4$, and ammonium tartarate solutions of vary-

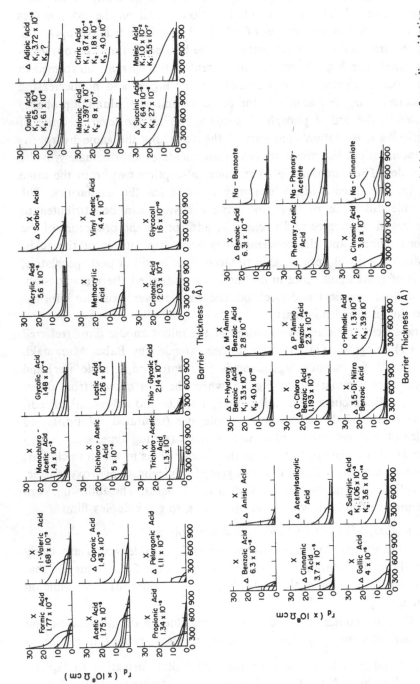

Fig. 45. Voltage–time transient (differential specific resistance) at constant current during the initial stage of anodic oxidation in aliphatic and aromatic acids. (After Baba, Tajima, and Muramaki.[421])

ing pH by measuring current-density transients at constant voltage, and showed in every case that the film in its earliest stage is of a compact barrier nature without pore nuclei. However, after the forming current has fallen to a particular value, dependent on the anodizing solution and, in particular, on its pH, it increases considerably, after passing through a maximum it finally reaches a steady value corresponding to the formation of the porous layer and the dissolution of Al^{3+} at the base of each pore. They interpret the rise in forming current density as caused by pore initiation (this corresponds to the critical voltage break by constant-current anodizing observed earlier) and the near hexagonal pattern is fully established by the time the final steady current is reached.

The possible charge carriers within the barrier layer are Al^{3+}, O^{2-}, and OH^- ions. Considering that more Al^{3+} than the equivalent of O^{2-} or OH^- must be transported through the barrier layer because some of the Al^{3+} dissolves into solution in the pore, it is unlikely that the values of various cation and anion mobilities are close to each other. If the Al^{3+} mobility is greater than that of O^{2-} and OH^- in the steady state, there will be a positive ionic space charge in the barrier layer. This will produce an inhibiting effect near the growing nucleus and will discourage the further concentration of current near the pore. Thus, the probability of initiation of a second pore will increase with distance from the first. These opposing effects will lead to a maximum probability of further nucleation at a particular distance from the center of the first nucleus. The distance will be approximately twice the barrier layer thickness (Fig. 46). A third nucleus may be expected at the intersection of the similar circles about the first and second. Repetition of this process leads to a hexagonal pattern.

Barrett[220,261,263] applied ellipsometry to the rather complicated porous layer (Fig. 47). An independent method of measuring the amount of barrier layer in the presence of an overlying porous film is based on the difference

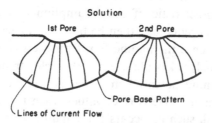

Fig. 46. Schematic picture demonstrating the spreading of pore nuclei on the surface of the initial barrier film. (After Hoar and Yahalom.[328])

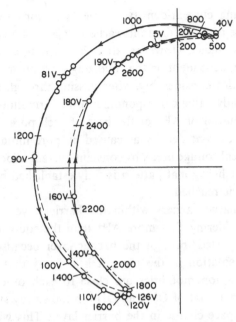

Fig. 47. Polar diagram of the reflection coefficients plotted for a growth curve of aluminum anodized in 3% ammonium tartarate. The dashed curve was computed for a substrate with optical constants 1.0–6.7j and a transparent film with refractive index $n_2 = 1.66$ immersed in a transparent medium with refractive index $n_1 = 1.346$. (After Barrett.[261,329,331,332])

in electrical properties of the two oxide types. The method of Hunter and Fowle[246] involves the study of current flow when anodically polarizing the surface in a suitable electrolyte. They studied ammonium tartarate at gradually increasing voltages until a sudden increase of current occurred. This increase is characteristic of the resumption of film formation. The voltage at which this occurs may then be taken as a measure of the barrier-layer film thickness since the anodic film formation is dependent on ionic conduction, which only takes place when the field strength is above about 10^7 V/cm and the entire potential drop lies across the barrier layer. This method, however, cannot give correct values when the leakage (electronic) current is substantial, such as occurs for impure Al or for alloys. Ellipsometry detects the onset of film growth, thus giving the correct value of the voltage at the resumption of film formation. In addition, ellipsometry can be performed *in situ* during anodizing. Considerations involved in the use of the ellipsometer for film-thickness measurements are described by Barrett and Winterbottom.[330] The refractive indices of both metal and film are necessary for the evaluation of instrument readings in terms of thickness. To a first approximation, the thickness values obtained are optical ones. Values quoted for films containing a porous layer should be considered as the thickness of an equivalent compact film and will represent an underestimate of the actual thickness. The sensitivity of the ellipsometer varies markedly in different thickness ranges. A good range exists from 0–250 Å, an excellent one between 900–1200 Å, followed by another good range between 1900–2000 Å. Ellipsometry provides a means of measuring the barrier thickness under a porous or hydrated layer in cases where purely electrical measurements are difficult. Together with the total thickness measurements, one is also able to determine the porous-layer thickness, with a certain amount of uncertainty stemming from the refractive index. Barrett's measurements indicated that for a porous film formed in ammonium tartarate at 70 V, the refractive index is not below 1.54. However, the refractive index is expected to increase with forming voltage since, by electron microscopy, the pore volume increases somewhat with formation voltage.

Barrett further states that the rate of pore formation on anodizing at constant voltage depends on the electrolyte, the pH, and the formation voltage. For ammonium tartarate the rate is the same between pH 2 and 6, but shows a marked increase at higher pH values. To judge by electron diffraction and micrographs, the porous film at high pH is similar to that in more acid electrolytes.

The anodizing behavior of specimens deliberately covered with a layer of boehmite resembles that of specimens covered with porous anodic films. The boehmite does not hinder anodic pore formation. The rate of pore formation increases with voltage, but the basis for this effect is not explained.

The experimental conditions fall naturally into three categories, according to whether the field strength across the barrier layer is below, equal to, or above that required to maintain a balance between the dissolution for pore formation and barrier film growth. At the higher field strengths, aluminum-ion migration explains the filling up of the pores. At the equilibrium field, barrier growth below the oxide–electrolyte interface balances dissolution, and as both processes have finite rates, the conclusion is that anion migration must occur. Thus, ionic current is in general composed of both anion and cation migration. At yet lower field strengths dissolution predominates over barrier growth. Dissolution of anodic films is field dependent, although the exact nature of the dependence has not been determined. Conditions seem to be more favorable for dissolution in the pores than at the surface of a uniform barrier-type film, but enhanced uniform dissolution can also be observed.

Theory of Dyeing Anodic Films

Giles[333] has presented an excellent review of the problem of dyeing anodic films. An outline of his concept follows: the anodized film is very porous and, therefore, has a high specific surface area, about 100 m^2/g for chromic acid film—comparable to that of charcoal. The sulfuric acid film contains more than 10% aluminum sulfate and some alumina monohydrate (boehmite). The surface state of anodic films is considered to be similar to the so-called "active alumina" used as adsorbent. The zeta potential of alumina is known to be positive in water. The origin of the charge is due to the ionization of the hydrated surface:

$$Al(OH)_3 \rightleftharpoons Al(OH)_2{}^+ + OH^-$$

$$AlOOH \rightleftharpoons AlO_2{}^- + H^+$$

Pretreatment of the material with an acid, H^+X^-, covers the surface with acid anions that may be either covalently bound or present as (hydrated) ions

$$>Al\!-\!OH + H^+X^- \rightleftharpoons Al\!-\!X + H_2O$$
$$\updownarrow$$
$$Al^+X^-$$

The anion may be replaced by dye anions in water and it is this reaction that is the basis of dyeing anodic films. Usually the dyes for anodic films are anionic, that is, they dissociate in water into a dye anion and a metal cation (usually Na^+). They are used in a hot solution containing about 1–2 g/liter dye with addition of a little acetic acid. The temperature must be below 70°C in order to prevent sealing, which is the postdyeing treatment. Sealing blocks the pores which secures the dye adsorbed on the pore wall and opening.

Both forms of alumina, film and powder, exert an interesting variety of forces in adsorbing solutes, as follows:

1. *A bond with aromatic hydrocarbons* (in nonaqueous solvents) which, it is suggested, may be a π-electron complex between the hydrocarbon and charged aluminum atoms exposed by mechanical damage.

2. *Hydrogen bonds*, formed by hydrogen acceptance from the solute or hydrogen donation by hydroxyl groups in the alumina surface. (Recent experiments by Giles showed that a proportion of the bonded solute may be covalently linked to the oxide).

3. *Chelate bonding*, with aromatic solutes containing acidic groups suitably oriented to form a solute–aluminum chelate complex, e.g., catechol or mordant dyes (*cf.* Alizarine Red S, above).

4. *Covalent bond* formation with anionic, e.g., sulfonate, groups.

5. *Ion exchange*, with anionic, e.g., sulfonate, groups.

The operation of 2–5 (above) has been diagnosed in adsorptions by anodic films. In normal dyeing processes, however, only 3, 4, and 5 seem to be important.

According to Mita,[204] the dyes having a good affinity to anodic films are those with the following structure or bonds:

1. Anthraquinone derivatives 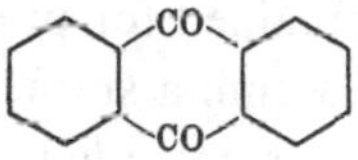
2. Nitroso —NO
3. Monoazo —N=N—
4. Diazo —N=N—, —N=N—
5. Xanthene 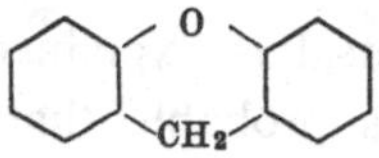
6. Dihydropyrazole or pyrazoline derivatives

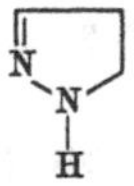

Light-fast dyes are:

1. Anthraquinone derivatives
2. Monoazo
3. Xanthene
4. Dihydropyrazole or pyrazoline derivatives

Anthraquinone derivatives are best in affinity and in light fastness and the monoazo dyes come next.

Morfopoulos and Parreira[334] performed electrokinetic studies on an oxidized aluminium surface by streaming potential measurements in NaCl,

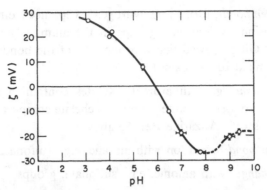

Fig. 48. The electrokinetic behavior of a pure alumina surface in contact with sodium chloride (1 ×
$10^{-3}N$ NaCl). (After Morfopoulos and Parreira.[334])

$NaNO_3$, and $Al_2(SO_4)_3$ solutions at constant ionic strength and variable
electrolyte concentration, respectively. Figure 48 shows zeta-potential values
as a function of pH obtained for pure alumina in contact with NaCl solutions. At low pH the alumina surface exhibits a positive character, and at
pH = 3.1, +26 mV were recorded. With decreasing acidity, the zeta potential decreases, and at about pH = 5.9 the isoelectric point is reached.
At high pH values a charge reversal is observed and a minimum of −27 mV
on the curve is observed at a pH of about 8. Then the potential changes
slightly and levels off beyond pH = 9 at about −20 mV.

The mechanism of charge reversal is explained by the following:
when alumina is heated in air, it is probable that AlOOH is formed
which dissociates

at low pH, $AlOOH + H^+ \rightarrow H_2O + AlO^-$

at high pH, $AlOOH + OH^- \rightarrow AlOO^- + H_2O$

When the alumina is in water for periods of time shorter than 8 h, the same
mechanism will be valid, but when it is in water longer than 10 h, $Al(OH)_3$
will be formed on the surface of the alumina and, thus,

at low pH, $Al(OH)_3 + H^+ \rightarrow Al(OH)_2^+ + H_2O$

at high pH, $Al(OH)_3 \rightarrow AlO_3H_2^- + H_2O$

Therefore, protons or hydroxyl ions are the potential-determining ions
for the alumina surface.

Kuroda and Uji[335] investigated the relationship between the zeta potential of anodic films and dye affinity. They point out that the remarkable difference in the dye affinities of textiles and anodized coatings is that the dye affinity of the former makes no appreciable change with elapsed time, while that of the latter begins to decrease just after anodizing. Furthermore, for anodic coatings, a characteristic phenomenon of "sealing" exists. The zeta potential of the sealed sulfuric acid coating was nearly constant (+30 mV), while that of the unsealed coating fluctuated and varied with the amount of rinsing approaching the sealed value and exhibiting a decrease in dye affinity with longer rinsing. When the sample was kept in a desiccator, the adsorption power remained unchanged for a long time (Fig. 49).

They also measured the isoelectric point of sulfuric acid coatings to be at pH = 6. Below pH 6, the zeta potential in HCl solutions was positive, and above pH 6, the value in NaOH solutions was negative. The data, together with the values in acid dye solutions, are shown in Fig. 50.

They obtained analogous results when comparing the dye affinity and mechanism of wool and anodized coatings, but state that the mechanism in the latter is more complicated due to the instability of the surface electric charge.

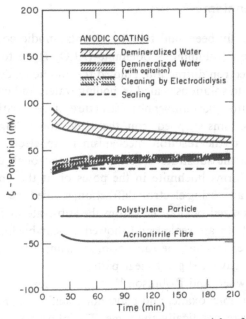

Fig. 49. Time dependence of zeta potentials of variously treated anodic coatings. (After Kuroda and Uji.[335])

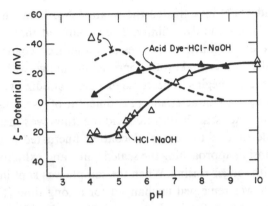

Fig. 50. Zeta potential of sulfuric acid film in acid
dye, HCl, and NaOH solutions of varying pH. (After
Kuroda and Uji.[336])

Yokota[337] reported that the isoelectric point of the sealed oxalic acid
film was at pH $= 9.2$ which is identical with the value for γ $Al(OH)_3$ reported by Parks.

Sealing Mechanism

It has generally been said that by sealing, anodic coatings are partly
converted to more voluminous boehmite ($Al_2O_3 \cdot H_2O$) to plug the pores.
Hart,[338] using electron diffraction, studied the anodic oxide films formed in
ammonium borate solutions sealed in boiling water, and confirmed that the
sealing changes the quasi-amorphous structure of the coating to boehmite,
whereas with the films stripped from the metal, this change did not occur.
He suggested that the reaction mechanism is an electrochemical one in
which Al^{3+} from the substrate moves outward thorough the oxide film to
meet OH^- and forms boehmite in the pores or at the openings.

Spooner,[329] also using electron diffraction, studied the sealing of sulfuric
acid films as-anodized or detached from the substrate in $HgCl_2$–methanol,
and noted that both are sealed and showed a clear boehmite pattern. The
weight gain by sealing was the same also. A steam-sealed ($115°C$, 30 min),
stripped coating gave a slightly clear pattern.

Hunter, Towner, and Robinson[340] compared coatings formed in dilute
sulfuric acid, chromic, oxalic, and phosphoric acids. The hydration rates of
the first three were practically the same. The phosphoric acid coating hydrated at a somewhat slower rate. Inasmuch as this coating is known to
have larger pores than the others, and consequently might be expected to

hydrate more readily, the pore structure would not be responsible for the degree of hydration. With barrier-type films about 500 Å thick formed in ammonium tartarate solution, which would be completely hydrated in boiling water for about 10 min, no hydration occurred after 30 min of boiling in a dilute phosphoric acid (pH = 3.5), or even in a solution containing 9 ppm NaH_2PO_4. They concluded that in the presence of phosphate ion in the solution, hydration is almost completely prevented. Richaud[341] also noted that phosphate and silicate in low concentration reduced the sealing quality.

Hoar and Wood[273] studied the sealing of porous films in sulfuric acid films by heating them in a nickel acetate solution, in dichromate and chromate solutions, and in distilled water. They performed ac-bridge measurements of the electrical impedance of the film during sealing over a wide range of frequencies. They assumed the equivalent circuits of unsealed and partially sealed porous films as shown in Fig. 51, and have given each component of resistance and capacitance the corresponding values from the values observed when balancing resistance R_b and capacitance C_b by trial and error (Table 20). The mechanism of sealing in a dilute nickel acetate solution was examined in detail. They suggest that changes in the complex impedance during sealing are best interpreted by supposing that the pores become filled mainly by inward movement of their sides, with some plugging contribution at their outer end.

Chromate and especially dichromate solutions of the concentrations often employed in practice appear to give more rapid sealing than the commonly used nickel acetate solution, whereas plain water gives much slower sealing. The temperature dependence of the overall process is appreciable for nickel acetate sealing and small for dichromate sealing.

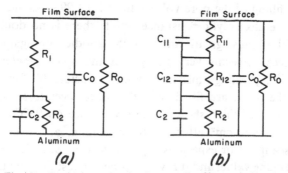

Fig. 51. Equivalent circuits for anodic films. a) Simple electric analogue for unsealed film (Jason and Wood). b) Modified electric analogue for partially sealed porous film. (After Hoar and Wood.[342])

Table 20. Values of Components in Analogues According to Fig. 51 (After Hoar and Wood[342])

	Analogue for 30-min sealed film	Analogue for 24-h sealed film
C_e (μF/cm^2)	0.001*	0.00094
$R_{1.1}$ (k$\Omega \cdot$cm^2)	2.34	83
$C_{1.1}$ (μF/cm^2)	0.195	0.0167
$R_{1.2}$ (k$\Omega \cdot$cm^2)	1.66	17
$C_{1.2}$ (μF/cm^2)	0.005	0.0003
C_2 (μF/cm^2)	0.45*	0.32

R_0. 5 $\Omega \cdot$cm^2 (estimated; precise value immaterial).

R_2. 700 k$\Omega \cdot$cm^2 (experimental, from simple analogue for unsealed film; precise value immaterial).

* Values held fixed. All other values adjusted by trial and error for best balance with R_b and C_b previously obtained by balancing with film impedance during sealing. Film thickness, 15×10^{-4} cm. Sealing in nickel acetate solution, 95°C.

An atomistic mechanism is proposed for the partial hydration of the original Al_2O_3 films to AlO(OH) involving lattice transport of OH$^-$ ions under a high field produced by H$^+$ ions that have first "advanced" by thermal diffusion. This model is extended to account for CrO_4 uptake into the film in dichromate and chromate sealing.

Hoar and Wood emphasize that the corrosion-preventing properties of an anodic oxide film depend on the degree of sealing. The practical success of sealing conducted for periods on the order of 30 min seems to be due to the filling of the pore volume to at least 99%, as indicated by the increase of pore electrolytic resistance. While there is no doubt that such a degree of filling must greatly improve the anodic film against most corrosive influences, it seems very likely that the much greater resistances observed over long sealing times may indicate considerable further improvement. When the pores are filled with the nickel acetate sealing solution, a film 1.5×10^{-3} cm thick shows a pore resistance (at 95°C) of 10 $\Omega \cdot$cm^2 when unsealed, 4 k$\Omega \cdot$cm^2 after a 30-min sealing, and over 100 k$\Omega \cdot$cm^2 after a 24-h sealing. For especially high protection, such long-time sealing might have genuine value, and it is worth practical testing from the corrosion point of view; it would also be interesting to see whether merely maintaining a 30-min sealed film at 95°C, without further immersion in the sealing solution, produces a further improvement in its protective value.

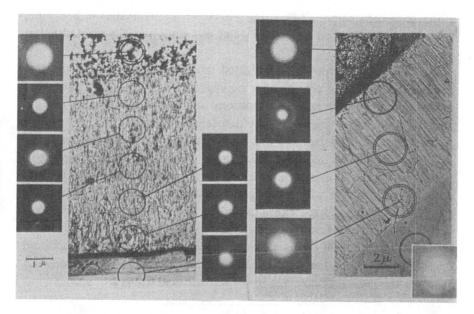

Fig. 52. Electron diffraction of a selected area and corresponding electron micrograph. a) Steam-sealed sulfuric acid film. b) Boiling-water-sealed oxalic acid film. (After Akahori and Fukushima.[343])

The observations that films dried and aged after anodizing seal less quickly, whereas partly sealed films can be dried and aged without impairing the rate of further sealing, have an obvious practical interest. If further work should show that these phenomena are general, it should be simple to arrange practical schedules that avoid any drying or ageing of material at the stage immediately after anodizing.

The influence, if any, of dyeing conducted after anodizing on the rate and completeness of the subsequent sealing requires further examination.

Akahori and Fukushima[274] anodized as-rolled aluminum foil of 99.99% purity in sulfuric and oxalic acids to form anodic coatings of 5–10 μ and sealed them with 5 kg/cm^2 steam (150°C) and boiling demineralized and distilled water (pH = 5.9–6.2 at 20°C), and examined their surfaces and cross-sectional structures, comparing them with unsealed coatings. Two typical pictures from electron micrograph and electron microdiffraction (Hitachi HU-11A) are shown in Fig. 52. They concluded that the powdery substance which was produced on the steam-sealed surface consists of quasi-boehmite needles; this powdery substance is not formed on surfaces sealed in boiling water. For some years the depth of boehmitization into the pores during sealing has been unresolved. They confirmed that the entire pore structure of the steam-sealed coating was converted to crystals of boehmite, while the coating sealed by boiling water was only partially crystallized at an outer region.

Amore and Murphy[344] investigated methods of preventing the formation of powderable coating after boiling water sealing. After careful examination of the pretreatment conditions, such as bright dipping, rinsing, anodizing, etc., they found that the powderable-coating effect was related to contamination and assumed that some necessary contaminant was missing from the process solutions. They finally found that some phosphate in the sealing water prevents the formation of powderable products. The optimum condition is with a phosphate concentration of 3–10 ppm and a pH 5.5–7.0. The anodized panels sealed in this way passed the dye-stain test, CASS test, and Ford Fact test. The reaction between Al^{3+} and phosphate ions in solution forms a stable, insoluble precipitate of aluminum phosphate. They assume that the exchange reaction of phosphate with sulfate is the most probable one in the sulfuric acid coating.

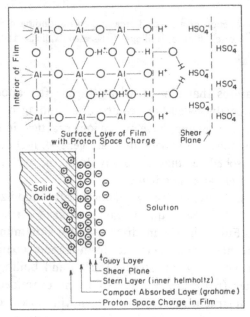

Fig. 53. Probable isoelectric arrangement of ions in the anodic film–electrolyte interphase for a sealed film. Top) Chemical diagram showing balance of anions and cations in the film and inner adsorbed layers. Bottom) Schematic diagram showing charge balance. (After Murphy.[50,51])

Spooner[226] early reported that the actual weight gain (water) by sealing is not enough to plug pores. To explain this result Murphy[50,51] recently proposed a model (Fig. 53) showing the sealed surface structure corresponding to the unsealed model referred to in Fig. 43. According to him, the model will explain Spooner's observation as well. By sealing, the proton space charge assumed by Murphy is neutralized by OH^- ions to form molecular water or to attract protons and assist their migration out of the coating. High-pH solutions with high OH^- concentrations favor rapid replacement of sulfate and neutralization of the proton space charge. With this mechanism, there is no need for the pores to be plugged. Sealing only makes the pore wall inert.

Barkman[235] comprehensively studied sealing and defined macrosealing, which is a direct reaction of water with substrate aluminum facilitated by various macrodiscontinuities during normal sealing. The papers by Pearlstein[345] and Neunzing and Röhrig[346] are illuminating, and the literature on sealing, until 1959, is listed and reviewed by Wood.[349]

To understand the mechanism of sealing with steam or hot water, the series of papers on aluminum–water reactions by Altenpohl,[347,348] Draley and coworkers,[395–397] and Kalpan[398] will be of great help.

PROPERTIES OF ANODIC OXIDE FILMS ON ALUMINUM

In the foregoing discussion most of the properties of anodic oxide films on aluminum have been described so that here they will be only briefly reviewed.

The density of α Al_2O_3 is 4.0, that of γ Al_2O_3 is 3.4. Barrier films are compact and duplex films with pores contain appreciable amounts of anions. Thus, Burgers, Classen, and Zernike[277] gave 3.1–3.3. Mason[351] obtained the densities of sealed and unsealed films of various electrolytes, and pore volumes, as shown in Table 21.

The hardness of the oxide itself is very high, usually similar to that of sapphire. However, this value can not be obtained unless the film is thick enough not to be influenced by the soft substrate. Moreover, in the case of porous films, the hardness is maximum near the metal–oxide interface and becomes softer outward. The hardness is dependent upon the condition of electrolysis. Hardness ranges from VHN 300–1400 (see the section on hard anodizing).

The harder films are more brittle and less ductile. Crazing is often experienced in heavy films, particularly in hard coatings. According to Hill and Mason,[352] cracks are produced by elongations higher than 0.5%.

Table 21. Density and Porosity of Anodic Coatings Made in Various Electrolytes (After Mason[351])

Electrolyte	Temperature		Density, g/cm³		Pore volume, %
	°F	°C	unsealed	sealed	
15% H_2SO_4	70	21.1	2.96	2.65	15.8
	80	26.7	2.93	2.56	19.2
7% H_2SO_4	70	21.1	2.97	2.70	13.7
	80	26.7	2.95	2.65	15.4
9% $(COOH)_2 \cdot 2H_2O$	70	21.1	2.89	2.68	11.2
	80	26.7	2.91	2.68	12.0
20% $NaHSO_4$*	91	32.8	3.00	2.58	21.0
8% CrO_3†	95	35.0	3.14	2.71	20.0

* Coated for 20 min at 0.65 A/dm² (6 A/ft²).
† Coated for 30 min at 40 V.

Crazing is also due to the differences in the thermal expansion coefficients of the substrate and the anodic film. The value of the latter is a fifth of that of the former.[3] It is said that crazing in hard coating is partly due to the rapid change of temperature when samples are withdrawn from a rather cold bath. It is also partially attributed to the shrinkage of the film associated with dehydration, which has been clearly confirmed by the DTA and thermobalance measurement. Since the bonding of the metal and oxide is very strong, crazing often extends to the soft substrate, thereby damaging the material subjected to repeated stress or mechanical working.

For continuous anodizing of wire and strips for the electric and canning industries, special attention is paid to the electrolyte and electrolytic conditions in order to avoid the troubles caused by subsequent mechanical working. The electrolyte usually is kept warmer, often contains chlorides, and ac is employed in order to improve the ductility of the film (see the section on continuous an rapid anodizing).

The thermal conductivity of anodic film is about one tenth of that of aluminum, and its thermal emissivity is 3–6 times higher depending on the radiation wavelength. The lowest values exist in the region of 2–5 μ. According to Phillips,[353] the emissivity increase is of the greater practical importance. Thus, anodized aluminum is more favorable than bare aluminum where maximum radiation heat transfer is desired. The reflectivity, in this region of the spectrum, shows a well defined absorption band at about 3 μ which Gwyer and Pullen[354] ascribe to water combined or adsorbed in the film. They show that it can be almost eliminated by sealing the film by heating and waxing, instead of by boiling.

In the visual region, the percentage of light reflected specularly from a polished and anodized specimen falls off linearly with the film thickness, by about 0.25%/μ. The percentage diffusely reflected falls off more rapidly, being on the order of 1%/μ. Reflectivity obtainable by electropolishing or by chemical polishing is greatly affected by the purity and alloying of the metal and also by the method employed. An alkaline bath needs aluminum of a higher purity than acid baths. In the case of the Brytal process, Pullen[52] reports that with super-purity aluminum almost 90% of the incident light is reflected, but with 99.5% pure metal, the figure falls to about 70%. In the visual region, the percentage of light reflected varies slightly with the wavelength, decreasing towards the red end of the spectrum. In the ultraviolet, reflectivity is high, amounting to about 90% for super-purity metal, and is almost independent of the wave length; there are no absorption bands.[355] Khan, Leachi, and Ilkes[356] studied the thickness and optical properties of anodic films on Al.

Kelleher[357] described the temperature control of a spacecraft's interior environment by sulfuric and chromic acid coatings to make the ratios of solar energy absorption to emittance ($S/\alpha\varepsilon_{TN}$) less than 1 (0.17–0.51). Weaver,[104] Clauss,[358] and Janssen[359] also reported on the thermal control properties of anodic coatings where ETN represents radiation normal to the surface being measured. Solar degradation of anodic coatings was tested under high-intensity ultraviolet radiation 10 times that of the sun, since 20000–40000 Å range ultraviolet radiation, usually absorbed in the earth's atmosphere, induces rupture of many chemical bonds in space. All high-reflective alloys exhibited an increase in solar absorption of 31.4–42%. She also proposed a repair process by anodizing for postassembly scratches. She states that the "passive" method of anodic coatings reduces the weight of spacecraft considerably as compared with the "active" temperature-control approach by other cooling devices. Further, anodic coating is lighter and more corrosion and abrasion resistant than many organic coatings. Clarke, Hamilton, and Beck established a barrier-layer coating process for use for spacecrafts.[422]

Optical properties of refractive index and double refraction were described earlier.

Alumina is an electric insulator and, therefore, anodic films are good insulators, the dielectric constant being 8–10 according the the method of preparation. As-anodized films are very hygroscopic. This property may be used as a hygroscopic meter by measuring the electrostatic capacity change.

Franckenstein[360] showed that the electrical resistance of the anodized aluminum is a little better than porcelain in moist air, but inferior to it in dry air. The specific resistance of a dry film is of the order of $4 \times 10^{15} \ \Omega \cdot cm$ at room temperature, falling to $9 \times 10^{12} \ \Omega \cdot cm$ at 300°C. It is thus rather greater than that of rubber ($2 \times 10^{15} \ \Omega \cdot cm$ at room temperature) and less than that of mica ($9 \times 10^{15} \ \Omega \cdot cm$).

Barrier layers are used in the electrical industry where their thinness and dielectric properties permit their use as electrolytic capacitors. The lower the forming voltage and the thinner the film, the higher the capacitance per unit area. For example, Hermann[361] reports that a borate film formed at 40 V may be 0.04 μ thick and have a capacitance of about 0.16 mF/cm^2, whereas one formed at 400 V may be 0.54 μ thick with a capacitance of 0.015 mF/cm^2. Miyata[362] found the following relationship between breakdown voltage and electrostatic capacitance:

$$C \times V = 6.54 \quad (C \text{ in } F/cm^2) \tag{28}$$

Tihol and Hull[363] used a low-pressure gas plasma of oxygen ions to anodize evaporated aluminum films on glass substrate. Capacitors anodized at 50 V have shown excellent electrical and life performance when operated at 20 V, even at 150°C.

Mechanical properties of the anodic films have been extensively investigated in connection with the development of "hard coating," already described.

From the scientific point of view, one of the most interesting mechanical properties in recent years is the stress associated with the oxide growth. Bradhurst and Llwelyn[364] anodized foil and wire of 99.99% Al in ammonium borate and citrate solutions, and measured the stress produced in the oxide films. They found that the current density is important in determining the nature of the stress, while Davis and coworkers[265,266] have shown that during anodizing in the same electrolyte, the relative amounts of the two ionic species diffusing in the electrolyte are dependent on the rate of oxidation (current density), which Hoar also pointed out.

The results of Bradhurst and Llwelyn and Davis and coworkers are shown in Fig. 54. At a low current density (0.1 mA/cm^2), oxygen migration through the film is predominant, while at a higher current density (about 0.5 mA/cm^2) cationic movement becomes predominant. Vertical lines on the stress axis represent the total range of stress observed in each experiment

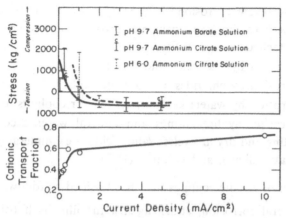

Fig. 54. Curve 1) Stress in anodic oxide films on aluminum as a function of current density of formation (in ammonium borate and ammonium citrate solutions). Curve 2) Fraction of cationic transport in oxide films on aluminum as a function of current density of formation (determined by Davis et al.[265,266]) in ammonium citrate solution. (After Bradhurst and Leach.[364])

for oxide films of 150–3000 Å thick. At a low current density, there is a greater tendency for compressive stress to develop, the stress reversal occurs at higher current densities, which is in qualitative agreement with Vermilyea.[365]

CORROSION PROBLEMS IN ANODIZED ALUMINUM

Corrosion of anodic coatings occur in various environments and atmospheres. Even under the same corrosive conditions and with an equal coating thickness, the protective power or the rate of corrosion of the coatings differ greatly depending upon the nature of the oxide films, the quality of sealing, and, when films are colored, the fastness of the dyes. Therefore, the resolution of problems associated with corrosion of anodic films are too complicated and varied to be covered in this review.

From the viewpoint of corrosive media, corrosion of anodic coatings is classified according to the following important environmental situations:

1. Corrosion by SO_2 and Cl^-: atmospheric corrosion.[366,367]
2. Corrosion by soil and chloride on the highways.[368]
3. Corrosion in alkaline media: concrete cement, mortars, and plasters frequently containing chloride as solidification promotors

cause severe corrosion.[369-371] Stray currents promote corrosion. Sometimes contact or galvanic corrosion occurs when aluminum is attached to steel.[372,373]

4. Corrosion by chemicals and foods.
5. Corrosion by chlorides and organic acids.
6. Corrosion by water: kettles, jars, reactor coolants, etc.[374,375]
7. Corrosion by hot springs and mineral water accompanied by wetting and drying cycles: saline (chloride), mineral (sulfate, carbonate, alkali), and volcanic (H_2S).

Corrosion of anodic coatings may be classified as follows:

1. General corrosion, attack against the film itself (dissolution of the oxide film).
2. Selective corrosion of substrate aluminum, including intergranular corrosion through the defects of the oxide film, for example, by heavy metals included in the film or in the corrosive media, or by halide ions in the environments. Pitting corrosion is of this category.
3. Corrosion accelerated by the thinning of the coating by weathering during prolonged exposures as described by Cochran,[376] which comes partly from erosive action of the wind and partly from chemical attack by the atmosphere (e.g., SO_2).

The defects of the coating come from inhomogeneity of the substrate, poor sealing, and from the craze produced, during anodizing, by volume change (particularly in the case of a thick coating) and, after anodizing, by thermal expansion or shrinkage, or by dehydration of the coating by heating or by mechanical and physical effects.

It is to be noted that inhibitors which are effective for preventing corrosion of aluminum, such as acridine, have no effect upon dissolution of anodic films, since these inhibitors, except passivators (oxidants), protect aluminum but not aluminum oxide.

In order to evaluate the corrosion resistance of anodic coatings, various test methods have been proposed and specified or are under active discussion among major countries interested in anodizing.

In the case of Japan, the Japan Industrial Standard (JIS)[377,378] for anodized aluminum has been effective since 1952 and is specified solely as an alkaline-corrosion test (10% NaOH at 35°C), which is clearly the "film-dissolution" test. At that time, it seemed, no comprehensive specifications existed among major countries. This specification was mainly for house wares, including kettles, and particularly for oxalic acid films with steam sealing, though the lower grades were specified for sulfuric acid film too. In the meantime, the problems of corrosion of anodized kettles by tap water and well water arose, and the accelerated copper sulfate–sodium chloride corrosion test was adopted to check the "defects" of the anodic coating.[374,375] In other countries, specifications for anodic coatings came into existence around 1960 or later and were applied mainly to sulfuric acid coating with boiling water sealing, mainly for architectural use. The different emphasis in specifications is no doubt due to the historical development in the respective countries.

According to a recent paper by Sacchi,[379] "CIDA Recommendations on Anodized Architectural Aluminum" defines the protective value of anodic coating by considering it as a direct function of film thickness, with the additional caution that the film must be properly sealed, and if dyed, that the dye must be light fast. They omitted the "overall corrosion" tests such as the following:

acidified salt-spray test (ASTM, BS, DIN, UNI)
CASS (copper-accelerated acetic acid–salt-spray test) (ASTM, JIS(?))
immersion test
sulfur dioxide–humidity test (BS, DIN)
electrolytic test, FACT (Ford Anodized Aluminum Corrosion Test),
    etc.
alkali resistance test (JIS)

They seem to pay more attention to the checking of sealing quality:

dye stain (ASTM, MIL, BS, DIN, UNI, ABE, VSM)
acid stain
artificial sweat (BS)
sulfite blooming (BS)
acetic weight loss (DIN)
sulfur dioxide–humidity (BS)
Kesternich or HCl (UNI)
AZTAC (impedance)

Sealing and coating thickness are of importance in predicting service performance. Cochran[376] states that sealing markedly improves the resistance of the anodic coating to staining and blooming and improves the light-fastness of colored coatings. In industrial atmosphere exposure tests, un-sealed or poorly sealed coatings invariably develop severe blooming within a few months.

Table 22. Comparison of Tests on Anodic Coatings Sealed Under Different Conditions (After Cochran[76])

Sealing treatment	Dye-stain test	AZTAC impedance, kΩ	6-h CASS test			Blooming after 6 months' atmospheric exposure
			area stained, %	number of breakdowns	grid rating	
Distilled water	passed	120	0	4	92	nil
Nickel acetate	passed	174	0	1	98	nil
Potassium phosphate	passed	2.7	0	1	98	severe
None—not sealed	failed	3.3	20	15	84	severe
Effect of silicate contaminant in distilled water and nickel acetate sealants						
Distilled water						
10 ppm SiO_3^-	passed	71	<1	25	60	nil
25 ppm SiO_3^-	passed	31	<1	40	60	nil
50 ppm SiO_3^-	failed	3.6	5	50	60	slight
100 ppm SiO_3^-	failed	3.3	10	10	80	severe
250 ppm SiO_3^-	failed	3.4	10	19	60	severe
500 ppm SiO_3^-	failed	3.5	10	30	60	severe
1000 ppm SiO_3^-	failed	3.4	10	17	60	severe
Nickel acetate						
10 ppm SiO_3^-	passed	180	<1	5	92	nil
25 ppm SiO_3^-	passed	149	0	1	98	nil
50 ppm SiO_3^-	passed	136	0	1	98	nil
100 ppm SiO_3^-	passed	113	0	0	100	slight

All coatings were 0.3 mil thick, formed on chemically brightened 3×6 in. panels of alloy 5457-H25. Anodizing conditions: 20 min in 16 wt. % H_2SO_4 at 70°F, 12 A/ft². All sealing treatments were for 15 min at boiling temperature. Dye-stain test by ASTM Method B136. Nickel acetate seal was 2 g/liter conc. with 1 g/liter Eloxan salt, a proprietary compound added to minimize sealing smudge. pH = 5.7. Phosphate treatment was in 2 g/liter KH_2PO_4. pH = 4.6. Silicate contaminant was introduced as Na_2SiO_3. pH values of contaminated sealing solutions were adjusted to 6.0 (distilled water) and 5.7 (nickel acetate) using acetic acid.

Of rather recent origin is the ALCOA AZTAC test[376,392] which makes use of the sensitivity of impedance to quality of sealing and to coating thickness. Anodized trim may be specified to have a minimum AZTAC value of 90 kΩ. When the coating thickness is known, AZTAC can be used for determining quality of sealing as a function of time, temperature, and pH. When a coating has not been sealed, very low AZTAC values (<4 kΩ) are obtained regardless of the coating thickness. Table 22 shows the comparison of various tests by Cochran.[376]

There is an extensive literature concerning the corrosion problems and protective properties of anodic coatings. Readers should refer to the comprehensive and rather recent papers by Cochran and Engelhart,[376,391] Vanden Berg,[381] Sacchi and Prati,[382,389] Kutzelnigg,[383,394] Kape and Whittaker,[384,385] Henley and Porter,[386,390] Darnault,[387] CIDA/EWAA Recommendations,[388] and Kesternich[393] in addition to the literature described above.

ACKNOWLEDGMENT

The author is grateful to the Light Metal Educational Foundations (Keikinzoku Shogaku Kai), Osaka, Japan for the long-standing financial and bibliographical assistance for the researches on anodic oxidation of aluminum, and to his colleagues Dr. N. Baba, Dr. T. Mori, and Mrs. M. Shimura, and also to Mr. T. Fukushima (presently at the National Institute for Metals, Tokyo), for their cooperation in research programs over many years.

REFERENCES

1. A. Güntherschulze and H. Betz, *Die Elektrolyt Kondensatoren*, Verlag M. Krain, Berlin (1937).
2. A. Jenny, *Die Elektrolytische Oxydation des Aluminiums und seine Legierungen*, Verlag von Steinkopf, Dresden and Leipzig (1938); *Anodic Oxidation of Aluminium and its Alloys* (transl. by W. Lewis), C. Griffin (1940) (no references).
3. Max Schenk, *Werkstoffe Aluminium und seine Anodische Oxydation*, A. Francke, Bern (1948).
4. A. M. Georgiev, *The Electrolytic Capacitor*, Murray Hill Books, Inc., New York (1945).
5. A. Miyata, *Anodic Oxidation* (in Japanese), Nikkann Kogyo Press, Tokyo (1953).
6. W. W. G. Hübner and A. Schiltknecht, *Die Praxis der Anodischen Oxydation*, Aluminium Verlag, Düsseldorf (1956) (no references); *The Practical Anodizing of Aluminium* (transl. by W. Lewis), MacDonald and Evans, London (1960) (with selected bibliography).

7. C. Etienne and F. Flusin, *Le Traitement de Surface de l'aluminium et des Alliages*, Aluminium Français, Paris (1957).

8. S. Wernick and R. Pinner, *The Surface Treatment and Finishing of Aluminium and its Alloys*, 3rd. Ed., Robert Draper, Teddington (1964).

9. T. P. Hoar, Anodic Behaviour of Metals, in *Modern Aspects of Electrochemistry*, Vol. 1, J. O'M. Bockris, ed., Butterworth Publications, London (1959), pp. 262–342.

10. L. Young, *Anodic Oxide Films*, Academic Press, New York (1961).

11. W. S. Goruk, L. Young, and F. G. R. Zobel, Anodic and Electronic Currents at High Field in Oxide Films, in *Modern Aspects of Electrochemistry*, Vol. 4, J. O'M. Bockris, ed., Plenum Press, New York (1966), pp. 176–250.

12. D. A. Vermilyea, Anodic Films, in *Advances in Electrochemistry and Electrochemical Engineering*, Vol. 3, P. Delahay, ed., Interscience, New York (1963), pp. 211–286.

13. A. V. Shreider, Oksidirovanie Alyuminiya i ego Spravov (Oxidation of Aluminum and Its Alloys), Metallurgizdat, Moscow (1960).

14. A. N. Gorbev, Anodnoe Okislenie Alyuminevykh Spravov (Anodic Oxidation of Aluminum Alloys), Akademiya Nauk SSSR (1961).

15. Fedot'ev and Grilke, *Electropolishing, Anodizing and Electrolytic Pickling of Metals* (Moscow, 1957) (transl. by A. Behr), Robert Draper, Teddington (1959).

16. G. H. Kissin, ed., *The Finishing of Aluminum*, Rheinhold, New York (1963).

17. ASTM, *Anodized Aluminum* (Symposium, Cleveland, 1965), ASTM Special Technical Publications No. 388 (1965).

18. S. Tajima, ed., *Metal Finishing Handbook*, revised Ed., Sangyo Toshio Publishing Co., Tokyo (1956).

19. *Textbook of Alumite* (in 9 Vol., in Japanese), Aluminium Sheet Products Society, Tokyo (1967).

20. Proceedings of the Conference on Anodizing (Nottingham, 1961), Aluminium Federation, London (1962).

21. Proceedings of the International Symposium on Anodizing (Birmingham, 1967), Aluminium Federation, London (1967).

22. H. Buff, Über das elektrische Verhalten des Aluminiums, *Liebigs Annalen* **102**, 265 (1857).

23. F. Wöhler und H. Buff, Über eine Verbindung von Silicium mit Wasserstoff, *Liebigs Annalen* **103**, 218 (1857).

24. E. Ducretet, La résistance électrochimique, offerte pour l'aluminium employé comme électrode positive dans un voltamètre, *Compt. Rend.* **80**, 280 (1875).

25. C. Pollak, DRP. 92,564 (1896).

26. L. Graetz, Über ein elektrochemisches Verfahren, um Wechselströme in Gleichströme zu verwandeln, *Z. Elektrochem.* **4**, 67 (1897).

27. K. Norden, Über den Vorgang an der Aluminiumanode (Ein Beitrag zur electrochemischen Umformung von Wechselstrom in Gleichstrom), *Z. Elektrochem.* **6**, 159 (1899).

28. C. E. Skinner and L. W. Chubb, The Electrolytic Insulation of Aluminium Wire, *Trans. Electrochem. Soc.* **26**, 137 (1914).

29. T. Kujirai and S. Ueki, Jap. P. 61,920 (Aug. 15, 1924, appl. Dec. 28, 1923).

30. S. Ueki, Jap. P. 62,278 (Feb. 5, 1924, appl. Dec. 20, 1923).

31. T. Kujirai and S. Ueki, Brit. P. 226,536 (1925).

32. T. Kujirai and S. Ueki, US P. 1,735,286 (1926).

33. S. Setoh and S. Ueki, US P. 1,735,609 (Nov. 12, 1926).

34. G. Bengough and J. M. Stuart, Brit. P. 223,994 (Nov. 3, 1924, appl. Aug. 2, 1923;) US P. 1,771,910 (July 29, 1930).

35. G. Bengough and J. M. Stuart, Brit. P., 223,995 (Appl. Aug. 2, 1923).

36. S. Setoh and A. Miyata, Electrolytic Oxydation of Aluminium and Its Industrial Applications, Proc. World Engineering Congress, Vol. 22, Tokyo, 1929 (1931), pp. 73–100.

37. H. Röhrig, Elektrolytisch erzeugte oxydische Schutzschichten auf Aluminium, Z. Elektrochem. 37, 721 (1931).

38. H. Röhrig, Brit. P. 406,998 (1934).

39. C. H. R. Gower and S. O'Brien, Brit. P. 290,901 (1927).

40. C. H. R. Gower, Brit. P. 290,921 (1928), DR P. 600,387 (1928).

41. E. W. Küttner, DR P. 620,898 (May 22, 1928).

42. H. Ginsberg, H. Neunzig und W. Sautter, Die Entwicklung der Overflächenschutzverfahren fur Aluminium aus der Sicht der VAW, Metall 21, 570 (1967).

43. W. Wetzki, Aus der Praxis der Anodischen Oxydation von Aluminium, Galvanotechnik 57, 843 (1966).

44. S. Setoh and A. Miyata, Researches on Anodic Film of Aluminium II, Anodic Behaviours of Aluminium in Aq. Solutions of Oxalic Acid, Sci. Pap. Inst. Phys. Chem. Res., Tokyo 19 (397), 237 (1932).

45. Th. Rummel, Über Wachstum und Aufbau elektrolytisch erzeugter Aluminium Oxydschichten, Z. Physik 99, 518 (1936).

46. W. Baumann, Entstehung und Struktur elektrolytisch erzeugter Aluminiumoxydschichten, Z. Physik 111, 708 (1939).

47. J. W. Cuthbertson, Anodic Oxidation of Aluminium, J. Inst. Metals 65, 95 (1939).

48. F. Keller, M. S. Hunter, and D. L. Robinson, Structural Features of Anodic Oxide Films on Aluminum, J. Electrochem. Soc. 100, 411 (1953).

49. J. F. Murphy and C. E. Michelson, A Theory for the Formation of Anodic Oxide Coatings on Aluminium, Proc. Conf. Anodizing Aluminium (Nottingham, 1961), Aluminium Federation, London (1962), pp. 83–95.

50. J. F. Murphy, Chemical and Electrochemical Factors in the Mechanism of Formation and Properties of Anodic Coatings, Proc. Symposium on Anodizing Aluminium (Birmingham, 1967), Aluminium Federation, London (1967), pp. 3–16.

51. J. F. Murphy, Practical Implications of Research on Anodic Coatings on Aluminum, Plating 54, 1241 (1967).

52. N. D. Pullen, An Anodic Treatment for the Production of Aluminium Reflectors, J. Inst. Metals, 151 (1936).

53. N. D. Pullen, Brit. P. 49,162 (Dec. 17, 1934).

54. R. B. Mason and M. Tosterud, US P. 2,040,618 (May 12, 1936); 2,108,603 (1938).

55. T. Nakayama, Jap. P., 128,891 (1939).

56. N. D. Tomashov, Anodic Oxidation and Its Application in the Construction of Engines, Vestnik Inzhenerov i Tekhniskov 59 (1946).

57. N. D. Tomashov, Anodizing and Its Uses in Engine Construction, Light Metals 9, 429 (1946).

58. B. E. Diel and L. Swanson (to Kaiser Aluminum and Chemical Corp.), US P. 3,031,387 (April 24, 1962, appl. Dec. 7, 1959).

59. S. Tajima, M. Soda, and T. Mori, Properties and Mechanism of Formation of α-Alumina (Corundum) Films on Aluminium by Anodic Oxidation in Bisulphate Melts, Electrochim. Acta 1, 205 (1959).

60. S. Tajima and N. Baba, Properties and Mechanism of Formation of Thick Anodic Oxide Films on Aluminium from the Nonaqueous System of Boric Acid and Formamide, *Electrochim. Acta* **9**, 1509 (1964); *Light Metals, Japan* **14**, 320 (1964).

61. W. Walkenhorst, Ein einfaches Verfahren zur Herstellung strukturloser Trägerschichten aus Aluminiumoxyd, *Naturwiss.* **34**, 373 (1947).

62. S. Tajima, Y. Itoh, and T. Fukushima, Anodic Behavior of Highly Pure Aluminum in Organic Acid Bath. I. Direct Current Electrolysis in Monocarbonic Acids, **23**, 296 (1955); II. Direct Current Electrolysis in Oxycarbonic Acid, *Denki Kagaku* **23**, 342 (1955); III. Direct Current Electrolysis in Oxycarbonic Acids and Maleic Acid, *Denki Kagaku* **23**, 395 (1955).

63. S. Tajima, Beitrag zur Theorie der anodisch gebildeten Oxydschichten auf Aluminium, *Metall* **18**, 581 (1964).

64. L. Bosdorf, A Review of Anodic Oxidation and New Processes in this Field, *Metal Finish. J.* 104 (1966).

65. R. Gardam, Die Entwicklung der Anodisation des Aluminiums und seiner Legierungen, *Aluminium* **40**, 165 (1964); Aluminium Conference, Budapest (1963).

66. S. Tajima and M. Shimura, Statistical Analysis of Formation and Dissolution of Anodic Oxide Films on Aluminium, *Denki Kagaku* **32**, 883 (1964); *Electroplating and Metal Finishing* **17**, 144 (1964).

67. R. B. Mason and C. J. Slunder, Anodic Reactions of Aluminum and Its Alloys in Sulfuric and Oxalic Acid Electrolytes, *Ind. Eng. Chem.* **39**, 1602 (1947).

68. J. Walton, Beitrag zur Technik der anodischen Oxydation von Aluminium unter besonderer Berücksichtigung des Alumilite Verfahrens, Aluminium-Archiv., Vol. 1, Aluminium Zentrale, Berlin (1935).

69. H. Bengston and Pettit, The Alumilite Process for Decorating and Protecting Aluminum Products, *Machinist* (Europ. Ed.) **77**, 76 (1933).

70. A. Prati, F. Sacchi, and G. Paolini, Effect of Operating Conditions on the Properties of Sulphuric Acid Anodic Oxide Coatings, *Electroplating and Metal Finishing* **2**, 44 (1963).

71. G. R. Darrow, Engineering the Sulfuric Anodic Processes, *Anodized Aluminum*, ASTM STP 388 (1965).

72. Anodisch Oxydiertes Aluminium für Dekorative Zwecke, Aluminium-Merkblatt O_4, Aluminium Zentrale E.V., Düsseldorf (5, Auflage, 1967).

73. G. H. Kissin, B. E. Deal, and R. V. Paulson, Anodizing Characteristics of Commercial Aluminum Alloys in Sulfuric Acid, *Finishing of Aluminum*, G. H. Kissin, ed., Reinhold, New York (1963), pp. 13–31.

74. M. N. Tyukina, Investigations of the Anodic Oxidation of Al–Cu Alloys in Sulfuric Acid, *Zh. Prik. Khim.* **36**, 338 (1963).

75. A. F. Bogozavlevskij, Anodic Oxidation of Aluminum in an Ultrasonic Field, *Zh. Prik. Khim.* **37**, 2256 (1964).

76. S. Setoh and A. Miyata, Researches on the Anodic Oxide Film on Aluminium. I. Effect of Concentration of the Electrolyte on the Formation of Anodic Film, *Sci. Pap. Inst. Phys. Chem. Res., Tokyo* **19**, 189 (1932).

77. S. Tajima, Anodizing Aluminum with Oxalic Acid, *Products Finishing* **17**, 43 (1952).

78. W. W. Hübner, Die anodische Oxydation des Aluminiums in verschiedenen zusammengesetzten Oxalsäurelösungen, Dissertation, Eidgenössische Technische Hochschule, Zürich (1948).

79. Max Schenk, Swiss P. 182,415, 188,228, 221,938, and 232,613.

80. E. Zurbrügg, Bewährung der GS-Oxydschichten von Aluminium, *Aluminium* **39**, Aluminium Anwendungen in Bauwesen (1963), pp. 419–421.
81. H. Neunzig und V. Röhrig, Anodisch erzeugte Oxydschichten in Schwefel-Oxal Säure Elektrolyten, *Metall* **18**, 590 (1964).
82. W. Sautter, Einfluss der Verfahrenstechnik auf die Eigenschaften anodisch erzeugter Oxydschichten von Aluminium, *Aluminium* **42**, 636 (1966).
83. G. D. Bengough and H. Sutton, The Protection of Aluminium and Its Alloys Against Corrosion by Anodic Oxidation, *Engineering* **122**, 274 (1926); *Metal Industry* (London) **29**, 153 ,175 (1926).
84. B. C. Lewsey, Anodic Oxidation and Colouring of Aluminium, *Electroplating and Metal Spraying* 8 (1952).
85. A. W. Brace and R. Peek, Production and Properties of Opaque Coating by Anodic Oxidation in Chromic Acid, *Trans. Inst. Metal Finish.* **34**, Advance copy No. 3 (1957).
86. F. Modic, Neuere Erfahrungen auf dem Gebiet der anodischen Oxydation von Aluminium in Chromsäure, *Aluminium* **39**, 169 (1963).
87. W. E. Gorden and M. E. Cupery, Sulfamic Acid, Industrial Application, *Ind. Eng. Chem.* **31**, 1237 (1939).
88. R. Piontelli, Un nuovo bagno per l'ossidazione anodica dell'alluminio, *La Ricerca Scientifica* **12**, 1195 (1941); Ital. P. 388,932 (Mar. 16, 1941).
89. R. Piontelli, Über die Anwendung sulfaminsäure Bäder bei der Abscheidung galvanischer Überzüge und bei anodischer Oxydation, *Korr. u. Metallschutz* **19**, 110 (1943).
90. S. Tajima, T. Fukushima, and Y. Kimura, Anodizing Aluminum with Sulfamic Acid, Comparison with Oxalic and Sulfuric Acid Processes, *Metal Finishing* (Oct.) **67**, (Nov.) 65, 74 (1952).
91. E. W. Schweikher (to E. I. du Pont et Nemours & Co.) US P. 2,430,304 (1947).
92. T. Fukushima, Anodizing and Electropolishing with Sulfamic Acid, *Chimica Metallica, Tokyo* **2**, 234 (1965).
93. J. Fischer, US P. 2,036,962 (1936).
94. J. Fischer, US P. 2,095,519 (1937).
95. R. C. Spooner and D. P. Seraphim, Phosphoric Acid Anodizing of Aluminium and Its Application to Electroplating, *Trans. Inst. Metal Finish.* **31** (3), (1953).
96. ALCOA, *Electroplating on Aluminum and Its Alloys* (1946).
97. H. J. Wittrock, Nickel–Chromium Plating upon Anodized Aluminum, *Finishing of Aluminum*, G. H. Kissin, ed., Reihold, New York (1963), pp. 163–191.
98. S. Tajima, Besondere Vorbehandlungen für Metalle und Nichtleiter, in *Handbuch der Galvanotechnik*, Vols. 1 and 2, Dettner–Elze, ed., Carl Hanser Verlag, München (1965), pp. 1033–1049.
99. H. J. Wittrock, Nickel–Chromium Plating upon Anodized Aluminum, paper read at the 48th Ann. Convention of the American Electroplaters' Soc. (1961).
100. A. F. Dickinson, Tests on Alzak Aluminium Reflectors, *Trans. Illum. Eng. Soc.* **29**, 368 (1934).
101. J. D. Edwards, Aluminum for Reflectors, *Trans. Illum. Eng. Soc.* **29**, 351 (1934).
102. S. G. Frederichsen, Manufacturing of Bright Anodized Parts for the Car Industry, *Metallen* (Sweden) **22**, 131 (1966).
103. E. Hutgren and O. Jakobson, Bright Anodizing of Aluminum, *Metallen* (Sweden) **22**, 151 (1966).

104. J. H. Weaver, Bright Anodized Coatings for Temperature Control of Space Vehicles, *Plating* (Dec.), 1165 (1964).
105. J. M. Andrus and R. Pettit, Chemical Preparation of Aluminum for Chemical, Electrochemical Brightening and Anodic Coating, in *Anodized Aluminum*, ASTM STP 388 (1965), pp. 1–20.
106. E. Zurbrügg and H. Hug, Choice of Material for Anodizing, Electrolytic and Chemical Polishing, *Aluminium Suisse* 1, 1 (1951).
107. E. Zurbrügg and H. Hug, Electrolytic and Chemical Polishing, *Aluminium Suisse* 2, 11 (1952).
108. E. Baumann and H. Neunzig, Chemically Brightened Anodized Aluminium and Its Employment in Automobile Manufacture, *Trans. Inst. Metal Finish.* 33, Advance copy No. 4 (1956).
109. G. E. Gardam and R. Peek, Studies in Bright Anodizing by the Ammonium Bifluoride–Nitric Acid Process, *Trans. Inst. Metal Finish.* 33, Advance copy No. 3 (1956).
110. A. W. Brace, The Anodizing and Brightening of Aluminium and Its Alloys, Influence of Composition and Structure, *Metallurgia* 55, 173 (1957).
111. N. Kawashima, Effect of Alloying Elements and Structure of Aluminum and Its Alloys on Electropolishing and Anodizing, Aluminum Finishing Research Report., Metropolitan Industrial Research Inst., Tokyo (1957), No. 26, pp. 1–18.
112. R. Peek and A. W. Brace, Bright Anodizing by the Modified Erftwerk Process, *Electroplating and Metal Finishing* (March), 1 (1958).
113. R. Lattey, Der heutige Stand des Glanzeloxierens von Aluminium, *Aluminium* 34, 382 (1958).
114. A. W. Brace, European Bright Anodizing Practice, *Proc. Am. Electroplaters' Soc.* 46, 216, 379 (1959).
115. A. W. Brace, Anodized Aluminium Body Trim, *Bulletin of Body Engineers* (May), 3 (1958).
116. W. E. Cooke, Factors Affecting Loss of Brightness and Image Clarity during Anodizing of Bright Trim Aluminum Alloys in Sulfuric Acid Electrolyte, *Plating* (Nov.), 1157 (1962).
117. B. A. Shenoi, K. S. Indira, and K. Vijayalakshimi, Bright Finishing Super-Purity Aluminum, *Metal Finishing* (April), 44 (May), 49 (1962).
118. B. A. Scott and H. M. Bigford, Bright Anodized Aluminium Surfaces, Proc. Conf. on Anodizing (Nottingham, 1961), Aluminium Federation, London (1962), pp. 55–65.
119. *Anonym.*, Automatically Brightening and Anodizing Aluminium, *Automotive Engineer* (March), 2 (1962).
120. A. Reuter, Betriebliche Fehlerquellen beim Glänzen und Anodisieren, *Aluminium* 39, 421 (1963).
121. S. Terai and H. Nishimura, Application of Bright Anodized Aluminum Alloys for Automobile, *Sumitomo Light Metal Technical Reports* 6, 135 (1966).
122. R. Sundberg and L. O. Samuelson, Aluminium Alloys for Bright Anodizing, *Metallen* (Sweden) 22, 135 (1966).
123. A. Teubler, Metallurgical Aspects of Bright Anodizing, Proc. Intern. Symp. on Anodising (Birmingham, 1967), Aluminium Federation, London (1967), pp. 71–95.
124. S. Setoh and A. Miyata, Effect of Concentration of the Electrolyte on the Formation of Anodic Film, *Sci. Pap. Inst. Phys. Chem. Res., Tokyo* 19 (396), 189 (1932).

125. Th. Rummel, Über Wachstum und Aufbau elektrolytisch erzeugter Aluminium-oxydschichten, *Z. Physik* **99**, 518 (1936).

126. G. Elssner, Schichtwachstum und Massenveränderung bei der elektrolytische Oxydation des Aluminiums, *Aluminium* **25**, 310 (1943).

127. G. Elssner und A. Breyer, Das Dickenwachstum anodisch erzeugter Oxydschichten auf Aluminium und seine Grenzen, *Archiv. für Metallkunde* **4**, 120 (1948).

128. N. D. Tomashov, Anodic Oxidation and Its Application in the Construction of Engines, *Vestnik Inzhenerov i Teknikov* (2), 59 (1946).

129. N. D. Tomashov, Anodizing and Its Application in Engine Construction, *Light Metals* **9**, 429 (1948).

130. G. L. Martin, US P. 2,692,851, 2,692,852 (1950), Brit. P. 701,390 (1951).

131. W. J. Campbell, Hard Surfacing Light Alloys by Anodizing, *Light Metals* **15**, 46 (1952).

132. W. J. Campbell, Anodized Aluminium Surfaces for Wear Resistance, *J. Electrodep. Tech. Soc.* **28**, 273 (1952).

133. W. J. Campbell, Anodic Finishes for Wear Resistance, *Metal Finishing J.* (Jan.), 5 (1964).

134. F. Gillig, Study of Hard Coatings for Aluminum, WADC (Wright Air Development Center) Technical Report 53-151 (May, 1953), p. 113.

135. W. J. Campbell, Anodic Finishes for Wear Resistance, Proc. Conf. on Anodizing (Nottingham, 1961), Aluminium Federation, London (1962), pp. 137–149.

136. Sanford Process Co., US P. 2,897,125, 2,692,852.

137. J. M. Kape, Thick Oxide Films on Aluminium and Its Alloys, *Met. Ind.* **91** (4) 63; (5) 90; (6) 109; (7) 129; (8) 148; (9) 171; (10) 198; (11) 217, 222; (12) 239 (1957). W. M. Doyle and L. T. Harris (to High Duty Allog Ltd.), Brit. P. 727,749.

138. P. Csokán, Beiträge zur Kenntnis der anodischen Oxydation von Aluminium in verdünnter, kalter Schwefelsäure, *Metalloberfläche* **15**, 113 (1961).

139. P. Csokán, Beitrag zur Frage des Bildungsmechanisms von anodische erzeugten Hartoxydschichten, *Werk. u. Korr.* **12**, 288 (1961).

140. P. Csokán, Einfluss der Oberflächenbeschaffenheit des Aluminiums auf die anodische Hartoxydbildung, *Schleif- und Poliertechnik* **1**, 25 (1961).

141. P. Csokán, Hard Anodizing, *Electroplating and Metal Finishing* **15**, 75 (1962).

142. P. Csokán und G. Sinay, Untersuchung der mechanischen Eigenschaften von harteloxierten Alumniumblechen, *Werk. u. Korr.* **11**, 224 (1960).

143. P. Csokán, Réalisation d'un revêtment d'oxyde dur sur l'aluminium par oxydation anodique, *Corrosion et Anticorrosion* **8**, 158 (1960).

144. P. Csokán, Anwendung von harteloxierten Aluminiumbestandteilen für den Maschienenbau, *Dechema-Monographien* **45**, 319 (1962).

145. P. Csokán, Über die Prüfmethoden von anodisch erzeugten Hartoxydschichten, *Elektrie* **8**, 245 (1962).

146. P. Csokán, Kemény oxiddal bevont aluminiumkokillak gyakorlati felhasznalasa, *GEP* **12**, 489 (1960).

147. P. Csokán, Festigkeitseigenschaften der mit Hartoxydschichten überzogenen Aluminiumoberfläche, *Bleche* (12), 649 (1962).

148. K. Isawa, A. Furuichi, and K. Takamura, Studies on Hard Anodic Films on Al and Its Alloys, *Sci. Pap. Inst. Phys. Chem. Res., Tokyo* **36**, 685 (1960).

149. S. Tajima, Formation and Properties of Hard Anodic Films on Aluminum, *Chimica Metallica*, Tokyo **1**, 61 (1964).

150. M. Stalzer, Entwicklungen auf dem Gebiet des Harteloxierens, *Metalloberfläche* **16**, 11 (1962).

151. H. J. Wiesner and H. A. Meers, Hard Anodizing of Fuel Metering Component, discussion, 45th Ann. Techn. Proc., Am. Electroplaters' Soc. (1958).

152. A. Chappuis, Anodische Oxydation der Federgehäuse automatischer Uhren, *Aluminium Suisse* **12**, 130 (1962).

153. Lelong, P. Segonds, and J. Herenguel, An Anodising Process to Produce Very Thick Oxide Films, Fr. P. 1,005,592 (1952); *Rev. Aluminium* **37**, 67 (1960).

154. A. Jogarao and B. Sambamurty, Studies on Hard Anodizing of Aluminium I, Proc. Symp. on Electrodeposition and Metal Finishing (Karaikudi, 1957), India Section, Electrochem. Soc., Inc. (1960), pp. 147–155.

155. A. Jogarao and B. Sambamurty, Electrolytic Formation of Abrasion and Heat-Resistant Coatings on Aluminium, *J. Sci. Ind. Res.* **19B** (2), 82 (1960); (Ind. P. Appli. No. 67,572).

156. V. Lakshminarasimhan, K. R. Narasimhan, and B. A. Shenoi, Hard Anodizing Aluminum and Its Alloys, *Metal Finishing* (March), 60 (1967).

157. R. V. Vanden Berg, Hard Aluminum Finishes Resist Wear and Abrasion, *Iron Age* **170** (18), 81 (1952).

158. H. A. Johnson, Mechanical Properties of Aluminum Hard Coatings, *Product Engineering* (Sept.), 136 (1954).

159. F. Sacchi, L'ossidazione anodica a spessore dell'alluminio e delle sue leghe, *Alluminio* 5 (1955).

160. L. Bosdorf und A. Breyer, Die Harteloxierung, *Aluminium* (7/8), 321 (1955).

161. A. W. Sweet, Practical Aspects of Aluminum Hard Coating, *Plating* 1191 (1957).

162. E. P. Owens, How to Judge the Quality of Aluminum Hard Coatings, *Iron Age* (Feb. 12), 108 (1959).

163. K. Tsunoda, Hard Anodizing, Aluminum Finishing Research Rept., Metropolitan Ind. Res. Inst., Tokyo (1959), No. 38, pp. 1–22.

164. E. Hermann, Harteloxierung, *Galvanotechnik* (8), 387 (1960).

165. D. R. Johnson, Hard Coat as a Metal Finishing Process, *Plating* 986 (1963).

166. T. L. Rama Char and J. Sundrarajan, Recent Development in Hard anodizing, Proc. Symp. on Electrodeposition and Metal Finishing (Karaikudi, 1957), India Section, Electrochem. Soc., Inc. 1960), pp. 160–164.

167. E. Lichtenberger and M. Holló, Influence du traitement préliminaire sur la structure des péllicules d'oxyde anodique formée en acide oxalique–acide formique, *Métaux, Corrosion-Industrie* (449), 1 (1963).

168. L. Bosdorf, Die anodische Oxydation und Ihre neuen Verfahren auf diesem Gebiete, *Galvanotechnik* **54**, 66 (1963); *Metal Finishing J.* (March), 104 (1966).

169. G. E. Gardam, Verfahren zur Erzeugung anodisch oxydierter Schichten mit Eigenfärbung auf Aluminium für Architekturzwecke, *Aluminium* **41**, 423 (1965).

170. E. A. Bloch und E. Zurbrügg, Neue Möglichkeiten zur Erzeugung farbiger Oxidschichten auf Aluminium, *Aluminium* **42**, 101–103 (1966).

171. J. M. Kape, The Use of Integral Colour Processes for Architectural Anodizing, Proc. Intern. Symposium on Anodizing (Birmingham, 1967), Aluminium Federation (1967), pp. 123–132; *Light Metals* **25** (307), 26 (1963).

172. J. M. Kape, Unusual Anodizing Processes and Their Practical Applications, *Electroplating and Metal Finishing* **14**, 407 (1961).

173. I. Mita, Coloring of Anodic Oxide Films on Aluminum, *Chimica Metallica*, Tokyo **2**, 308 (1965).

174. G. Paolini, Integral Color Anodizing for Architectural Use, *Aluminio* **35**, 435 (1966).

175. P. O. Aronson, Color for Anodized Aluminum, *Metallen* (Sweden) **21**, 67 (1965).

176. F. C. Porter, Anodizing Practice, Recent Advances, *Corros. Techn.* **10**, 5, 40 (1963).

177. Korpiun (to Schering–Kahlbaum AG) DRP. 657,902 (appl. Dec. 12, 1935), 664,240 (appl. Dec. 10, 1935).

178. J. M. Kape, Some Thoughts on the Treatment of Surface by Anodic Oxidation, *Metal Finishing J.* (Feb., 7 (1958).

179. F. L. Church, Light-Fast Colors for Anodized Architectural Aluminum, *Modern Metals* (Sept.), 36 (1962).

180. P. Brenner and H. U. von Vogel, Einfluss von Struktur auf die Farbe der Eloxalschichen der Al–Mg Legierungen, *Aluminium* **19**, 696 (1937).

181. F. Keller and G. W. Wilcox, Anodically Oxidized Aluminum Alloys—Metallographic Examination, *Metals and Alloys* (June), 8 (1939).

182. F. Keller, G. W. Wilcox, M. Tosterud, and C. J. Slunder, Anodic Coatings on Aluminum, Behavior of Alloy Contents, *Metals and Alloys* (July), 7 (1939).

183. R. Richuad, L'Oxydation anodique des alliages d'aluminium contenant du silicium, *Memoirs Scientifiques de la Rev. de Métallurgie* **56**, 30 (1959).

184. H. Vosskühler and H. Zeiger, Grau anodisierbare Aluminiumknetlegierungen, *Aluminium* **38**, 291 (1962).

185. G. Kappel, "Grinatal," eine Grautonlegierung des Typs AlSi für dekorative Anwendung, *Aluminium* **39**, 428 (1963).

186. E. Zurbrügg, Grinatal und Grinacolor, *Aluminium Suisse* **14**, 8 (1964).

187. "AlSi" mit grauer Eigenfärbung bei der anodischen Oxydation, Aluminium-Merkblatt W 16, Aluminium Zentrale E.V., Düsseldorf (1967).

188. T. Nakayama, Can. Pat. 602,194 (1961).

189. S. Tajima, F. Satoh, N. Baba, and T. Fukushima, Influence of Current Forms on Anodic Oxidation of Aluminum in Sulfuric Acid and Oxalic Acid, *J. Electrochem. Soc., Japan*, Overseas edition E262 (1958).

190. F. B. Flick, US P. 1,526,127 (appl. July 10, 1923).

191. H. Bengston, US P. 1,869,041 (appl. June 11, 1930).

192. R. Shirotsuka, Jap. P. 62,278 (appl. Jan. 24, 1930).

193. M. Tosterud, US P. 1,900,472 (Sept. 9, 1931).

194. *Weather Resistant Aluminium Dyestuffs for Architecture*, Basel, Durand, & Hugenin S.A. (1959).

195. C. T. Speiser, La Résistance au soleil et aux intempéries des colorants organiques, *Rev. Aluminium* (Feb.), 225 (1962).

196. R. C. Spooner, Light-Fastness of Organic Dyes on Anodized Aluminum, *Metal Finishing J.* (Nov.), 1 (1958).

197. R. C. Spooner, Light-Fast Organic Dyes for Anodizing, *Met. Ind.* **94** (4), 68; (6), 107 (1959).

198. C. T. Speiser, Theory and Practice in Dyeing and Sealing Anodized Aluminium, *Electroplating and Metal Finishing* **9**, 106, 128 (1956).

199. *Multi-Color Effects on Anodized Aluminium*, Basel, Durand, & Hugenin S.A. (1957).

200. Multi-Color Effects on Anodized Aluminium, *Trans. Inst. Metal. Fin.* **35**, 91 (1958).

201. C. T. Speiser, Neuere Entwicklung auf dem Gebiet des Färbens von anodisch oxydiertem Aluminium, *Aluminium* **42**, 421 (1966).

202. Hübner–Schiltknecht, *Die Praxis der Anodischen Oxydation*, Aluminium Verlag, Düsseldorf (1956); *The Practical Anodizing of Aluminium* (transl. by W. Lewis), MacDonald & Evans, London (1960), p. 334.

203. S. Shono and I. Mita, Dyeing Anodic Oxide Films on Aluminium, *J. Met. Fin. Soc.*, Japan **10**, 238, 306 (1956).

204. I. Mita, Coloring Anodic Oxide Films on Aluminium, *Chimica Metallica*, Tokyo, **2**, 308 (1965).

205. R. V. Vanden Berg, Color and Textures for Aluminum, *Products Engineering* (Sept.), 101 (1957).

206. R. V. Vanden Berg, Technique for Coloring Anodized Aluminum, *Products Finishing* (Feb.), 2 (1960).

207. E. Hermann and W. Hübner, Färben von anodisch oxydierten Aluminium mittels anorganischer Pigments, *Aluminium Suisse* **5**, 134 (1955).

208. J. M. Kape and H. Mills, The Prdduction of Coloured Anodic Film without the use of Dyestuffs, Advance copy of the Spring Conf., Brighton, Inst. Metal Fin. (1959), p. 37.

209. I. Mita and H. Kawase, Coloring Architectural Aluminum with Inorganic Pigments and the Light-Fastness and Outdoor Exposure Tests, Aluminum Finishing Research Report, Metropolitan Industrial Res. Inst., Tokyo (1963), No. 65, pp. 1–20.

210. Asada, Jap. P. Prov. No. Sho 38-1715 (1963); Brit. P. 1,022,927.

211. H. Ginsberg (to Vereinigte Aluminium Werke AG), DR P. 670,903 (1938).

212. O. Schaaber (to S. Junghans), DR P. Anm. 16/10 P350,87D (Dec. 9, 1949); US P. 2,666,023 (1954).

213. *Tatsuta Technical Review* (Tatsuta Electric Wire and Cable Co., Japan) (1), 88 (1967).

214. T. Yasuda and H. Hamaguchi, Application of α-Alumina Film Anodically Formed in Bisulfate Melts in Electrical Engineering Field, preprint No. 510 of the Joint Meeting of four electrical engineering societies, Japan (1961).

215. J. M. Kape, Anodizing in Non-Aqueous Solvent System, *Plating* (Jan.), 26 (1968).

216. D. Nickelson, Continuous Anodizing and Lacquering, *Metal. Ind.* (Jan.), 22 (1960).

217. J. M. Kape, Recent Development in Anodizing, *Light Metal Ind.* (March), 45 (1964).

218. P. Csokán, Technologie der kontinuierlichen Anodisierung von Aluminiumbändern und drähten, *Werk. u. Korr.* **4**, 307 (1964).

219. H. N. Johnston, Continuously Anodized Aluminum Coil, *Ind. Finish.* (Feb.), 108 (1967).

220. F. L. Church, Continuous Process Anodizes One Side of Coiled Sheet at High Speed at Low Cost, *Modern Metals* **19**, 32 (1963).

221. E. Hermann, Eloxieren von Bändern und Drähten, *Aluminium* **29**, 465, 513 (1953).

222. H. Richaud, Continuous Electrolytic Oxidation of Aluminium Wire and Strip, *Metal Finishing J.* **3** (26), 71 (1957).

223. E. F. Barkman, New Techniques Expand Continuous Anodizing Applications, Proc. Intern. Symp. on Anodizing (Birmingham, 1967), Aluminium Federation, London (1967), pp. 115–121.

224. E. F. Barkman, Technological Advances in Electrochemical Finishing of Aluminum, *Plating* (May), 508 (1967).

225. E. Lichtenberger–Bajza and G. J. V. Jagandha–Raju, Rapid Anodizing, *Electroplating and Metal Finishing* **15**, 342 (1962).

226. R. C. Spooner, The Sealing of Sulfuric Acid Anodic Films on Aluminum, 44th Annual Techn. Proc., Am. Electroplaters' Soc. (1957), pp. 132–142.

227. F. Sacchi and G. Paolini, Controllo del fissaggio degli strati dei ossido anodico mediante la prova alla goccia, *Alluminio* **30**, 9 (1961).

228. H. Richaud, The Influence of Sealing Water Impurities on the Quality of the Oxide Film, Proc. Conf. on Anodizing (Nottingham, 1961), Aluminium Federation (1962), pp. 181–185.

229. N. Neunzig and V. Röhrig, Über die Abhängigkeit der Güte der Eloxalschicht von ihrer Nachverdichtung, *Aluminium* **38**, 150 (1962).

230. W. E. Cooke and R. C. Spooner, Sealing Anodized Aluminum, *Metal Finishing J.* (July) (1960).

231. J. Elze, Prüfung der Nachverdichtung anodisch erzeugter Oxydschichten auf Aluminium, *Aluminium* **38**, 161 (1962).

232. G. Paolini and F. Sacchi, Controllo della prova di fissaggio ISML su strati ossido anodico ottenuti in diverse condizioni operative, *Alluminio* **31**, 411 (1962).

233. G. Darnault, Die Bedeutung der Verdichtung und deren Prüfung für das Verhalten von anodisch erzeugten Oxydschichten auf Aluminium und seinen Legierungen, *Aluminium* **40**, 290 (1964).

234. H. Birtel und W. Leute, Die Kontrolle der Verdichtung anodische erzeugter Oxydschichten auf Aluminium, *Aluminium* **41**, 52 (1965).

235. E. F. Barkman, Sealing and Post Anodic Treatments, in *Aodized Aluminum*, ASTM STP 388 (1965), pp. 85–120.

236. M. Pourbaix, *Atlas of Electrochemical Equilibrium in Aqueous Solutions*, Pergamon Press, Oxford (1966).

237. E. Deltombe and M. Pourbaix, The Electrochemical Behavior of Aluminum, Potential pH Diagram of the System Al–H_2O at 25°C, *Corrosion* **14**, 496t (1958); Rapport technique RT 42 of CEBELCOR, December 1955.

238. J. E. Lewis and R. C. Plumb, Studies of Anodic Behavior of Aluminum, I. The Direction of Ionic Movement, *J. Electrochem. Soc.* **105**, 495 (1958).

239. R. C. Plumb, Studies of the Anodic Behavior of Aluminum, II. Coulometry of Barrier Layer Production, *J. Electrochem. Soc.* **105**, 498 (1958).

240. E. J. W. Verwey, Incomplete Atomic Arrangement in Crystals, *J. Chem. Phys.* **3**, 592 (1935).

241. E. J. W. Verwey, The Structure of the Electrolytic Oxide Layers on Aluminum, *Z. Krist.* **91**, 317 (1935).

242. C. S. Taylor, C. M. Tucker, and J. D. Edwards, Anodic Coatings with Crystalline Structure on Aluminum, *Trans. Electrochem. Soc.* **88**, 325 (1945).

243. N. D. Pullen, Oxide Films on Aluminium, *Met. Ind.*, London **54**, 327 (1939).

244. J. D. Edwards and F. Keller, Formation of Anodic Coatings on Aluminum, *Trans. Electrochem. Soc.* **79**, 135 (1941).

245. A. Domony und E. Lichtenberger–Bajza, Untersuchung des Bildungsmechanismus und der Eigenschaften der durch verschiedene äussere Energieverhältnisse auf Aluminiumoberflächen entstehenden Schutzschichten, *Metalloberfläche* **15**, 134 (1961).

246. M. S. Hunter and P. Fowle, Determination of Barrier Layer Thickness of Anodic Oxide Coatings, *J. Electrochem. Soc.* **103**, 481 (1956).

247. H. Ginsberg and K. Wefers, Zur Struktur der anodischen Deckschichten auf Aluminiumoberflächen, *Metall* **16**, 173, 202 (1963).
248. G. A. Dorsey, Jr., The Characterization of Anodic Aluminas, I. Composition of Films from Anodic Aluminas, *J. Electrochem. Soc.* **113**, 169 (1966); II. Effect of Anodizing Temperature on Coatings from Aliphatic Dicarboxilic Electrolytes, **113**, 173 (1966); III. Barrier Layer Composition and Structure, **113**, 284 (1966).
249. F. Liechti and W. D. Treadwell, Zur Kenntnis elektrolytisch erzeugter Oxydschichten auf Aluminium, *Helv. Chim. Acta* **30**, 1204 (1947).
250. R. B. Mason, Factors Affecting the Formation of Anodic Oxide Coatings in Sulfuric Acid Electrolytes, *J. Electrochem. Soc.* **102**, 671 (1965).
251. E. Lichtenberger–Bajza, A. Domony, and P. Csokán, Untersuchung der Struktur und andere Eigenschaften von durch anodische Oxydation auf Aluminium erzeugten Hartoxydschichten, *Werk. und Korr.* **11**, 701 (1960).
252. A. W. Brace and H. Baker, The Incorporation of Sulfate into Anodic Coatings Using Radioactive Sulfur as Tracer, *Trans. Inst. Met. Fin.* **40**, 31 (1963).
253. E. Raub, H. Kawase, and A. von Krusenstjern, Untersuchungen zur anodischen Oxydation von Aluminium in Schwefelsäure Elektrolyten mit Hilfe der Radioindikatormethode, *Metalloberfläche* **21**, 144 (1967).
254. W. McNeil and L. L. Guss, Anodic Film Growth by Anion Deposition in Aluminate, Tungstate and Phosphate Solutions, *J. Electrochem. Soc.* **110**, 853 (1963).
255. S. Tajima, N. Baba, and M. Shimura, Einfluss von Anionen und Inhibitoren auf die Primärenwachstumsvorgänge der anodischen Oxydschichten auf Aluminium, Extended Abstracts, 17th Meeting of CITCE (1966), pp. 27–29; *Electrochim. Acta* **12**, 955 (1967).
256. S. Tajima, N. Baba, T. Mori, and M. Shimura, Influence of Anions on the Initial Growth of Anodic Oxide Films on Aluminium Including Integrally Coloured Films, Proc. Intern. Symp. on Anodizing (Birmingham, 1967), Aluminium Federation, London (1967), pp. 37–42.
257. S. Tajima, T. Mori, and Y. Shibata, Optical Studies on Anodic Films on Aluminum I, *Light Metals*, Japan **14** (66), 239 (1964).
258. S. Tajima, T. Mori, N. Morita, and S. Yamao, Optical Studies on Anodic Oxide Films on Aluminum II, *Light Metals*, Japan **14**, 245 (1964).
259. M. J. Pryor and D. S. Keir, A Method for the Isolation of Surface Films from Aluminum Alloys and the Mechanism of the Reactions Involved, *J. Electrochem. Soc.* **102**, 370 (1955).
260. E. J. W. Verwey, Electrolytic Conduction in a Solid Insulator at High Field—The Formation of the Anodic Oxide Film on Aluminum, *Physica* **2**, 1059 (1935).
261. M. A. Barrett, *Studies with Ellipsometry, II. Optical Constants of Aluminium and of Aluminium Oxide Films*, Metallurgisk Institutt, Norges Tekniske Hogskole, Trondheim (1967).
262. K. Hüber, Die anodische Glänzung und ihre Beziehung zur Anodischen Passivierung, *Z. Electrochem.* **55**, 165 (1951).
263. S. Anderson, Mechanism of Electrolytic Oxidation of Aluminum, *J. Appl. Phys.* **15**, 477 (1944).
264. G. Hass and A. P. Bradford, Anodically Produced Multiple Oxide Films for Increasing the Reflectance of Evaporated Aluminum, *J. Opt. Soc., Am.* **44**, 810 (1954).
265. J. J. Davis, J. P. S. Pringle, A. L. Graham, and F. Brown, A Tracer Study of Anodic Oxidation, *J. Electrochem. Soc.* **109**, 999 (1962).

266. J. A. Davis, J. Domeij, J. P. S. Pringle, and F. Brown, *J. Electrochem. Soc.* **112**, 675 (1965).

267. P. E. Doherty and R. S. Davis, The Formation of Surface Pits by the Condensation of Vacancies, *Acta Met.* **7**, 118 (1959).

268. J. Whitton, Measurement of Ionic Mobilities in the Anodic Oxides of Tantalum and Zirconium by a Precision Section Technique. *J. Electrochem. Soc.* **115**, 58 (1968).

269. H. A. Francis, Direct Observation of the Anodic Oxidation of Aluminum, *J. Electrochem. Soc.* **112**, 1234 (1965).

270. D. Vermilyea, The Formation of Anodic Oxide Films on Tantalum in Non-Aqueous Solutions, *Acta Met.* **2**, 482 (1954).

271. A. J. Dekker and W. Vh. van Geel, On the Amorphous and Crystalline Oxide Layers on Aluminium, *Philips Rep.* **2**, 313 (1947).

272. G. Amsel and D. Samuel, The Mechanism of Anodic Oxidation, *J. Phys. Chem. Solids.* **23**, 1707 (1962).

273. W. J. Bernard, Ionic Movement during the Growth of Anodic Oxide Films on Aluminum, *J. Electrochem. Soc.* **109**, 1082 (1962).

274. R. W. Berry, B. L. Kennedy, and H. Waggener, An Investigation of the Diffusing Species in the Anodic Growth of Tantalum and Aluminum Oxides, presented at the spring meeting of the Electrochem. Soc., Pittsburgh, Pa. (1963).

275. J. J. McMullen and M. J. Pryor, The Relation between Passivation, Corrosion and the Electrical Characteristics of Aluminium Oxide Films, Proc. 1st Intern. Congr. on Metallic Corrosion, Butterworth, London (1961), pp. 149–156.

276. J. C. Banter, Incorporation of Ions in Anodic Oxide Films on Zirconium and the Effect on Film Behavior, *J. Electrochem. Soc.* **114**, 508 (1967).

277. T. P. Hoar and N. F. Mott, Mechanism of Porous Anodic Oxide Films on Aluminium, *J. Phys. Chem. Solids* **9**, 97 (1959).

278. M. A. Heine and M. J. Pryor, The Distribution of A.C. Resistance in Oxide Film on Aluminum, *J. Electrochem. Soc.* **110**, 1205 (1963).

279. A. J. Brock and G. C. Wood, Hydroxyl Ion and Proton Mobility during Anodic Oxidation of Aluminium, *Electrochim. Acta* **12**, 395 (1967).

280. N. F. Mott, Theory of the Formation of Protective Oxide Films on Metals, III *Trans. Farad. Soc.* **43**, 429 (1947).

281. N. Cabrera and N. F. Mott, *Rept. Progr. in Physics* **12**, 163 (1948).

282. A. Charlesby, Electron Currents in Thin Oxide Layers on Aluminum, *Proc. Phys. Soc.* **B66**, 317, 533 (1953).

283. S. Tajima, Y. Tanabe, M. Shimura, and T. Mori, Electronmicroscopic and Optical Studies on the Electrocrystallization of Alpha-Alumina (Corundum) Films on Aluminium, *Electrochim. Acta* **6**, 127 (1962).

284. M. J. Dignam, Oxide Films on Aluminum, l. Ionic Conduction and Structure, *J. Electrochem. Soc.* **109**, 184 (1962).

285. C. P. Bean, J. C. Fischer, and D. A. Vermilyea, Ionic Conductivity of Tantalum Oxide at Very High Field, *Phys. Rev.* **101**, 551 (1956).

286. J. D. Dewald, Transient Effects in the Ionic Conduction of Anodic Oxide Films at High Field, *J. Phys. Chem. Solids.* **2**, 55 (1957).

287. M. J. Dignam, Oxide Films on Aluminum II. Kinetics of Formation in Oxygen, *J. Electrochem. Soc.* **109**, 192 (1962).

288. P. J. Ryan and M. J. Dignam, High-Field Kinetics Process in Anodic Oxide Films on Aluminum, *Canad. J. Chem.* **40**, 1875 (1962).

289. M. J. Dignam, Field Dependence and Anomolous Temperature Dependence of Tafel Slopes for Processes Occuring at High Electrostatic Fields, *Canad. J. Chem.* **42**, 1155 (1964).

290. M. J. Dignam, Conduction Properties of Valve Metal–Oxide Systems I. A New Theory, *J. Electrochem. Soc.* **112**, 722 (1965).

292. M. J. Dignam, D. Goad, and M. Sole, Determination of the Field Dependence of the Tafel Slope for the Steady-State Anodic Oxidation of Aluminum, *Canad. J. Chem.* **43**, 800 (1965).

293. M. J. Dignam and W. R. Fawcett, The Kinetics and Mechanism of Oxidation of Superpurity Aluminum in Dry Oxygen, I. Apparatus Description and the Growth of "Amorphous" Oxide, *J. Electrochem. Soc.* **113**, 656 (1966).

294. M. J. Dignam and W. R. Fawcett, The Kinetics and Mechanism of Oxidation of Superpurity Aluminum in Dry Oxygen, II. The Growth of Crystals of γ-Alumina, *J. Electrochem. Soc.* **113**, 663 (1966).

295. K. Nagase, On the Thickness of Aluminium Film in Electrolytic Capacitor, *Mem. Inst. Sci. Ind. Res., Osaka Univ.* **10**, 66 (1953).

296. K. Nagase, *Mem. Inst. Sci. Ind. Res., Osaka Univ.* **12**, 67 (1955).

297. D. A. Vermilyea, The Increasing Rate of Voltage and the Relation between Residual Current and Formation Time at Formation of Aluminum Oxide Film. Ionic Conductivity of Anodic Films at High Field Strength: Transient Behavior, *J. Electrochem. Soc.* **104**, 427 (1957).

298. W. J. Bernard and J. W. Cook, The Growth of Barrier Oxide Films on Aluminum, *J. Electrochem. Soc.* **106**, 643 (1959).

299. H. A. Johansen, G. B. Adams, and P. Van Rysselberghe, Anodic Oxidation of Aluminum, Chromium, Hafnium, Niobium, Tantalum, Vanadiam and Zirconium at Very Low Current Densities, *J. Electrochem. Soc.* **104**, 339 (1957).

300. K. Videm, Behaviour of Aluminium in Barrier Layer Anodising Electrolytes, Extended Abstracts of 17th Meeting of CITCE, Tokyo (1966), pp. 30–32.

301. H. Fischer and F. Kurz, Über mikroskopisches Bild anodischer Oxydfilme auf Aluminium und ihr Wachstum, *Korros. und Metallschutz.* **18**, 42 (1942).

302. P. Lacombe and L. Beaujard, in *Etudes sur les aspects des pellicules d'oxydation anodique formés sur l'aluminium et ses alliages*, Ed. comité gén. d'Org. des ind. mec., Paris (1944).

303. K. Huber, Pore-volume of electrolytically produced protective coatings on aluminum, *J. Colloid Sci.* **3**, 197 (1948).

304. K. Huber, Studien zur Chemie und zur Struktur anodisch erzeugter Niederschläge und Deckschichten IV. Die polarisationsoptische Analyse der dispersen Struktur oxydischer Deckschichten auf Aluminium, *Helv. Chim. Acta* **28**, 1416 (1945).

305. K. Huber and A. Gaugler, Über optische Untersuchung an Deckschichten auf Aluminium, *Experimentia* **111** (7), 1 (1947).

306. K. Huber, Pore volume of electrolitically produced protective coatings on aluminum, *J. Colloid. Sci.* **3**, 197 (1948).

307. R. L. Burwell, Jr., P. A. Smudski, and T. P. May, Ethylene Adsorption Isotherm at—183°, *J. Am. Chem. Soc.* **69**, 1525 (1947).

308. F. P. Zalivalov, M. N. Tyukina, and N. D. Tomaschov, Effect of Electrolysis Conditions on the Formation and Growth of Anodic Oxide Films on Aluminium, *J. Phys. Chem. USSR* **35**, 879 (1961).

309. C. J. L. Booker, J. L. Wood, and A. Walsh, Electronmicrographs from Thick Oxide Layers on Aluminum, *Nature* **176**, 222 (1955); *Brit. J. App. Phys.* **8**, 347 (1957).

310. G. C. Wood, Proc. Conf. on Anodizing (Nottingham, 1961), Aluminium Federation, London (1962), pp. 109–110.

311. H. Grubitsch, W. Geymer, and E. Burik, Eloxalschichten mit definierter Porengrösse, *Aluminium* **37**, 569 (1961).

312. L. A. Cosgrove, Porosity of Anodic Oxide Coatings on Aluminum. Comparison of n-Butane and Krypton Sorption, *J. Phys. Chem.* **60**, 385 (1956).

313. G. Paolini, M. Masoero, F. Sacchi, and M. Paganelli, An Investigation of Porous Anodic Oxide Films on Aluminium by Comparative Adsorption, Gravimetric and Electron Optical Measurements, *J. Electrochem. Soc.* **112**, 32 (1965).

314. M. S. Hunter and P. Fowle, Factors Affecting the Formation of Anodic Oxide Coatings, *J. Electrochem. Soc.* **101**, 514 (1954).

315. Th. Skulikidis, S. Karalis, and P. Mentojiannis, Untersuchung der Sekundärstruktur von anodischen Deckschichten auf Aluminium durch Farbstoffaufnahmen I. Über zwei aufeinander liegende Hauptdeckschichten, *Kolloid-Z.* **149**, 6 (1956).

316. Th. Skulikidis, Ch. Papathanasiu, and J. Marangosis, Untersuchung der Sekundärstruktur von anodischen Deckschichten auf Aluminium durch Farbstoffaufnahme, II. Kinetik der Färbung, *Kolloid-Z.* **150**, 54 (1957).

317. Th. Skulikidis, N. Mallios, and K. Kazantzis, Kinetik der Alterung von anodischen Deckschichten auf Aluminium durch Farbstoffaufnahmen, *Kolloid-Z.* **159**, 130 (1958).

318. Th. Skulikidis, G. Paraskevopoulis, and D. Argyriou, Beitrag zur Kinetik der anodischen Oxydatyon von Aluminium, *Kolloid-Z.* **168**, 154 (1960).

319. Th. Skulikidis and P. Bekiaroglou, Kinetik der Alterung von anodischen Deckschichten auf Aluminium durch Farbstoffaufnahmen, II. Höhere Temperaturen, *Kolloid-Z.* **178**, 45 (1961).

320. H. Akahori, Electronmicroscopic Study of Growing Mechanism of Aluminium Oxide Films, *J. Electronmicroscopy* **10**, 175 (1961).

321. W. Kaden, Beitrag zu den Wachstumsvorgängen von Oxydschichten auf Aluminium, *Aluminium* **39**, 33 (1963).

322. W. Kaden, Neuere Erkenntnisse und Erfahrungen über verschiedene Variationen der anodischen Oxydation des Aluminiums, *Aluminium* **39**, 424 (1963).

323. T. A. Renshaw, A Study of Pore Structure of Anodized Aluminum, *J. Electrochem. Soc.* **108**, 185 (1961).

324. E. F. Barkman, The Structure, Composition and Mechanism of Formation of Anodic Oxides on Aluminum, Proc. Conf. on Anodizing. (Birmingham, 1967), Aluminium Federation, London (1967), pp. 27–36.

325. R. W. Franklin, Characteristics of Barrier Coatings for Electrolytic Capacitor, Proc. Conf. on Anodizing (Nottingham, 1961), Aluminium Federation, London (1962), pp. 96–107.

326. D. J. Stirland and R. W. Bicknell, Studies of the Structure of Anodic Oxide Films on Aluminum, *J. Electrochem. Soc.* **106**, 481 (1959).

327. R. L. Burwell and T. P. May, Pittsburgh Inten. Conf. on Surface Reactions, Vol. 10, Corrosion Publishing Co., Pittsburgh (1948).

328. T. P. Hoar and J. Yahalom, The Initiation of Pores in Anodic Oxide Films Formed on Aluminum in Acid Solutions, *J. Electrochem. Soc.* **110**, 614 (1963).

329. M. A. Barrett and A. B. Winterbottom, Film Formation on Aluminium under Various Electrochemical Conditions in Aqueous Electrolytes. First Intern. Congr. on Metallic Corrosion (London, 1961), Butterworth, London (1963), pp. 657–661.

330. A. B. Winterbottom, Increased Scope of Ellipsometric Studies of Surface Film Formation, Proc. Symp. on the Ellipsometric Studies of Surface Film Formation, Proc. Symp. on the Ellipsometer and Its Use in the Measurement of Surfaces and Thin Films, NBS, Washington, D.C. (1963), pp. 97–112.

331. M. A. Barrett, Optical Study of the Formation and Stability of Anodic Films on Aluminum, Proc. Symp. on the Ellipsometer and Its Use in the Measurement of Surfaces and Thin Films, NBS, Washington, D.C. (1963), pp. 213–228.

332. M. A. Barrett, Some Observations on the Mechanism of Anodic Oxide Formation on Aluminium, Scandinavian Corrosion Congress, Helsinki (1964), p. 6.

333. G. H. Giles, The Theory of Dyeing Anodic Finishes, Proc. Conf. on Anodizing (Nottingham, 1961), Aluminium Federation, London (1962), pp. 174–180.

334. V. C. P. Morfopoulos and H. C. Parreira, Elektrokinetic Studies on Oxidized Aluminium Surfaces, *Corro. Sci.* **7**, 241 (1967).

335. K. Kuroda and K. Uji, On Zeta-Potential and Dyeing Affinity of Anodic Oxide Coatings on Aluminum, *Chimica Metallica*, Tokyo **2**, 303 (1965); *J. Metal Finish. Soc.*, Japan **18**, 169 (1967).

336. K. Kuroda and K. Uji, Relation between pH as well as Concentration of Anodic Dyestuff Solution and Dyeability of Anodic Oxide Coating on Aluminum, *J. Metal Finish. Soc.*, Japan **18**, 481 (1967).

337. H. Yokota, Zeta-Potential of Anodic Oxide Films on Aluminum, *Denki Kagaku* **35**, 14 (1967).

338. K. R. Hart, A Study of Boehmite Formation on Aluminium Surfaces by Electron Diffraction, *Trans. Farad. Soc.* **50**, 269 (1954).

339. R. C. Spooner, Water Sealing of Detached Aluminium Oxide Anodic Film, *Nature* **178**, 1113 (1956).

340. M. S. Hunter, P. E. Towner, and D. L. Robinson, Hydration of Anodic Oxide Films, 46th Ann. Techn. Proc., Am. Electroplaters' Soc. (1959), p. 6.

341. H. Richaud, The Influence of Sealing Water Impurities on the Quality of the Oxide Films, Proc. Conf. on Anodizing (Nottingham, 1961), Aluminium Federation, London (1962), pp. 181–185.

342. T. P. Hoar and G. C. Wood, The Sealing of Porous Anodic Oxide Films on Aluminium, Proc. Conf. on Anodizing (Nottingham, 1961), Aluminium Federation, London (1962), pp. 186–220; *Electrochim. Acta* **7**, 333 (1962).

343. H. Akahori and T. Fukushima, Study on the Hydration of Alumite by Electron Microscopy and Electron Micro-Diffraction, *J. Electron Microscopy* **13**, 162 (1964).

344. C. A. Amore and J. F. Murphy, Sealing of Anodized Aluminum, *Metal Finishing* (Nov.), 50 (1965).

345. F. Pearlstein, Sealing Anodized Aluminum, *Metal Finishing J.* (Aug.), 40 (1960).

346. H. Neunzig and V. Röhrig, Über die Abhängizkeit der Güte der Eloxalschicht von Ihrer Nachverdichtung, *Aluminium* **38**, 150 (1962).

347. D. G. Altenpohl, Zur Frage des Auftretens von Böhmit bei der Reaktion zwischen Aluminium und kochendem oder überhitztem Wasser, *Aluminium* **36**, 438 (1960).

348. D. G. Altenpohl, Use of Boehmite films for corrosion protection of aluminium, *Corrosion* **18**, 143t (1962); *Aluminium* **31**, 10, 62 (1955).

349. G. C. Wood, Sealing Anodic Oxide Films on Aluminium, *Trans. Inst. Metal Finish.* **36**, winter. Part 6 (1959), ADA Reprint No. 87 (1959).

350. G. Bürgers, A. Classen, and J. Zernicke, Über die chemische Natur der Oxidschichten, die sich bei anodischer Polarisation auf den Metallen, Aluminium, Zirkon, Titan und Tantal bilden, *Z. Physik* **74**, 593 (1932).

351. R. B. Mason, Density and Porosity of Anodic Coatings on Aluminum, *Metal Finishing J.* (Aug.), 55, 67 (1957).

352. H. N. Hill and R. B. Mason, Anodic Coatings on Aluminum, Their Brittleness, *Metals and Alloys* **14**, 972 (1942).

353. H. W. L. Phillips, The Nature and Properties of the Anodic Film on Aluminium and Its Alloys, in *Properties of Metallic Surfaces*, The Inst. of Metals, London (1953), pp. 237–252.

354. N. D. Pullen, Oxide Films on Aluminium. Some Physical Properties, *Met. Ind.* London, **54**, 327 (1939); *Metallurgia* **21**, 57 (1939).

355. L. Tassara, *La Lucidatura Elettrolitica dei Metalli*, Antonio Cordani S.A., Milano (1945).

356. I. H. Khan, J. S. L. Leachi, and N. J. M. Ilkes, The Thickness and Optical Properties of Films on Anodic Aluminium Oxide, *Corr. Sci.* **6**, 483 (1966).

357. T. M. Kelleher, Thermal Control Properties of Immersion and Repair Anodic Coatings, Proc. Conf. on Anodizing (Birmingham, 1967), Aluminium Federation, London (1967), pp. 71–77.

358. E. J. Clauss, ed., *Surface Effects on Space Materials*, J. Wiley and Sons, New York (1960).

359. J. E. Janssen, Normal Spectral Reflectance of Anodized Coatings on Aluminum, Magnesium, Titanium and Berylium, ASD TR 61-147, Sept. (1961).

360. G. Franckenstein, Gleichspannungsmessungen an elektrolytisch erzeugtem Aluminiumoxyd, *Ann. Physik* **26**, 17 (1936).

361. W. Hermann, Zum Mechanismus der Oxydschichtbildung auf Aluminium-Anoden von Elektrolytkondensatoren, *Siemens Wiss. Veröffent., Werkstoffsonderheft* 182 (1940).

362. A. Miyata, Dielectric Characteristics of Aluminum Anode during Formation, *Rept. Inst. Phys. Chem. Res., Tokyo* **19**, 369 (1940).

363. G. J. Tibol and R. W. Hull, Plasma Anodized Aluminum Oxide Films, *J. Electrochem. Soc.* **111**, 1368 (1964).

364. D. H. Bradhurst and J. S. Llwelyn, The Mechanical Properties of Thin Anodic Films on Aluminum, *J. Electrochem. Soc.* **113**, 1245 (1966).

365. D. A. Vermilyea, Stresses in Anodic Films, *J. Electrochem. Soc.* **110**, 345 (1963).

366. P. C. Sheasby, Factors Affecting the Performance of Externally Exposed Anodized Aluminium, Proc. Conf. on Anodizing (Birmingham, 1967), Aluminium Federation, London (1967), pp. 133–144.

367. R. C. Spooner, Outdoor Exposure of Anodic Coatings on Aluminum: Effect of Sealing, *Finishing of Aluminum*, H. G. Kissin, ed., Reinhold, New York (1963), pp. 127–162.

368. R. F. Stratful, Highway Corrosion Problems, *Materials Protection* (Sept.), 8 (1963).

369. R. E. Brooks, Compatibility of Aluminum with Alkaline Building Materials, *Materials Protection* (July), 44 (1962).

370. C. J. Walton and F. L. McCeary, The Compatibility of Aluminum with Alkaline Building Products, paper presented at the 13th Ann. Conf. NACE, St. Louis, 1957, Alcoa Research Lab. (1957).

371. F. Endtinger, Beständigkeit von Aluminium gegen alkalische Baustoffe, *Aluminium Suisse* **17** (April), 75 (1967).

372. M. J. Pryor and D. S. Keir, Galvanic Corrosion I. Current Flow and Polarization Characteristics of the Aluminum–Steel and Zinc–Steel Couples in Sodium Chloride Solution, *J. Electrochem. Soc.* **104**, 269 (1957); II, Effect of pH and Dissolved Oxygen Concentration of the Aluminum–Steel Couple, *J. Electrochem. Soc.* **105**, 629 (1958).

373. S. Tajima, T. Mori, and M. Komiya, Contact Corrosion of Bimetallic Couples of Aluminum or Anodized Aluminum and Mild Steel or Stainless Steel in Inorganic Acid Media, *Boshoku Gijutsu* **14**, 109 (1965).

374. Corrosion Prevention Committee (S. Tajima, Chief), Studies on Prevention of Corrosion of Aluminum kettles, Light Metal Sheet Products Society, Tokyo (1963), p. 57.

375. S. Tajima, Azione dell'acqua sui bollitori di alluminio anodizato, *Alluminio* (3), 135 (1962).

376. W. C. Cochran, Test Methods for Anodized Aluminum and Their Significance, Proc. Conf. on Anodizing (Birmingham, 1967), Aluminium Federation, London (1967), pp. 145–163.

377. Anodizing Aluminum, Japan Industrial Standard, JIS H-8601 (July, 1952).

378. J. M. Kape, The Deterioration of Anodised Aluminium Exposed to Corrosive Environments, *Electroplating and Metal Finishing* (Feb.), 7 (1959).

379. F. Sacchi, Quality Control Scheme for Anodized Aluminium, Proc. Conf. on Anodizing (Birmingham, 1967), Aluminium Federation, London (1967), pp. 167–176.

380. J. Stone, H. A. Tuttle, and H. N. Bogart, Ford Anodized Aluminum Corrosion Test –FACT–, Plating, July (1966), pp. 877–888.

381. R. V. Vanden Berg, Specifications and Nomenclature for the Commercial Use of Anodic Coatings in the U.S.A., UNIPREA (International Symposium for the Standardization of Surface Treatments of Metal Materials), Torino (Oct., 1961), p. 13.

382. F. Sacchi and A. Prati, Standardization of Anodic Films on Aluminium in Italy, UNIPREA, Torino (Oct., 1961), p. 7.

383. A. Kutzelnigg, A New Way of Determining the Aluminium Oxide Film Thickness of Anodized Light Alloy, UNIPREA, Torino (Oct., 1961), p. 2.

384. J. M. Kape, Testing of Anodic Coatings for Outdoor Exposure Service, UNIPREA, Torino (Oct., 1961), p. 7.

385. J. M. Kape and J. A. Whittaker, Five Year Exposure Test on Anodized Aluminium in a Severe Industrial Atmosphere, *Trans. Inst. Metal. Finish.* **43**, 106 (1965).

386. V. F. Henley and F. C. Porter, Standardisation of Anodic Oxidation Coatings on Aluminium in the United Kingdom, UNIPREA, Torino (Oct., 1961), p. 5.

387. G. Darnault, Critical Analysis of Present Standardization in France and Future Programs in the Field of Anodizing Treatments, UNIPREA, Torino (Oct., 1961), p. 7.

388. CIDA/EWAA, Recommendations on Anodized Aluminium for Architectural Purposes (Centre International de Developpement de l'Aluminium, ed.), 2nd Ed., Paris (August, 1965).

389. E. Di Russo, P. Lenzi, A. Prati, and F. Sacchi, An Investigation of the Mechanism of Pitting Corrosion on Anodized Al–Si–Mg Architectural Alloys, paper presented at 28th Corrosion Week, CEBELCOR, Bruxelles (June, 1965).

390. F. C. Porter, Anodized Aluminium, Specification and Quality Control, *Metal Finishing J.* (Feb.), 65 (1964).

391. W. C. Cochran, E. T. Engelhart, and D. J. George, Improved Methods, New Tests Upgrade Anodized Auto Trim, *Modern Metals* 16, 82 (1960).

392. E. T. Engelhart and D. J. George, Evaluation of Anodic Coatings by Impedance Measurements, *Materials Protection* 3 (11), 24 (1964).

393. W. Kesternich, Prüfung metallischer und nichtmetallischer Schutzüberzüge auf Korrosionsbeständigkeit in einer künstlichen Industrieatmosphäre, *Stahl und Eisen* 11, 687 (1951).

394. A. Kutzelnigg, Vergleich zwischen dem Ergebnis der Freibewitterung in Industrieluft und der Kurzbeanspruchung in Schwefeldioxyd enthaltender Atmosphäre, *Werk. u. Korr.* 9, 429 (1958).

395. J. E. Draley and W. E. Ruther, Aqueous Corrosion of Aluminum, *Corrosion* 10, 12 (1957).

396. S. Mori and J. E. Draley, Oxide Dissolution and Its Effect on the Corrosion of 1100 Aluminum, in Water at 70°C, *J. Electrochem. Soc.* 114, 352 (1967).

397. J. E. Draley, S. Mori, and R. E. Loess, The Corrosion of 1100 Aluminum in Water from 50° to 95°C, *J. Electrochem. Soc.* 114, 353 (1967).

398. L. H. Kalpan, Aluminum Oxide Films from the Reaction of Aluminum and Water Vapor, *Electrochem. Technology* 3, 335 (1965).

399. K. Yamamoto, N. Baba, and S. Tajima, Mechanism of Formation of Conversion Films on Aluminium from Inhibited Alkaline Solution, Compt. Rend. 2éme Symp. Européen sur les Inhibiteurs de Corrosion (Ferrara, 1966), Università di Ferrara (1966).

400. S. Setoh and A. Miyata, Anodic Films of Aluminum and Their Applications, *Denki Kagaku* 1, 15 (1933).

401. R. B. Mason, Effect of Aluminum Sulfate in the Sulfuric Acid Electrolyte on Anodic Polarization, *J. Electrochem. Soc.* 103, 425 (1956).

402. J. M. Kape, The Use of Malonic Acid as an Anodizing Electrolyte, *Metallurgia* (Nov.), 181 (1959).

403. J. Pake, Control of Hot Water in Sealing Anodic Coatings, *Prod. Fin.* 30 (1), 66, 170 (1965).

404. M. Karsulin, ed., Symposium sur les Buxites, Oxydes et Hydroxydes d'Aluminium, Vol. 1, II (Zagreb, 1963), Acad. Yougoslavie des Sciences et des Arts, Zagreb (1963).

405. R. B. Mason and P. E. Fowle, Anodic Behavior of Aluminum and Its Alloys in Sulfuric Acid Electrolytes, *J. Electrochem. Soc.* 101, 53 (1954).

406. J. G. Morris, Anodizing of Aluminum Alloys–Metallurgical Structure Factors, *Anodized Aluminum*, ASTM STP 388(1965), pp. 21–33.

407. P. F. Schmidt, On the Mechanisms of Electrolytic Rectification, *J. Electrochem. Soc.* 115, 167 (1968).

408. D. A. Vermilyea, Anodic Films, Protons and Electrolytic Rectification, *J. Electrochem. Soc.* 115, 177 (1968).

409. R. H. Doremus, Ionic Transport in Amorphous Oxides, *J. Electrochem. Soc.* 115, 181 (1968).

410. D. M. Smith and G. A. Shirn, Conduction and Stoichiometry in Heat-Treated, Anodic Oxide Films, *J. Electrochem. Soc.* 115, 186 (1968).

411. K. Lehovec, Impedance for Tunnel Exchange of Electrons Across the Annealed Ta/Ta_2O_5 Interface, *J. Electrochem. Soc.* 115, 192 (1968).

412. P. J. Boddy, Oxygen Evolution on Semiconducting TiO_2, *J. Electrochem. Soc.* **115**, 199 (1968).

413. F. Huber, On the Rectification of Anodic Oxide Films of Titanium, *J. Electrochem. Soc.* **115**, 203 (1968).

414. J. M. Hale, The Insulator–Electrolyte Interface, *J. Electrochem. Soc.* **115**, 208 (1968).

415. C. E. Michelson, The Current–Voltage Characteristics of Porous Anodic Oxides on Aluminum, *J. Electrochem. Soc.* **115**, 213 (1968).

416. C. G. Dunn, Information on Anodic Oxides on Valve Metals: Oxide Growth at Constant Rate of Voltage Increase, *J. Electrochem. Soc.* **115**, 219 (1968).

417. A. W. Smith, The Impedance, Rectification and Electroluminescence of Anodic Oxide Films on Aluminium, *Canad. J. Phys.* **37**, 591 (1959).

418. Z. Ruziewicz, Luminescence of Oxide Films Produced during Anodic Oxidation of Aluminium, *Bull. Acad. Polonaise des Sciences* Cl. III-4, 537 (1956).

419. M. Hollo, Electronoptical Investigations on the Structure and Mechanism of Formation of Anodic Oxide Layers on Aluminium, Advance copy of the 1st. Intern. Congr. on Metallic Corrosion (London, 1961), Butterworth, London (1961), pp. 449–454.

420. E. Raub and N. Baba, Uber einige Eigenschaften der beim chemischen und anodischen Glänzen entstehenden Deckschichten auf Aluminium, *Metalloberfläche* **19**, 285 (1965).

421. S. Tajima, N. Baba, and T. Muramaki, On Voltage–Time Transient in Anodic Oxidation of Aluminum in Aliphatic and Aromatic Acids, Extended Abstracts, Spring Meeting of the Electrochem. Soc., Japan (April, 1968), B-72 (1968).

422. D. R. Clarke, D. W. Hamilton, and T. R. Beck, Development of a Barrier-Layer Anodic Coating for Aluminum, *Plating* **54**, 1342 (1967).

423. A. La Vecchia, R. Piontelli, F. Siniscalco, and A. Valengo, Anodizing with Sulfamate Bath, Proc. Symp. on Sulfamic Acid and Its Metallurgical Applications (Milan, 1966), Associazione Italiana di Metallurgia (Milan, 1966), pp. 401–410; *Electrochim. Metal.* **2**, 85 (1967).

424. A. La Vecchia, G. P. Piazzesi, and F. Sinscalco, Preliminary Investigations on Al Anodic Layers from Sulfamate Bath, Proc. Symp. on Sulfamic Acid and its Metallurgical Applications (Milan, 1966), Associazione Italiana di Metallurgia, Milano (1967), pp. 411–424; *Electrochim. Metal.* **2**, 85 (1967).

INDEX

Cyclohexylamine, to adjust pH, 190

D

Damage, Zn monocrystals, influence on embrittlement, 90
Debris layer
Silicon carbide impaction influence, 66
Surface model, 61, 62
Thickness estimate, Au, Fe, Al, 64
Decylamine inhibitors for stainless steels in acids, 198
Deuterium oxide conductance measurements, 40
Dibenzylsulfoxide inhibitors for stainless steels in acids, 198
Dicyclohexylamine nitrite vs Zn, Pb, Cd, 187
Dicyclohexylamine, oil-soluble salts inhibitors, 187
Dielectric constant, water, temperature dependence, 3
Diisopropylamine nitrate vs Zn, Pb, and Cd, 187
Dinonylnaphthalene sulfonic acid inhibitors for gasoline, 201
Di-o-tolylthiourea, gelatin inhibitors for Zn, Cd, Pb, 200
Dislocations
Chlorate layer barrier to, 121
Creep phenomena, 113
Density measurements, 68
Dislocation-rich layers, influence on, 76
Emergence, stress required for, 129
Equilibrium film thickness, 129
Films as barriers to emergence, 122
Generation mechanism, 68
Glide path edge vs flow stress distribution on Cu crystals, 78
Lattice spacing, influence on, 77
Mobility, influence of 7.2 N CaCl, 113
Mobility, influence of DMF and DMSO, 111
Mobility, influence of environment sensitivity on drilling nonmetallics, 116
Mobility, influence on drilling efficiency, 117
Multiplication by cross slip, 106
Oxide ceramic crystals, influence on, 103
Pile up, rupture of films, and current density surges due to, 122
Pinning theory, 60
Point defect, influence of impurity atom interactions, 111
Reduction of dibenzylsulfoxide to dibenzylsulfide, influence on, 165
Residual, at interface, influence, 134
Screw, influence by image forces, 70
Solute atom reaction in hydrogen charging, 127
Surface conditions effect on, 58
Surface vs interior interaction, 127
Torsional pileups, influence when released, 122
Transfer from substrate, 132

Dislocations (cont.)
Transfer, influence of a ±3% lattice misfit, 134
Vacuum, entrapment by, 129
Work-hardening, influence on character of, 76
Double layer, AgCl, point-defect hardening in, 108
Double layers, influence of potential difference across, 81
Drilling, efficiency vs environment, 117
Ductility, anodic oxide films on Al, 335
Duranodic anodizing, 260
Dye absorption by sulfuric acid film, 239
Dyes
Absorption, oxide layer properties favoring, 314
Anodizing for color, 251, 265
Chromic acid anodizing, for, 245
Sealing, influence on, 333
Structure with affinity for anodic films, 327
Sulfamic acid anodizing, for, 247
Theory for coloring anodized films by, 326

E

Electrocapillary curves, influence of surface-active additions, 163
Electrocapillary effects, 81
Electrochemical double layer, influence of metal charge, 162
Electrodeposition of metals <300°C, 41
Electrodeposits of Au and rhodium, variations in ductility, 135
Electrodes
Antimony/antimony oxide, <300°C, 23
Calibration in sulfuric acid, <250°C, vs hydroiodide, <200°C, 23
Carbon in hydrogen–oxygen fuel cells, 43
Cesium vs Pt black, <200°C, 43
Dropping Me, <200°C, for double-layer study, 42
External reference, 23
Glass, <30,000 psig, 19
Hydrogen, <275°C, 16
Hydrogen diffusion in <200°C fuel cells, 43
Me, dropping, use of, 163, 177
Me/MeCl, <263°C, > 1 M HCl, 22
Me/Me sulfate vs <0.2 M sulfuric acid, <80°C, 22
Niboride, <200°C, in 50% KOH fuel cells, 43
Palladium, 19
pH measurements in acids and alkalies, 40
Platinum, 23
Pressure-balanced, 25
Pt-black in hydrocarbon–O$_2$ fuel cells, 43
Ru carbonate and bicarbonate vs Pt-black, 43
Silver/silver bromide to measure pH, <200°C, 21
Silver/silver chloride, 150°C, 21
Silver/silver sulfate vs <0.05 M sulfuric acid, <250°C, 22